VOLCANIC ROCKS AND SOILS

PROCEEDINGS OF THE INTERNATIONAL WORKSHOP ON VOLCANIC ROCKS AND SOILS
LACCO AMENO, ISCHIA ISLAND, ITALY, 24–25 SEPTEMBER 2016

Volcanic Rocks and Soils

Editors

Tatiana Rotonda
Sapienza University of Rome, Rome, Italy

Manuela Cecconi
University of Perugia, Perugia, Italy

Francesco Silvestri
University of Napoli Federico II, Naples, Italy

Paolo Tommasi
CNR, Institute for Environmental Geology and Geo-Engineering, Rome, Italy

CRC Press
Taylor & Francis Group
Boca Raton London New York Leiden

CRC Press is an imprint of the
Taylor & Francis Group, an **informa** business

A BALKEMA BOOK

Organized by

Under the auspices of

ISRM

SIMSG | ISSMGE

Cover photo: (front) Erosional forms in the Pizzi Bianchi tuffs, Ischia Island (courtesy of www.prontoischia.it) (back) Cliff on the left bank of the Cava Scura canyon (Pizzi Bianchi tuffs), Ischia Island (by Paolo Tommasi)

CRC Press/Balkema is an imprint of the Taylor & Francis Group, an informa business

© 2016 Taylor & Francis Group, London, UK

Typeset by MPS Limited, Chennai, India
Printed and bound in Great Britain by CPI Group (UK) Ltd, Croydon, CR0 4YY

Published by: CRC Press/Balkema
                 P.O. Box 11320, 2301 EH Leiden, The Netherlands
                 e-mail: Pub.NL@taylorandfrancis.com
                 www.crcpress.com – www.taylorandfrancis.com

ISBN: 978-1-138-02886-9 (Hbk)
ISBN: 978-1-315-64791-3 (eBook PDF)

Volcanic Rocks and Soils – Rotonda et al. (eds)
© 2016 Taylor & Francis Group, London, ISBN 978-1-138-02886-9

Table of contents

Session 3: Mechanical behaviour of volcanic soils

Session 4: Geotechnical aspects of natural hazards

Session 5: Geotechnical problems of engineering structures

Volcanic Rocks and Soils – Rotonda et al. (eds)
© *2016 Taylor & Francis Group, London, ISBN 978-1-138-02886-9*

Preface

According to tradition, after the first three Workshops on Volcanic Rocks were held respectively on Madeira, on the Azores and on Tenerife, also the fourth edition of this event is hosted on a volcanic island, Ischia.

The island of Ischia is very famous worldwide for the beauty of its nature (it is known as "the Green Island") and for its thermal springs. It is located in a wide volcanic area surrounded by two main active volcanic districts. The first one, known all over the world, is the Somma-Vesuvius, located East of the Bay of Naples. The second volcanic district is sited to the West of Naples and it is formed by some tens of volcanic craters, of different size and age. This area is called Phlegraean Fields, whose name derives from the Greek word, φλέγω, which means "to burn". It is covered by rocks (Campanian ignimbrite, Neapolitan Yellow tuff, etc.) and soils (pozzolanas, ashes, pumices), which outcrop in the city of Naples and are spread throughout large parts of the Campania Region. Another famous site is the small city of Pozzuoli, that gave the name "Pozzolana" to the well-graded pyroclastic soil, very diffused in the area and used, since the Roman period, as a component of cements.

For this Workshop the Organising Committee decided to consider both volcanic rocks and soils, which are widespread throughout Central and Southern Italy, and particularly in the Campanian volcanic district. The Organising Committee is grateful to the Associazione Geotecnica Italiana (AGI) for the organization of the event and both the International Society for Rock Mechanics (ISRM) and the International Society for Soil Mechanics and Geotechnical Engineering (ISSMGE) for having co-sponsored the event.

The aim of the Workshop is to bring together geotechnical engineers, geologists, volcanologists, structural and hydraulic engineers, etc., interested in both research and practical problems regarding volcanic rocks and soils.

The Workshop is divided in the following five sessions:

- Structural features of volcanic materials
- Mechanical behaviour of volcanic rocks
- Mechanical behaviour of volcanic soils
- Geotechnical aspects of natural hazards
- Geotechnical problems of engineering structures

Each session is opened by a Keynote Lecture, delivered by an internationally recognised expert, followed by the presentation of selected contributions and an open discussion. A Special Lecture is dedicated to pyroclastic soils from the Campania region.

The technical visits to the island of Ischia and to the Phlegraean Fields, which for this kind of event are as important as the scientific sessions, will include several sites of both archaeological, natural and technical interest.

As the Chairman of the Organising Committee, I would like to thank all the members of the Organising and Scientific Committees, and particularly the Editors of the Proceedings.

Stefano Aversa

Volcanic Rocks and Soils – Rotonda et al. (eds)
© 2016 Taylor & Francis Group, London, ISBN 978-1-138-02886-9

Committees

ORGANISING COMMITTEE

Stefano Aversa	Italy
Manuela Cecconi	Italy
Settimio Ferlisi	Italy
Nicola Nocilla	Italy
Lucio Olivares	Italy
Tatiana Rotonda	Italy
Francesco Silvestri	Italy
Claudio Soccodato	Italy
Paolo Tommasi	Italy

INTERNATIONAL SCIENTIFIC COMMITTEE

Lucia Capra	Mexico
Carlos Dinis da Gama	Portugal
Derek Elsworth	USA
Aldo Evangelista	Italy
John Hudson	UK
Seiichi Miura	Japan
Charles W.W. Ng	Hong Kong
Claudio Olalla	Spain
Mick Pender	New Zealand
Luciano Picarelli	Italy
Tatiana Rotonda	Italy
Paolo Tommasi	Italy
Resat Ulusay	Turkey
Giulia Viggiani	Italy

Keynote Lectures

Volcanic Rocks and Soils – Rotonda et al. (eds)
© *2016 Taylor & Francis Group, London, ISBN 978-1-138-02886-9*

Sensitive pyroclastic-derived halloysitic soils in northern New Zealand: Interplay of microstructure, minerals, and geomechanics

Vicki G. Moon, David J. Lowe, Michael J. Cunningham, Justin B. Wyatt & Willem P. de Lange
School of Science, University of Waikato, Hamilton, New Zealand

G.J. (Jock) Churchman
School of Agriculture, Food, and Wine, University of Adelaide, South Australia, Australia

Tobias Mörz, Stefan Kreiter & Max O. Kluger
Marum – Center for Marine and Environmental Sciences, University of Bremen, Bremen, Germany

M. Ehsan Jorat
School of Civil Engineering and Geosciences, Newcastle University, Newcastle Upon Tyne, UK

ABSTRACT: Sensitive soils in the Bay of Plenty in North Island occur within weathered, rhyolitic pyroclastic and volcaniclastic deposits, with hydrated halloysite (not allophane) as the principal clay mineral. We evaluate the development of sensitivity and characteristic geomechnical behaviours for sequences of the silt-rich, halloysitic soils. Morphologically the halloysite comprises short tubes, spheroids, plates, and, uniquely, books. Key findings include (i) the varied morphologies of halloysite minerals within the microstructure create an open network with small pores and predominantly point contacts between clay particles; (ii) low plasticity, high natural water contents, low cohesion, low CPT tip resistance, and low permeability are attributable to the dominance of halloysite; (iii) boundary effects between pyroclastic units amplify Earth tide effects; and (iv) large spikes in pore water pressures follow rainfall events. The regular deposition since c. 0.93 Ma of siliceous pyroclastic deposits from ongoing explosive rhyolitic volcanism in TVZ, together with high natural water content and low permeability, have created a locally wet environment in the stratigraphic sequences that generates Si-enriched pore water from the weathering mainly of rhyolitic volcanic glass shards and plagioclase, providing conditions suitable for halloysite formation. Initial hydrolysis of glass shards also releases cations that promote cohesion between clay minerals. Eventual enleaching of these cations reduces cohesion between clay minerals, resulting in sensitive behavior.

1 INTRODUCTION

In the Bay of Plenty region of North Island, New Zealand (Figure 1), sensitive soil failures cause considerable infrastructure damage. A history of large landslides over the past 40 years includes:

- a large failure at Bramley Drive, Omokoroa in 1979, which led to the removal of five houses (Gulliver and Houghton, 1980);
- the collapse of the Ruahihi Canal in 1981 which resulted in more than 1 million m^3 of material being eroded and transported into nearby rivers and estuaries (Hatrick, 1982); and
- a series of landslides in various parts of Tauranga City and its surroundings in May 2005.

The Bramley Drive scarp, which had remained inactive for 30 years and had developed an extensive vegetation cover, was reactivated in 2011, and continued regression of the scarp face occurred throughout 2012.

Characteristically these landslides display a long runout distance of associated debris flows. This long runout is associated with, and evidence of, the sensitive nature of many materials in the Tauranga region (Keam, 2008; Wyatt, 2009; Arthurs, 2010; Cunningham, 2013; Jorat *et al.*, 2014a). Sensitivity is defined as a loss of strength upon remoulding, and is quantified as the ratio of undisturbed to remoulded undrained strength where both strengths are determined at the same moisture content. Values of <2 are insensitive, 4–8 are considered sensitive, 8–16 are extra sensitive, and >16 are referred to as "quick clays" (New Zealand Geotechnical Society, 2005).

Sensitive soil behaviour is classically described from glacial outwash deposits in Norway and Canada, where leaching of salts from an open, flocculated structure containing low-activity illitic clay minerals results in a loss of cohesion of the soils, which are then prone to failure in response to a weak trigger (Lundström *et al.*, 2009). Characteristic failures are

3

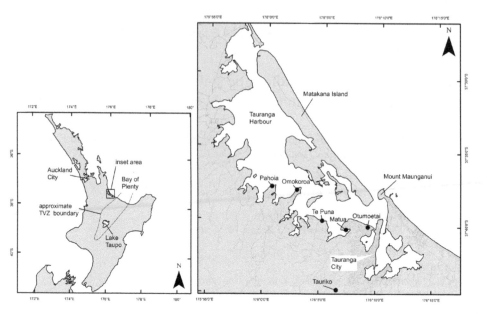

Figure 1. Location map of Bay of Plenty in North Island, New Zealand, showing sampling sites. The Bramley Drive landslide referred to in the text is located at the sampling position shown for Omokoroa. The main source volcanoes for the sequences of pyroclastic deposits are in the central Taupo Volcanic Zone (TVZ).

spreads and flows as large quantities of water are released upon remoulding, generating very fluid debris flows. In the Bay of Plenty, however, we do not have glacial clays, and the sensitivity is developed in a sequence of rhyolitic (silica-rich) pyroclastic materials that range from non- to strongly weathered. The term "pyroclastic" encompasses all the clastic or fragmental materials explosively erupted from a volcanic vent (Lowe, 2011, Lowe and Alloway, 2015). The sequence is complex, and includes primary pyroclastic fall deposits (tephra-fallout), pyroclastic density-current deposits – which are emplaced by gravity-controlled, laterally moving mixtures of pyroclasts and gas that include pyroclastic flows, which have high particle concentrations (generating ignimbrites), and pyroclastic surges, which have low particle concentrations – and a wide variety of reworked pyroclastic materials including slope wash, fluvial, and aeolian variants (Briggs *et al.*, 1996, 2005, 2006). Clearly, sensitivity in these materials is developed through a different mechanism than is the case for the Northern Hemisphere glaciogenic examples. In this paper we present mineralogical, microstructural, and geomechanical data from six sites in the Bay of Plenty, and examine the influence of microstructure on the behavior of the pyroclastic-derived soil materials on slopes.

2 DEVELOPMENT OF UNDERSTANDING OF SENSITIVITY IN NEW ZEALAND PYROCLASTIC SOILS

2.1 Sensitive New Zealand soils

Early work on sensitive soils in New Zealand reported sensitivities up to 140 in rhyolitic deposits at Bramley

Drive, Omokoroa, and attributed the high sensitivity to hydrated halloysite clays (Smalley *et al.*, 1980). The materials investigated had high clay contents compared with those of sensitive soils of the Northern Hemisphere, and high porosity and high natural water contents (many above the liquid limit) were measured. In the samples investigated, the halloysite was seen to have a spherical morphology and the authors inferred that this gave minimal interparticle interactions, particularly long-range interactions.

Jacquet (1990) undertook a systematic study of sensitivity in a range of tephra-derived soils in New Zealand, including andesitic tephra-fall beds containing a high proportion of allophane, along with a few samples containing mainly halloysite. Sensitivities ranged from 5–55, yet it was noted that the sensitivity of many of these materials was associated with a high undisturbed strength, rather than a particularly low remoulded strength. This high undisturbed strength was attributed to a relatively low moisture content (all soils were below the liquid limit). Sensitivity could not be seen to readily relate to mineralogical compositions or bulk properties of the materials. Jacquet (1990) observed fibrous webs of imogolite linking allophane particles, and attributed irreversible breakdown of electrostatic bonds between such clay minerals as explaining sensitivity in these tephra-derived soils. However, he noted that the halloysite mineral aggregates were larger than the allophane aggregations and hence had fewer contacts, making the halloysite aggregates less stable than those in the allophane-dominated soils.

Torrance (1992), in a discussion of Jacquet's (1990) paper, noted that for designation as a "quick clay", remoulded materials must behave as a liquid (shear

4

strength <0.5 kPa). Thus materials of high peak strength and relatively high remoulded strength, such as the sensitive soils studied in New Zealand, are not considered to be truly quick. Torrance (1992) also suggested that softening on remoulding is due to the release of free water from large pores, along with the breakdown of structural units (macrofabric or pedality) in the soil. A conceptual model for sensitivity development in tephra-derived soils was attempted: the model suggested that a tephra-fall origin produced a high void ratio deposit as a critical parameter, along with an allophanic mineralogy.

This early work recognized the importance of high porosity and high natural water contents in leading to low remoulded strength of the landslide debris. All authors recognized an association between halloysite or allophane, or both, in the soils and their concomitant sensitivity. However, following the Torrance (1992) paper, allophane appears to have taken the leading role as the "culprit" material for generating sensitivity, and this has been a pervasive view in New Zealand since this time, especially amongst the engineering community and also in the soil sciences (e.g., Allbrook, 1985; Lowe and Palmer, 2005; Neall, 2006; McDaniel et al., 2012). This perception persists despite Smalley et al. (1980) determining much greater sensitivities in halloysite-dominated soils. Torrance (1992) attributed the importance of allophane to its "non-crystalline" character, with allophane seen as an "amorphous" material that forms earlier in the weathering sequence than crystalline halloysite. However, these features and origins ascribed to allophane no longer pertain, as discussed below.

2.2 *Allophane and halloysite clay minerals*

Allophane and halloysite are both common aluminosilicate secondary minerals formed in soils developed mainly from unconsolidated pyroclastic (tephra) deposits (Churchman and Lowe, 2012; McDaniel et al., 2012).

Although previously described (erroneously) as amorphous because of the broad, low-intensity humps it generates on X-ray diffraction (XRD) traces (Lowe, 1995; Churchman and Lowe, 2012), allophane is now recognised as being "nanocrystalline", meaning that it has a structure (or short-range order) at the nanometer scale, i.e., in the 1–100 nm range (Theng and Yuan, 2008; Churchman and Lowe, 2012). Unit particles of allophane comprise tiny hollow spherules or "nanoballs" $\sim$3.5 to 5.0 nm in diameter with a chemical composition $(1-2)SiO_2 \cdot Al_2O_3 \cdot (2-3)H_2O$ (Abidin et al., 2007). The most-common, Al-rich, form is also referred to as proto-imogolite allophane (Al: Si $\sim$2: 1) (Parfitt, 1990, 2009).

Halloysite is a 1:1 kaolin-group clay mineral with a similar composition to kaolinite, but with interlayer water that can be driven off, giving hydrated and dehydrated forms or end members. The hydrated form has a 1.0 nm (10 Å) d-spacing, and the dehydrated form has a d-spacing of 0.7 nm (7 Å). Halloysite can adopt a continuous series of hydration states, from 2 to 0 molecules of H_2O per $Si_2Al_2O_5(OH)_4$ aluminosilicate layer, and these are interpreted as a type of interstratification of the two end member types (Churchman, 2015). This dehydration, and the associated d-spacing change, is one of the key characteristics distinguishing halloysites from kaolinite (Churchman and Lowe, 2012). Under normal environmental conditions, dehydration of halloysite micelles is an irreversible process (Joussein et al., 2005; Keeling, 2015).

Unlike allophane, halloysite has long-range order and is readily identifiable using XRD (Churchman et al., 1984; Joussein et al., 2005; Churchman and Lowe, 2012). Halloysite is known to occur in around 10 morphologies (Joussein et al., 2005), summarized into four main types by Churchman (2015): (i) tubular (perhaps most common), (ii) platy, (iii) spheroidal, and (iv) prismatic. An additional book-like morphology was recognized by Wyatt et al. (2010), as discussed further below.

Until the 1980s, halloysite was commonly considered to be a later stage of weathering of volcanic glass than allophane, with a weathering sequence (originally proposed by Fieldes, 1955) of glass → allophane → halloysite → kaolinite being envisaged (Kirkman, 1981; Wesley, 2010). The postulated mineralogical change from allophane to halloysite was considered to occur by a solid-phase transformation involving dehydration and "crystallization" from an "amorphous" material (Churchman and Lowe, 2012). An alternative view, now well established, envisages both allophane and halloysite being formed directly from the products of the dissolution mainly of volcanic glass (and feldspar) under different environmental conditions, with the concentration of Si in soil solution and the availability of Al being the main controlling factors (Joussein et al., 2005; Churchman and Lowe, 2012). Originally formulated by Parfitt et al. (1983, 1984) and supported by Lowe (1986, 1995) and others, this Si-leaching model shows that where Si is high in the soil solution (> $\sim$10 ppm), halloysite forms preferentially, whereas allophane is favoured where soil solutions are relatively low in Si (< $\sim$10 ppm) (Singleton et al., 1989; Churchman and Lowe, 2012). Thus halloysite forms preferentially on siliceous materials, in areas with relatively low precipitation, where drainage is impeded allowing buildup of Si in solution, or at depth in sequences where Si can accumulate by transfer from overlying siliceous deposits. Conversely, allophane forms on less siliceous (more Al-rich) materials, or in areas with high rates of leaching where precipitation is relatively high and the materials are freely drained, hence resulting in the loss of Si from uppermost soil materials (desilication) (Churchman et al., 2010; Churchman and Lowe, 2012).

For allophane to alter to halloysite (as proposed previously by Fieldes, 1955) would require a complete re-arrangement of the atomic structures and this could only occur by dissolution and re-precipitation processes because the allophane would need to "turn inside out" so that Si-tetrahedra are on the outside, not inside, of the curved Al-octahedral sheets (Parfitt, 1990; Lowe, 1995; Hiradate and Wada, 2005). The effect of time is clearly subordinate because glass can weather directly via dissolution either to allophane or halloysite depending on glass composition and both macro- and micro-environmental conditions, not time (Lowe, 1986). Allophane under certain conditions remains stable, being recorded in deposits c. 0.34 Ma and older (Stevens and Vucetich, 1985; Churchman and Lowe, 2012).

2.3 Geomechanical properties

Key publications regarding the geomechanics of allophanic and halloysitic soils are those of Wesley (1973, 1977, 2001, 2009, 2010). In a study of soils developed on andesitic tephras in Indonesia, Wesley (1973) noted Atterberg limits which consistently plotted below the A-line (high compressibility "silts"); soils dominated by allophane showed a very wide range of liquid limits (80–250), whereas halloysite-dominated soils showed a smaller range (60–120). Plasticity indices for both clay minerals (18–80) indicated low activity materials, and both the allophanic and halloysitic soils showed high moisture retention with saturation levels remaining close to 100% with little variation throughout the year. There was no indication of swelling, but shrinkage cracks occurred in the halloysite-dominated soils in the dry season. For the same soils, relatively high cohesion and friction angle values were obtained ($\phi' = 31$–$40°$, $c' = 13$–23 kPa), with allophanic soils having slightly higher shear strength values than the halloysite-dominated soils (Wesley, 1977). Permeabilities of approximately 10^{-7} to 10^{-8} m s^{-1} were reported (Wesley, 1977). These soils were not sensitive (Wesley, 1973, 1977).

In later publications, Wesley (2009, 2010) largely concentrated on allophanic soils derived from tephras of intermediate (andesitic) composition. He reiterated the very high natural water contents, liquid and plastic limits, and comparatively high effective strength parameters of allophanic clays, and also noted irreversible drying as a key characteristic of soils containing allophane. However, Wesley (2007) also reported on findings from a very limited study of a landslide profile in Tauranga for which the parent material is rhyolitic pyroclastic materials. He assumed the materials to be allophanic on the basis of observations of field characteristics (no mineralogical analyses were presented) and noted high to extremely high sensitivity along with extreme variation in geomechanical characteristics between individual layers within the landslide sequence.

2.4 Microstructure

Recent student theses have considered the geomechanics and microstructure of sensitive soils in rhyolitic materials from New Zealand.

Keam (2008) studied mass movement processes particularly on the Omokoroa Peninsula in Tauranga Harbour. He identified a sensitive silt layer at the base of slope failures and undertook microstructural analysis of this layer which he described as having an "… open network of predominantly silt grains that are very weakly connected due to the absence of framework connectors", together with "… an abundance of pore space." The silt was seen to have a very high porosity but low inferred permeability, together with high natural moisture content that often exceeded the liquid limit. He also reported low cohesion, relatively high friction angle, and "significantly lowered" residual values. Keam (2008) described the microstructure as skeletal and mixed skeletal-matrix structures comprised predominantly of crystalline silt grains in granular particle matrices, with few clays. Abundant pore spaces were observed, ranging from ultrapores to macropores with occasional fissures. An open structure with considerable pore space was interpreted as collapsible, leading to sensitivity (Keam, 2008).

Arthurs (2010) studied a range of sensitive pyroclastic materials from around the North Island of New Zealand and concluded that an originally low-density deposit derived from vesicular glassy material, weathering to low activity secondary minerals, a "delicate" microstructure with a generally skeletal to matrix-skeletal arrangement of grains and aggregates, and high natural water content, all predisposed a material to sensitive behavior. Arthurs (2010) also suggested that syn-eruptive reworking may be a significant contributor to ultimate sensitive behavior, and that such reworking accounts for the considerable spatial variability evident in the pyroclastic deposits compared with glacial-outwash-derived sensitive soils of the Northern Hemisphere.

3 SAMPLING AND METHODOLOGY

The stratigraphic sequences in the Tauranga area of the Bay of Plenty are complex, but an overall stratigraphy recognizes several broad units that are well exposed in the present scarp of the Bramley Drive failure at Omokoroa where the sequence is very thick (Figure 2). Most of the deposits derive from caldera volcanoes in the Taupo Volcanic Zone (see Briggs et al., 2005; Wilson et al., 2009) southeast of Tauranga (Figure 1). At the top are Holocene and Pleistocene tephras representing the most recent eruptives and modern pedological soil horizons; the base of this unit comprises the Rotoehu Ash deposited c. 50,000 years ago (Briggs et al., 1996; Danisik et al., 2012). The Rotoehu Ash lies on a very distinctive dark reddish-brown clay-rich paleosol formed on the Hamilton Ash beds. (Paleosols are defined here as pedogenic soils on a landscape, or of an environment, of the past.) These beds are composed of a series of weathered tephra deposits with intercalated paleosols ranging in age from c. 0.08 to 0.34 Ma (Lowe et al., 2001). At Bramley Drive the Hamilton Ash beds reach a total thickness of ~9 m; this thickness is variable around the region. The Hamilton Ash beds lie on top of another very well developed, dark brown clay-rich paleosol which marks the top of the so-called Pahoia Tephra sequence – a poorly defined composite sequence of primary pyroclastic and reworked rhyolitic volcaniclastic materials ranging in age from approximately 0.34 to 2.18 Ma (Briggs *et al.*, 1996). The Pahoia Tephras are part of the Matua Subgroup, a widespread, complex unit that occurs throughout the Bay of Plenty (Pullar *et al.*, 1973; Briggs *et al.*, 1996, 2006), and which includes pyroclastic deposits of both fall and flow origin, lacustrine, estuarine, and (rare) aeolian sedimentary deposits, lignites, and fluvially reworked volcanogenic materials. At the Bramley Drive site, the Pahoia Tephras include at least six units that attain a combined thickness of > 12 m. These units are underlain by a weakly-welded ignimbrite provisionally identified as Te Puna Ignimbrite (0.93 Ma) (Gulliver and Houghton, 1980; Briggs *et al.*, 1996, 2005). In turn, the ignimbrite is underlain by a lignite deposit at shore platform level.

It is the units making up the Pahoia Tephras that are associated with the sensitive soil failures observed in the Tauranga area. This sequence is in places very thick, but thinner in others, and it is difficult to correlate single layers across any significant distance. It is likely that many of the units are formed by local reworking of primary pyroclastic material, and hence may have limited lateral extents. However, a consistent pattern can be seen at Omokoroa in that immediately above the Te Puna Ignimbrite is a lower sequence of tephras and intermixed reworked materials, typically very pale coloured and likely near permanent saturation, and often containing dispersed blue-black MnO$_2$ (pyrolusite) concretions or redox segregations (~5 % abundance) indicative of occasional drying out (Wyatt *et al.*, 2010). This lower, pale sequence locally reaches up to 8.5 m in thickness at Bramley Drive, and is covered by an upper sequence of brown-coloured (weathered) tephra deposits making up the remainder of the Pahoia Tephras. It is the pale, partially reworked lower sequence of pyroclastic and volcaniclastic materials, which is at or near saturation much of the time, that displays high sensitivity.

In this paper, materials from individual units within the sensitive parts of the lower Pahoia Tephras at Omokoroa Peninsula, and from correlatives at five further sites at Pahoia Peninsula, Te Puna, Otumoetai, Tauriko, and Matua (Figure 1), were sampled and tested. Further details of site stratigraphy and sample locations can be found in Wyatt (2009) and Cunningham (2013).

Sampling was undertaken based on recognition of sensitive layers in the field from the use of vane shear tests following standard methods (New Zealand Geotechnical Society, 2001). Mineralogy was determined using (i) XRD analysis of both bulk samples and of clay separates treated systematically by heating (to 110° and 550°C) and by adding formamide, using both glass-slide and ceramic-tile mounts, (ii) by analysis of acid oxalate extractions (AOE), and (iii) by scanning

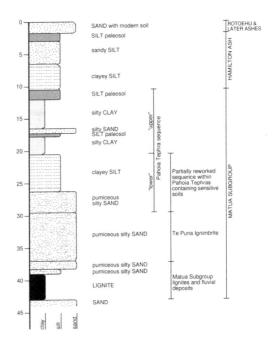

Figure 2. Simplified stratigraphy exposed in the landslide scarp at Bramley Drive, Omokoroa. The Pahoia Tephra sequence has been informally split into "upper" and "lower" portions for this paper. Te Puna Ignimbrite is c. 0.93 Ma and basal Hamilton Ash is c. 0.34 Ma in age. Ma, millions of years ago.

electron microscopy (SEM) equipped with energy-dispersive X-ray (EDX) capability (Wyatt *et al.*, 2010). Microstructures were examined on oven-dried samples using broken surfaces, powdered samples, and remoulded materials. Natural water content, dry bulk density, and Atterberg limits were determined following ISO standards (ISO/TS 17892-1:2004(E), ISO/TS 17892-2:2004(E), ISO/TS 17892-12:2004(E)), except that specimens were not allowed to dry when preparing for Atterberg limits tests (following Wesley, 1973); and particle density was determined using density bottles following the method outlined by Head (1992). Effective cohesion and friction angle from consolidated, undrained triaxial testing were measured following standard BS 1377-8:1990. Coefficients of consolidation (c_{vi}) and volume compressibility (m_{vi}) were determined for each applied loading in consolidation stages of the triaxial testing (BSI, 1999), and the coefficient of permeability was estimated using the method described by Head (1986). For one sample from Matua, the permeability was measured directly on a triaxial sample using two volume change devices.

A cone penetrometer test (CPTu) was undertaken at a site immediately behind the scarp of the Bramley Drive landslide at Omokoroa in February 2012 (Jorat *et al.*, 2014b). The instrument used (GOST) is an offshore CPT instrument developed at Bremen University (MARUM – Center for Marine Environmental Sciences) in Germany. GOST incorporates a small

Table 1. Measured field strength and bulk characteristics for sensitive materials from the Tauranga region.

Location	Peak vane strength (kN m^{-2})	Remoulded vane strength (kN m^{-2})	Sensitivity	Dry bulk density (kg m^{-3})	Particle density (kg m^{-3})	Porosity (%)	NMC[1] (%)	Saturation (%)
Omokoroa	83 ± 9	7 ± 2	13 ± 2	807 ± 28	2415 ± 5	67 ± 1	87 ± 3	105 ± 3
Te Puna	122	12	11	688	2220	69	109	109
Pahoia	68 ± 1	7 ± 1	10 ± 1	964 ± 12	2500 ± 110	61 ± 1	69 ± 1	109 ± 3
Tauriko	85 ± 33	14 ± 10	11 ± 5	740 ± 120	2560 ± 18	71 ± 5	96 ± 16	95 ± 6
Otumoetai	131 ± 34	17 ± 7	10 ± 2	803 ± 63	2663 ± 10	70 ± 2	81 ± 9	91 ± 1
Matua	60	6	10	1030	2780	63	64	105
range	45 – >227	2–36	5–20	589–975	2220–2686	60–77	64–115	83–111
average	110.3	12.5	10.8	804.3	2530.7	68.2	86.1	98.8
standard error	14.6	3.2	1.2	39.6	41.3	1.6	5.5	2.6

[1]NMC = Natural moisture content

(5 cm^2) piezocone, and thus gives high-resolution traces. GOST also has the capacity to undertake vibratory CPTu. At the time of this testing the vibratory capacity was still under development and exact control on the vibration characteristics had not been obtained: frequencies of approximately 15 Hz with vertical vibrations of a few millimetres amplitude were applied. Two separate CPTu runs were undertaken: a static run at 2 cm s^{-1} penetration speed, and a second vibratory run approximately 1 m away with the oscillation imposed on the same penetration rate. Jorat et al. (2014c) described the instrument design and modes of deployment.

A Digitilt borehole inclinometer from Slope Indicator™ has been used to obtain deformation measurements at Bramley Drive since June 2013. The inclinometer casing is located ~5 m behind the central part of the main landslide scarp and extends to a depth of 43 m. The A-axis is aligned at 320° T, parallel with the axis of the most recent movements of the landslide (2011–2012 regressions). Thus the A-axis is measuring predominantly a N-S component of any movement, and the B-axis is measuring predominantly E-W movement. Measurements were taken from 41.5 m to 0.5 m depth at 0.5 m intervals. Two runs were undertaken at each measurement time with the instrument turned through 180° between readings in order to cancel any instrument bias errors, and cumulative plots were derived from the difference between measured values for each point and those obtained from the first use of the instrument in June 2013.

Three pore pressure transducers at depths of 12 m, 21 m, and 27.5 m have been logging continuously since May 2013.

4 GRAIN SIZE AND MINERALOGY

Texturally, the samples are identified in the field as silts or clayey silts. This is supported by laboratory textural analysis which shows median values of clay: silt: sand of 6.5:71:22.5%. A dominance of silt is in keeping with a tephra-fall origin, but does not preclude reworking of initial tephra deposits as suggested by Arthurs (2010).

Clay contents are generally lower than those recorded for sensitive soils in the Northern Hemisphere, with similar silt contents. For example, Eilertsen et al. (2008) recorded ranges of 12–44% clay and 37–71% silt in Norwegian sensitive soil deposits, and Geertsma and Torrance (2005) noted average clay contents of 41.5% and silt of 58% in Canadian soils.

Mineralogically the samples are dominated by glass or alteration products of volcanic glass together with plagioclase, quartz, and subordinate mafic minerals comprising ferromagnesian minerals and Fe-Ti oxides. The clay mineral assemblages are in all cases dominated by hydrated halloysite, with minor kaolinite identified in just two samples from Otumoetai. The AOE analyses indicated that nanominerals (allophane, ferrihydrite) were absent (or negligible) from all samples. Sensitivity values (Table 1) show generally "sensitive" to "extra sensitive" materials (New Zealand Geotechnical Society, 2005). These values are comparable with others measured on pyroclastic soils in New Zealand, both in Tauranga (Smalley et al., 1980; Wesley, 2007; Keam, 2008; Arthurs, 2010) and more widely (Jacquet, 1990).

5 MICROSTRUCTURE

5.1 Components

The principal granular components of the materials are volcanic glass shards (Figure 3A–C). These show characteristic morphologies with sharp (angular) edges and bubble-rim textures representing fragmentation from a vesiculating magma. Shards are mostly crisp and clean (Figure 3A, B), although occasional shards show extensive surface degradation (Figure 3C). Rare feldspar and quartz crystals are seen (Figure 3D, E); these commonly show surface degradation representing weathering or damage during transport and deposition. One sample from Omokoroa included sparse diatom frustules indicating redeposition in a lacustrine or shallow marine environment (Figure 3F).

As noted earlier, halloysite is most commonly manifested as a tubular mineral (e.g., Churchman, 2015;

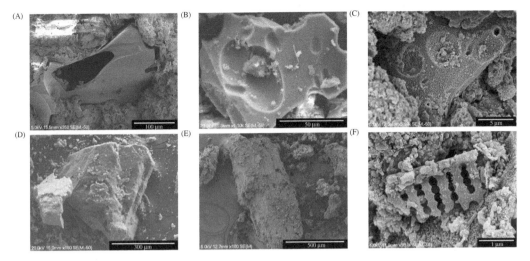

Figure 3. (A) Glass fragment from Tauriko showing vesicular texture and angular edges. (B) Clean glass fragment showing bubble-wall texture and minor adhering clay. (C) Glass fragment with altered surface. (D) Plagioclase feldspar crystal showing damage from transport and weathering. (E) Plagioclase feldspar crystal with ragged edges from weathering; adhering clay minerals are apparent. (F) Part of diatom frustule in Omokoroa sample. Photos: H. Turner.

Keeling, 2015), and typical "spiky" tubular halloysite is identified in some specimens (Figure 4A). However, a range of other morphologies is more common in the halloysites identified here. Most frequent are short, stubby tubes (Figure 4B) which range from ~0.3 and ~1 μm in length. These short tubes invariably show surface cracking which probably represent dessication cracks from sample preparation. Spheres or spheroids are another common form, as described earlier by Smalley *et al.* (1980) for the Omokoroa site. Spheroids (Figure 4C, D) are small (~0.1 to ~0.7 μm in diameter) (Wyatt *et al.*, 2010). Platy forms, typically with hexagonal plates, are also seen (Figure 4E, F). Most commonly the plates are small (< ~0.5 μm), but on occasion reach large diameters (>5 μm) Most interesting, however, is the tendency for the plates to coalesce or stack together into halloysite "books" identical to the characteristic morphology of kaolinite (Figure 4G). Such books comprised up to ~30% of bulk samples and ~10% of the clay fraction of samples examined via SEM. A variety of plate shapes making up the books occurred: irregular, quasi-hexagonal, elongated, and twisted-vermiform (Figure 4H). Most were curved. Plate widths ranged from ~1 to ~20 μm. Halloysite books of silt to sand size have not been previously reported (other than in the preliminary report by Wyatt *et al.*, 2010).

5.2 *Interactions between components*

Direct interaction between granular components is rarely seen, with virtually all contacts being mediated by clay minerals (Figure 5A, B). Where grain-to-grain contact is apparent, in all cases it exists across fractures in glass or crystals (Figure 5C), which may be due to specimen preparation or crystal fracture during weathering (or both). Clay mineral aggregates in

most cases drape loosely against the margins of granular components, with little evidence of bonding seen (Figure 5D), suggesting that the granular materials are weakly bound into the clay matrix.

Interactions between clay minerals vary depending on the dominant morphology of the halloysite minerals. Long tube morphologies have only been observed in a chaotic arrangement (Figure 4A) with largely point contacts and few edge-to-edge contacts. Equally, stubby tubes tend to have dominantly point contacts (Figure 4B), although their small size allows a closer arrangement than that seen in the longer tubes. Spheres/spheroids also contact at points (Figure 4D); their common association with stubby tubes gives a structure with considerable open space between individual clay crystals or aggregates (Figure 4B).

Platy clay crystals show the widest variety of interactions. Edge-to-edge or edge-to-face arrangements are common amongst individual plates, giving a typical "card-house" structure often associated with flocculation of clay minerals (Figure 5E). Face-to-face arrangements can form between a few individual crystals, but most commonly occur in association with the book morphology observed in these samples.

Interactions between different morphology clay minerals can vary from simple point contacts (Figure 5E), which are most commonly observed as a result of the non-conformable shape of the different clay components. In some instances closer contact between clay minerals can be seen, most obviously evidenced by halloysite spheroids or stubby tubes adhering to the margins of large books (Figure 5F).

5.3 *Structure*

Microstructure varies depending on the abundance of granular components in the material. Where there are

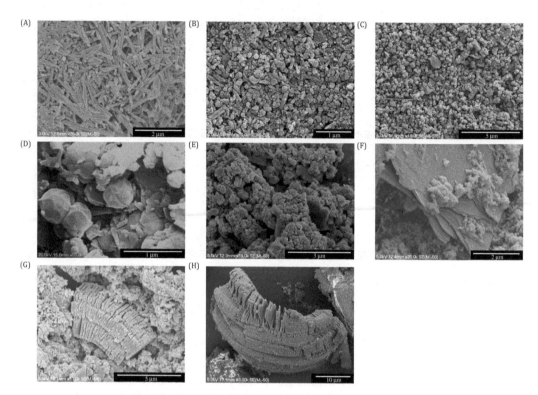

Figure 4. Halloysite clay morphologies observed in sensitive soil specimens in Tauranga. (A) Characteristic halloysite tubes forming spicules with considerable pore space incorporated within the chaotic arrangement of spicules. (B) Common stubby tubes showing surface desiccation cracks. (C) & (D) Spheres or spheroids. (E) Small plates. (F) Occasional large plates occur. (G) & (H) Plates coalesce to a "book" morphology usually considered typical of kaolinite (e.g., Dixon, 1989) but characterized here as comprising wholly hydrated (1.0 nm) halloysite. Photos: H. Turner.

few larger grains (Figure 6A), the clays form a matrix microstructure where any larger components "float" within a loosely-packed clay matrix. Arrangement of the matrix clays appears random, with a mixture of sizes and shapes forming a fragmentary matrix. As the proportion of granular materials increases (Figure 6B), the clays still maintain a matrix microstructure, but domains of clay aggregates begin to become evident. Where granular materials are abundant (Figure 6C), the interactions between individual grains are still mediated by clays, but in this case it is largely clay aggregates separating the grains. Whilst still a matrix microstructure in that the clay matrix dominates the way the different components can interact, this pattern is referred to as matrix-skeletal, indicating that there is a "skeleton" of point contacts between grains and clay aggregates.

5.4 Pore space

Pore space is ubiquitous in these samples. Pores occur in spaces created by loosely-packed clay minerals with only point contacts (Figure 4A). Pore spaces also occur within and around small clay aggregates (Figure 6D), and along fissures within the clay matrix (Figure 6E) or separating the clay matrix and granular components.

An overall image of a surface of these materials shows extensive pore space throughout (Figure 6F).

Interaction of clays in this way results in a highly porous structure (porosity 62–77%), but the pore space is hugely dominated by ultrapores (<0.1 μm) and micropores (< 0.5 μm) with dominant pore sizes being <1 μm. Whilst these micropores impart high porosity, we infer that they are poor at transmitting water.

6   GEOMECHANICS

6.1   Bulk material properties

Dry bulk density of all measured samples is typically very low (Table 1), and is associated with high porosity and void ratio values (Moon et al., 2013). Natural moisture content is high, meaning that the soils are characteristically at or close to saturation in their normal field conditions (Table 1), an observation supported by their pale, low chroma, colours indicative of reducing conditions and the presence of ~5% redox segregations of MnO_2 (Churchman and Lowe, 2012).

Atterberg limits (Table 2) are high. All samples plot below, but parallel to, the A-line; they are classed as high compressibility silts (MH). Effective strength parameters (c′, φ′) show averages of c′ = 15 kN m^{-2}

(A) (B) (C)

(D) (E) (F)

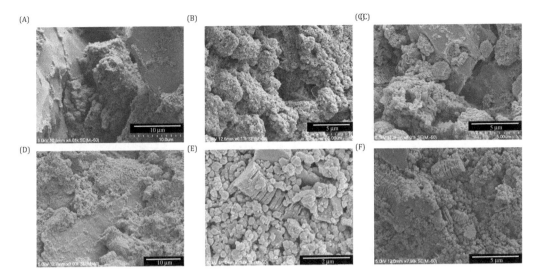

Figure 5. Component interactions. (A) Accumulated clay minerals forming bridge between two glass shards. (B) Clay minerals coat surfaces of coarser grains, meaning that most interactions are mediated by clays. (C) & (D) Glass shards and silt grains appear loosely wrapped in clay mineral aggregates. (E) Clay mineral grains generally only interact at point contacts. (F) With large book morphologies, halloysite spheres/spheroids or tubes are often seen adhering to the book surfaces. Photos: H. Turner.

and $\phi' = 33°$, indicating relatively low cohesion yet a high friction angle.

Permeability is low: estimated coefficients of permeability in the range 10^{-7} to 10^{-9} ms^{-1}, with the best measurements (those derived using two volume change devices) giving a very low average permeability for silty materials of 4×10^{-9} ms^{-1}.

6.2 Cone penetrometer

Traces for tip resistance and pore water pressure derived from the CPTu test are shown in Figure 7; only static tip resistance is shown, while both static and vibratory traces are included for pore water pressure.

In the static CPTu profile, the tip resistance responds to the effects of soil formation with increased tip resistance, most notably near the present ground surface and at each of the identified paleosols. In the intervening layers between the paleosols, the tip resistance of the materials is characteristically very low (<1.5 MPa). Just below the sequence exposed in the landslide scarp (about 30 m below ground surface), the tip resistance increases, and stays high but variable until the maximum depth of 38 m; this zone of increased tip resistance corresponds with the weakly-welded Te Puna Ignimbrite underlying the Pahoia Tephra sequence.

The excess pore water pressure trace indicates that the water table depth at the time was approximately 1.5 m. Below this the trace shows a steady rise in induced water pressures to a depth of 24 m. The induced pore water pressure then falls sharply at this point, corresponding with a spike in the tip resistance, indicating a thin coarser layer with increased permeability. A second pore water pressure peak

occurs at approximately 26 m, after which the induced pore water pressures fall in the ignimbrite, though still remain above hydrostatic. A zone from approximately 17–26 m depth shows particularly elevated induced pore water pressures, indicating low permeability through this sequence of materials. This depth range lies immediately above Te Puna Ignimbrite and coincides with the pale, partially reworked lower sequence of pyroclastic and volcanogenic materials within the Pahoia Tephras. It is interpreted to coincide with the assumed position of the initial failure zone for the 1979 failure at Bramley Drive (Gulliver and Houghton, 1980) which was inferred to be at the contact of the Te Puna Ignimbrite and the overlying materials.

The induced pore water pressure under vibratory CPTu shows shows particularly elevated pore water pressures in response to vibration developed across the entire Pahoia sequence, but most notably in the partially reworked lower Pahoia Tephras between 17 and 26 m. Through this zone the induced pore water pressures are up to three times greater than those developed in the static run. Pore water pressures developed in Hamilton Ash beds and recent tephras are very slightly elevated above those in the static run, whilst pressures in the Te Puna Ignimbrite are equivalent to those in the static run.

Despite the marked increase in pore water pressure during vibratory testing, only small changes in tip resistance are observed. Sasaki et al. in 1984 (Jorat et al., 2015) defined reduction ratio as:

$$RR = 1 - q_{cv}/q_{cs}$$

where: RR = reduction ratio; q_{cv} = vibratory tip resistance; q_{cs} = static tip resistance.

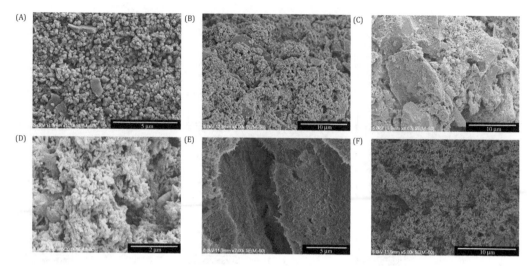

Figure 6. Microstructure and pore space. (A) With few silt-sized grains, the microstructure is matrix dominated with occasional grains separated by loosely-packed matrix clays. (B) As granular components increase, the microstructure remains matrix dominated but matrix clays form small aggregates surrounding silt-sized grains. (C) As granular components become significant, a skeletal-matrix microstructure is apparent. (D) Pore space most commonly occurs within matrix clays, both between individual clay crystals (Figure 4A) and amongst small clay aggregates. (E) The largest pores occur along fractures which generally separate grains and matrix materials. (F) An overall image shows extensive pore spaces. Photos: H. Turner.

Table 2. Measured geomechanical data for sensitive materials from the Tauranga region.

Location	Plastic limit	Liquid limit	Plasticity index	Liquidity index	Effective cohesion $(kN\ m^{-2})$	Effective friction angle $(°)$	Permeability $(*10^{-9}\ m\ s^{-1})$
Omokoroa	49 ± 3	73 ± 1	25 ± 3	1.6 ± 0.1	12 ± 4	33 ± 5	
Te Puna	46	89	44	1.5	16	41	
Pahoia	35 ± 1	54 ± 1	19 ± 1	1.9 ± 0.1	13 ± 3	34 ± 2	
Tauriko	48 ± 5	69 ± 9	21 ± 4	2.2 ± 0.2	18 ± 5	29 ± 2	3 ± 2[1]
Otumoetai	43 ± 5	79 ± 9	37 ± 4	1.1 ± 0.3	15 ± 7	32 ± 3	111 ± 30[1]
Matua	36	52	15	1.8			4.0 ± 0.3[2]
Range	32–57	52–96	13–44	0.3–2.4	5–35	26–41	0.5–160
Average	43.8	72.0	28.3	1.6	14.7	32.7	49.6
Standard error	2.3	4.4	3.0	0.2	2.4	1.5	16.3

[1] = permeability determined from consolidation; [2] = permeability determined using method of two volume change devices in the triaxial tester.

Reduction ratio values greater than 0.8 indicate a high liquefaction potential (Jorat et al., 2015). In the profile measured here, the reduction ratios remain less than 0.2, indicating little further loss of strength on vibration.

6.3 Borehole inclinometer

Borehole inclinometer results indicate that to date there is no clear shear surface developing in the inclinometer profile. However, wide fluctuations exist with apparent cumulative displacements at the top of the profile of up to 4 mm in the B axis and 1 mm in the A axis which initially appear to vary randomly in direction at different measuring times (Moon et al., 2015). While the small A axis displacements may be within the error of the instrument, the surprisingly larger displacements in the B axis (west–east direction) are beyond the estimates of instrumental error. When hourly records are measured for a single day the same magnitude fluctuations are recorded, but in this case a pattern can be seen in the direction of movement (Figure 8A), with the A axis of the casing swinging towards the positive direction in the early readings, then swinging back to negative values later in the day.

We infer the observed fluctuations to be the result of the solid Earth tides. Distortion of the Earth's mantle due to the gravitational attraction of the moon (primarily) and sun causes tidal effects that can be seen in displacement of the crust. Due to the Earth's rotation, there is a predominantly west–east variation, but other components are included in the total motion, giving an elliptical path for any point on or near the Earth's surface. At the latitude of Bramley Drive the solid

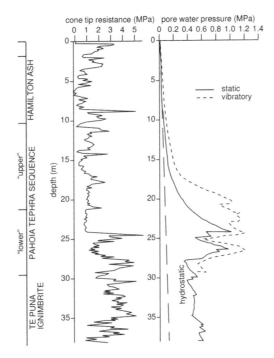

Figure 7. CPTu traces for static come tip resistance and induced pore water pressure in both static and vibratory modes at Bramley Drive. After Moon *et al.* (2015).

Earth tides cause a semi-diurnal rise and fall of the ground surface of up to approximately 16 cm. Well-established theoretical predictions of the solid Earth tides exist. To account for the influence of these tides on the cumulative profiles from the borehole inclinometer data we have fitted a linear regression line through the measured profile. The slope of this regression is our estimate of the displacement associated with the Earth tide. This measured slope is plotted against the predicted strain of the Earth tides (both normalized) in Figure 9. At the early stages of our measuring sequence (up to May 2014), the measured slope and predicted strain show remarkable phase agreement; this concordance seems to be confirmation that we are measuring Earth tide effects. It is notable however, that the measured variation is one to two orders of magnitude greater than predicted by the theoretical solution. Indeed, at the level of the predicted strain we would be unable to resolve the displacement at all with the simple inclinometer that we are using. Notably, the phase relationship largely disappears from the data during the austral winter of 2014.

By averaging the cumulative profiles seasonally (Figures 8B and C), the effects of Earth tides can largely be removed assuming that the offset caused by the Earth tides is randomly sampled by differing measuring times and days. The averages presented here are based on austral seasons: in general, winter is June to August; spring is September to November; summer is December to February; and autumn is March to May.

In the A axis (Figure 8B) there appears to be little obvious pattern until the end of spring 2014 when general movement in the positive (towards open face) direction is seen. However, the total displacement in this axis is very small and it is difficult to conclude that these measurements are beyond the error inherent in the instrument. Notably, these overall profiles curve steadily from the base, implying creep through the entire depth of the borehole.

Total movements determined in the B axis (Figure 8C) are larger. They also show relatively little overall displacement from winter 2013 to winter 2014, but accelerating displacement from spring 2014 to autumn 2015. In this case the lower portion of the profile remains largely vertical, with curvature increasing from a depth of approximately 26 m. This depth corresponds with the Pahoia Tephras, and in particular with the zone (lower Pahoia Tephras) that showed increased induced pore water pressures under vibratory CPTu.

Superimposed on the overall trends in the graphs of Figure 8 are small, sharp jumps in the curves. Many of these correspond with the major unit boundaries marked on the graphs (within the 0.5 m vertical measuring frequency of the instrument); others mostly coincide with boundaries representing the more detailed stratigraphy of the major units.

6.4 Piezometers

Piezometer records are plotted in Figure 10. The upper and middle piezometers respond directly with air pressure; the upper one shows a direct response whilst the middle one shows a damped response. The lower piezometer, however, does not obviously respond to

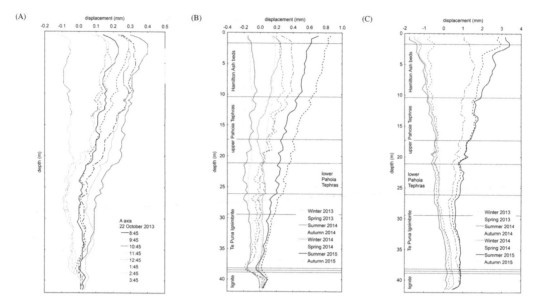

Figure 8.   Borehole inclinometer results from Bramley Drive, Omokoroa. (A) Cumulative plots of borehole inclinometer data from different times on 22 October 2013. (B and C) Average cumulative borehole inclinometer profiles for austral seasons from June 2013 to May 2015 for the A axis (B) and B axis (C).

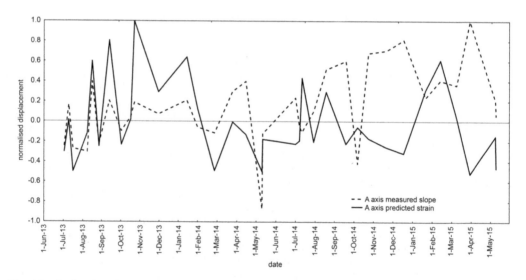

Figure 9.   Predicted Earth tide strain versus measured slope of A axis cumulative inclinometer plot for Bramley Drive, Omokoroa.

air pressure, but displays a lagged response to rainfall. We recognize two discrete aquifers: a shallow aquifer that includes the upper and middle piezometers; and a deep aquifer represented by the lower piezometer. The upper aquifer is based in the Pahoia Tephras and overlying materials and is open to the atmosphere. The lower aquifer is in the Te Puna Ignimbrite. Pore water pressures at the end of the austral summer in May 2013 were lower in the deep aquifer than in the shallow aquifer, and conversely for the austral winter in 2013. Since early summer 2014 the deep aquifer has

maintained a consistently lower pressure than that at the base of the overlying shallow aquifer.

During the austral winter and early spring of 2014 the trace is characterized by sharp increases in pore water pressure in the shallow aquifer; these increases correspond with intense rainfall events and are corroborated by episodes of suddenly rising water levels in standpipes monitored by the Western Bay of Plenty District Council approximately 300 m along the coast from this site. Only very minor transmission of these pressure spikes extended into the deep aquifer.

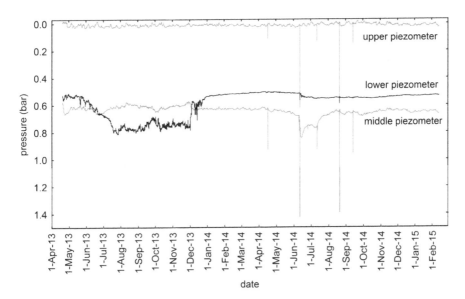

Figure 10.   Piezometer traces from May 2013 to February 2015 at Bramley Drive, Omokoroa.

Notably, the initiation of these spikes coincides with the loss of the clear phase relationship between measured and predicted Earth tide displacements, and it is immediately after this period that distinct displacement of the B axis cumulative borehole inclinometer trace was observed.

7   DISCUSSION

7.1   *Influence of halloysite*

The microstructure of these materials shows a variety of morphologies of the halloysite minerals, but most importantly, the clay sizes are mainly small, and their arrangements mean that the high porosity occurs almost entirely within very small, interconnected pore spaces. Thus, the materials can hold very large amounts of water, but that water cannot move readily within the soils (low permeability). Capillary effects in the narrow pore space mean that they remain close to saturated under normal field conditions.

Many of the key geomechanical characteristics of these materials are controlled by the dominance of halloysite in the clay mineral fraction. As noted earlier, halloysite is a 1:1 clay mineral with a repeating structure of one tetrahedral (silica) sheet and one octahedral (alumina) sheet bound with an interlayer space occupied (in hydrated form) by water molecules. As such, it exhibits relatively high plastic and liquid limits and a low activity (Wesley, 1973), as seen in the materials we examined in the Bay of Plenty region.

Halloysite also typically displays a low cation exchange capacity (CEC) (Joussein *et al.*, 2005; Churchman and Lowe, 2012). Cation exchange capacity is often related to the plasticity of clay minerals and soil. Because of their low CEC, low-activity clays such as kaolinite and halloysite have low values of plasticity

(Lal, 2006), and display a marked lack of cohesion when moulded (Bain, 1971). It is recognized that changes in the chemistry of the constituent pore water cause corresponding changes in sensitivity and residual shear strength (Moore, 1991; Andersson-Skold *et al.*, 2005; He *et al.*, 2015) of soil materials containing clays because of the importance of cation interaction with the charged clay surfaces. Hence weathering and water movement through the soil profile will be expected to impact on the cohesive characteristics of halloysite-dominated sensitive soils. Itami and Fujitani (2005) concluded that edge surfaces play a significant role in determining the flocculation behavior of halloysite clays, but noted that more investigation is needed to elucidate the charge characteristics of halloysite.

The low plasticity index measured for these sensitive soils means that the high porosity associated with the open structure results in the materials containing sufficient water for the liquidity index to be greater than one, hence remoulding to a fluid paste after failure is unsurprising.

For these materials, the presence of halloysite is seen as critical to the development of sensitivity. This conclusion contradicts previous work by Torrance (1992) who suggested that sensitivity in pyroclastic materials was associated with allophane because of its short-range order (nanocrystalline) structure, but is in keeping with the work of Smalley *et al.* (1980) who recognized very high sensitivities in halloysite-dominated soils.

7.1.1   *Development of halloysite*

As discussed earlier, the Si-leaching model pertaining to the formation of allophane and halloysite indicates that both can form directly from the synthesis of the products of dissolution of primary minerals and mineraloids (namely volcanic glass) via different pathways according to local conditions (Lowe, 1986). Halloysite formation is favoured by a Si-rich environment (Si concentration is $>\sim 10$ ppm) and a wet, even "stagnant", moisture regime; allophane, conversely, forms preferentially in soils where Si concentration is low in soil solution ($<\sim 10$ ppm), allowing development of Al-rich allophane (Churchman et al., 2010; Churchman and Lowe, 2012). Al-rich allophane occurs with good drainage where Si is able to be removed from the profile, and is favoured in andesitic and basaltic materials where the original Si content is lower in glass, and where Al availability is not limited by (for example) high contents of humic material that can form Al-humus complexes at the expense of (inorganic) allophane where pHs are low (Dahlgren et al., 2004; Churchman and Lowe, 2012; McDaniel et al., 2012). Thus, halloysite forms preferentially in the Tauranga region where the weathering and dissolution of the upper mainly rhyolitic (siliceous) pyroclastic deposits (including Hamilton Ash beds and upper Pahoia Tephra beds) have provided a ready source of Si that has migrated (leached) into the lower Pahoia Tephra beds; the high porosity yet low permeability of these lower units has resulted in consistently high natural moisture contents with limited water movement and so Si has effectively accumulated. A drier climate during cool glacial periods (which pertained ~ 80–90% of the time during the Quaternary) would also favour halloysite formation (Churchman and Lowe, 2012).

As reported above, we observed halloysite in a range of morphologies including tubes (both long and stubby), spheres and spheroids, and plates. However, the very large book forms seen in some of the samples are entirely novel, being nowhere recorded (to our knowledge) in the literature (other than by our research group in Wyatt et al. 2010). A suggested pathway for their development is described in brief below (a longer article is in preparation: Cunningham et al., in prep).

Fragmented siliceous glass shards, with a large surface area and often a vesicular character, are readily dissolved via hydrolysis (Gislason and Oelkers, 2003; Churchman and Lowe, 2012). The process liberates cations (which occupy intermolecular space amidst loosely linked SiO_4 tetrahedra in the glass) and Si into interstitial pore water, and leads to the rapid precipitation of secondary minerals from such solutions (Churchman and Lowe, 2012). In a locally very wet environment enriched in silica, formation of halloysite is favoured (Churchman et al., 2010). Some research indicates that spheroidal halloysite results from fast dissolution of glass and recrystallization from the resulting supersaturated solution, but others have found that spheroidal particles have higher Fe contents, thus suggesting a structural control. However, that spheroidal and tubular halloysite particles often occur together indicates there is probably no definitive distinction between the conditions that lead to the different particle shapes for halloysite (Churchman and Lowe, 2012). Nevertheless, it has been proposed that one form of halloysite may transform to another with ongoing weathering, and that

spheroidal halloysite may transform to a tubular form (Churchman, 2015). Thus the deposition of successive siliceous pyroclastic deposits in the Tauranga area, and their subsequent dissolution, provides an ongoing source of Si in soil solution that is leached down the profile to continue precipitation processes via Ostwald "ripening" (e.g., Dahlgren et al., 2004) at depth through long periods of time. Concomitantly, the weathering of glass, ferromagnesian minerals, and Fe-Ti oxides (from the original pyroclastic materials) enriches the profile with Fe, which is subsequently incorporated into the halloysite cell unit (Fe substitutes for Al in the octahedral sheet), leading to sheet unfurling and thus promoting the formation of plates – plates typically have relatively high Fe contents whereas tubes have much less (Papoulis et al. 2004; Joussein et al., 2005; Churchman and Lowe, 2012; Churchman, 2015). Our EDX analyses of flat surfaces of plates in halloysite books in a clay-fraction sample from Tauriko were compared with analyses of clusters of halloysite tubes (Wyatt et al., 2010). Although Si and Al contents were effectively identical for both morphologies (books: SiO_2 47.7 $\pm$ 1.1%, Al_2O_3 34.1 $\pm$ 0.5%; tubes: SiO_2 50.7 $\pm$ 2.1%, Al_2O_3 34.2 $\pm$ 1.3%), we found that the Fe content in the plates making up the books ($Fe_2O_3 = 5.2 \pm 0.2\%$) was significantly larger than in the tubes ($Fe_2O_3 = 3.2 \pm 0.3\%$). This enriched Fe content, consistent with ranges reported for plates in the literature (e.g., Joussein et al., 2005), indicates that Fe has replaced Al in octahedral positions, hence reducing the mismatch with the tetrahedral sheet, lessening layer curvature, and thus generating flat plates (Wyatt et al., 2010). At this point the Fe has either been all bound to the halloysite or is insoluble, leaving a relatively "clean" environment. This clean environment and onset of Ostwald ripening encourages large halloysite plates to form; the highly porous, open structure of the materials is required to give the space necessary for development of these large crystals.

Previously, halloysite tubes had been reported as forming on the edges of, and in between, kaolinite plates as a result of loss of structural rigidity (e.g., Robertson and Eggleton, 1991). However, Papoulis et al. (2004) invoked transformation from tubular halloysite to kaolinite via an unstable platy halloysite phase formed from "the interconnection of tubular halloysite to felted planar masses of halloysite" (p. 281). They suggested that resultant "halloysite-rich booklets" comprised both platy halloysite and newly-formed kaolinite together, and that eventually such halloysite-kaolinite booklets were "converted initially to a more stable but disordered kaolinite and finally to well-formed book type kaolinite" (p. 281). We suggest that this mechanism may apply in our study but, critically, that pure halloysite plates, thence pure halloysite books, are formed as an end point, i.e. without the coexistence of halloysite and kaolinite and without kaolinization. That the book-rich halloysitic material in Tauranga occurs at depth within permeable, siliceous pyroclastic and volcaniclastic materials would imply that site wetness conditions needed to maintain halloysite genesis, rather than kaolinite, have prevailed, as invoked by Churchman *et al.* (2010) in their Hong Kong study.

Coalescence of large plates to books appears to occur as the result of the partial dehydration of the profile that promotes shrinkage between the plates and thus, we propose, the amalgamation of the plates to form large books. This drying stage is evidenced by the scattered, fine MnO_2 redox segregations (concretions) in the sequence as described earlier. As the MnO_2 was not particularly concentrated in the profile (~5% abundance at most), partial (short-lived) dehydration is suspected (full dehydration would lead to kaolinite being thermodynamically favoured). This process of partial dehydration appears to be essential for the formation of books – at sites with no manganese concretions, plates but no books were observed.

In summary, the microenvironmental conditions within the lower Pahoia Tephra beds – Si rich, Fe^{2+} rich, permanently near or at saturation but with occasional or intermittent drying out – have kinetically favoured the transformation pathway halloysite tubes → halloysite plates → halloysite books. Potential transformation to kaolinite (cf. Papoulis *et al.*, 2004) is not occurring because the site remains wet most of the time (Churchman *et al.*, 2010). The Al sheet is positively charged while the silica sheet is negatively charged at the pH values expected in non-calcareous environments, viz. between 1.5 and 8.5 (Churchman *et al.*, 2015). When water is present in excess, the polar water molecules enter the interlayer and become associated with opposite charges in adjacent layers. As long as there is water, they will therefore be attracted within the interlayer position electrostatically, giving hydrated halloysite. Without water, adjacent layers move closer and rotation of silica tetrahedral occurs in order to align the layers, leading to platy kaolinite (Churchman and Lowe, 2012).

7.2 Other soil components

The other principal component of the materials is glass shards derived from vesiculation and comminution of the explosively erupting magma that gave rise to the pyroclastic deposits. These shards are of small size (silt or even in the clay-sized component) and hence have very large surface areas. As noted previously, they are composed of an amorphous solid comprising loosely linked SiO_4 tetrahedra with intermolecular spaces occupied by cations, and tend to have sharp, jagged morphologies representing the breakdown of larger vesicular clasts. Such materials are therefore very porous with low chemical stability and so break down very quickly at rates likely to be closely proportional to geometric surface areas (Dahlgren *et al.*, 2004; Wolff-Boenisch *et al.*, 2004; Churchman and Lowe, 2012), leading to rapid loss of surface cations through substitution with H^+ ions (hydrolysis) provided by carbonic acid in the soil solution. This process helps break Al-O bonds and liberates cations into

the soil solution, and leaves a weakened silica tetrahedral framework by removing adjoining Al atoms (Churchman and Lowe, 2012).

The shard morphology allows high frictional resistance to develop if shards are in contact.

The crystalline component of the materials is minor, and mainly consists of small (silt to clay sized) feldspar (plagioclase) crystals that show evidence of weathering. This weathering will also release cations into the soil solution (Churchman and Lowe, 2012).

The varied observed morphologies of the halloysite crystals means that most contacts between clay minerals occur at single points or over very short distances. This feature limits the opportunity to derive strength within the clay matrix as the contact surface areas are small. Similarly, weak grain/matrix contacts means ready disaggregation of coarse and fine components.

A largely skeletal microstructure also means that just as at the matrix scale there is limited surface area for interactions to develop strength, this is also true at the microstructural scale because we see dominantly point or small area contact points between aggregates of clay or clay plus grain mixes, or both.

A matrix dominated by point contacts and an overall skeletal microstructure has led to high porosity (low density) materials, which, combined with the low plasticity index of halloysite, allows for a liquidity index > 1 to be achieved quite readily. The overall small sizes of the components (glass shards, crystals, clays) means that the pores are dominantly very small (<0.5 mm) With this small pore size, capillary effects within the pore spaces allow for near-saturated conditions under most field situations, imparting high apparent cohesion. Small pores also mean that the permeability is low because capillary effects must be overcome to allow water movement, but there appears to be a high connectivity of the pore spaces.

7.3 Field characteristics

The low CEC of the halloysite means that it develops low cohesive strength, indicating that the materials develop strength largely from frictional resistance between grains. However, the materials are typically silts to clayey silts with very few coarser components; indeed, the measured laboratory effective strengths of the materials indicate surprisingly high angles of internal friction for a structure that is composed almost entirely of contacts mediated by clay minerals. This relatively high friction angle may reflect the consolidation state of the specimens during laboratory testing. In contrast, the strength measured *in situ* is low, as evidenced by low tip resistance through the lower Pahoia Tephra sequence in the CPTu testing. Low CPTu tip resistance has been reported for other pumiceous soils, for which crushing of pumice clasts is suggested as a cause of the low resistance (Orense *et al.*, 2012). In the case of the soils in our study, the mode of deposition means that the magmatic material was already shattered into largely discrete shards with very few showing an open, vesicular texture. Hence crushing of pumices is unlikely to account for the low tip resistance in these materials; weak bonding of the clay minerals and destruction of the resulting delicate structure is a more likely cause of the low strength on static CPTu testing. While large amounts of water can be contained within the open structure, the rate at which water can move within the small pore spaces is limited, so permeability is very low, and induced water pressures in CPTu testing rise to very high levels.

The tip resistance is only slightly reduced by cyclic CPTu testing; we suggest that as the structure of the material is already readily broken down by a static run of the CPTu probe, adding extra cyclic energy during the vibratory testing does not result in noticeably greater damage to the material. Conversely, cyclic CPTu testing shows a dramatic increase in pore water pressure compared with the static CPTu test throughout the sequence of materials identified as sensitive at the Bramley Drive site. This increase is related to the low permeability of these materials, and indicates that increased pore water pressures, and hence reduced effective stresses, will result from any seismic or other cyclic stresses imposed on the slope such as wind and wave impacts or even Earth tides. The potential for fracture under such imposed stresses may exist.

Magnification of Earth tide effects appears to occur throughout the stratigraphic sequence at Bramley Drive, including within the sensitive materials and overlying pyroclastic (tephra) layers and paleosols. However, the deformations appear be predominantly at boundaries, with key boundaries being the upper and lower portions of the sensitive units, and boundaries between individual tephra layers. We suggest that the observed enhancement of Earth tide motions is not directly associated with the microstructure of the materials themselves, but rather is a function of the roughness of the boundary surfaces where stick, slip, bonding, and rebonding may occur along the interfaces at each cycle of loading (as indicated by the sharp jumps in the inclinometer traces in Figure 9).

Pore water pressure measurements show two discrete aquifers, with the shallow aquifer based in the lower Pahoia Tephras that have low permeability measured on core samples. During periods of intense rainfall, infiltration into the upper horizons is rapid, leading to a dramatic spike in the pore water pressure trace in the shallow aquifer. The occurrence of these spikes in the austral winter of 2014 at the time when the phase relationship with the Earth tides was lost and some longer-term deformation is apparent in the borehole inclinometer data indicates permanent changes to the structure of the soils associated with the short-term high pore water pressures.

7.4 Development of sensitivity

A lack of cohesive interactions between clay minerals is central to the development of sensitivity in Northern Hemisphere soils. In that case, it is believed that fine-grained sediments, dominated by low-activity illitic clays derived from glacial outwash, were initially deposited in cation-rich saline waters where they developed a flocculated structure. Later uplift to sub-aerial conditions meant (i) no consolidation, and (ii) progressive leaching of Na^+ ions leading to loss of cohesion across the edge-to-face and edge-to-edge clay contacts. High water contents (LI > 1) and little cohesive strength lead to ready breakdown of the materials, which then form a very fluid debris as the platy clay minerals align, yet are separated and hydrated by water released from the abundant pore spaces.

The sensitive rhyolitic materials we examined in the Bay of Plenty originated in a pyroclastic (fragmental) environment that is "upwind" of prevailing winds and somewhat distal from the volcanic sources, giving small particle sizes. These small particles were deposited in a loose arrangement, either due to primary deposition by settling through air, or via secondary deposition in a low-energy fluvial, lacustrine, or estuarine environment. Arthurs (2010) noted the prevalence of reworking in the sensitive soils he examined. Reworking clearly does not preclude development of sensitivity, but it is unclear at present whether or not it is a requirement for sensitivity in these materials. Settling in a quiet environment allows development of an open structure that has not been subjected to significant loading since deposition, and hence the materials remain normally consolidated.

However, an open structure ensures high porosity, whereas small grain sizes mean pore spaces are small and maintain high water contents because of capillarity. Wyatt (2009) noted that materials which remained near the ground surface for some time and display evidence of weathering in an aerobic environment do not develop sensitivity to the same extent, and suggested that rapid burial by deposition of later materials is needed. Small pore spaces, high water contents, and little atmospheric exposure ensures that the local environment remains wet and with limited water movement.

Weathering of the silica-rich rhyolitic glasses (and also plagioclase) comprising both the sensitive materials and the overlying tephra deposits and paleosols means that soil solution is enriched in Si, providing conditions thermodynamically and kinetically favourable for halloysite (rather than allophane) to form from the dissolution products of the dissolving glass and plagioclase (Hodder et al., 1990, 1996). This weathering also provides cations that promote cohesion amongst clay minerals, allowing retention of the flocculated structure as secondary clay minerals precipitate. Likewise, rapid weathering of glass and mafic minerals provides Fe that promotes development of platy halloysite morphologies and, with intermittent dehydration, the formation of unique book morphologies in some instances.

Over time, the supply of cations and Fe gets used up (incorporated into clay crystal lattices or into other insoluble components) or enleached from the system. The low CEC of halloysite, in conjunction with low cation-concentration soil solutions, leads to progressive loss of cohesion across the contact points in the clays. Thus we are left with an open, loose structure of predominantly fine-grained materials that gain limited strength from true cohesion, additional strength from apparent cohesion associated with water films in the soil pores, and some strength from friction across point contacts between grains.

Upper horizons in the stratigraphic sequence are clearly highly permeable, and most water falling on the ground rapidly infiltrates directly into the surface soils. Lower permeability in the sensitive horizons restricts egress of water from the profile, resulting in very dramatic spikes in the pore water pressure signal following rainfall events. The loss of apparent cohesion and reduction in effective stress are thus likely triggers for landslide activity. However, we suggest that dilution of cation concentration in the pore waters associated with a large influx of fresh water, together with enleaching of cations over time as weathering progresses, reduces the true cohesion between clay minerals, resulting in lowered resistance to elevated pore water pressures as the materials weather.

Breakdown of the structure results in long runout because of the high water content. Complex interactions between the different clay morphologies means the runout is not nearly as fluid as in the equivalent landslides in the Northern Hemisphere because fully hydrated face-to-face arrangements do not develop. Interestingly, successive failures of the Bramley Drive slide over time have shown shorter runout distances and less fluid debris. These decreases will in part be due to smaller volumes involved in the later failures, but they may also indicate that progressive weathering near the face where water movement is greater (removing cations) is required for full development of sensitivity. When the time between events is shorter, a less sensitive response is observed.

Microstructure and mineralogy have revealed much about controls on the development of sensitivity in these materials; future work will concentrate on the chemistry of the pore waters and the interaction of cations with the halloysite component of the soils.

ACKNOWLEDGMENT

Access to the Bramley Drive site was provided by Western Bay of Plenty District Council. Special thanks to Wolfgang Schunn for managing instruments and operation of the CPTu unit, and Helen Turner for her help with the SEM and EDX analyses. We thank Ian Smalley for review comments.

REFERENCES

Abidin, Z., Matsue, N. and Henmi, T., 2007. Differential formation of allophane and imogolite: experimental and molecular orbital study. *Journal of Computer-aided Materials Design* 14: 5–18.

Allbrook, R. F., 1985. The effect of allophane on soil properties. *Applied Clay Science* 1: 65–69.

Andersson-Skold, Y., Torrance, J. K., Lind, B., Odén, K., Stevens, R. L. and Rankka, K., 2005. Quick clay – A case study of chemical perspective in Southwest Sweden. *Engineering Geology* 82: 107–118.

Arthurs, J. M., 2010. *The nature of sensitivity in rhyolitic pyroclastic soils from New Zealand.* PhD thesis, University of Auckland, New Zealand.

Bain, J. A., 1971. A plasticity chart as an aid to the identification and assessment of industrial clays. *Clay Minerals* 9: 1–17.

Briggs, R. M., Hall., G. J., Harmsworth., G. R., Hollis., A. G., Houghton., B. F., Hughes., G. R., Morgan. M. D. and Whitbread-Edwards, A. R., 1996. *Geology of the Tauranga Area – Sheet U14 1:50 000.* Department of Earth Sciences, University of Waikato, Occasional Report 22. 57 pp.+ map.

Briggs, R. M., Houghton, B. F., McWilliams, M. and Wilson, C. J. N., 2005. $^{40}Ar/^{39}Ar$ ages of silicic volcanic rocks in the Tauranga-Kaimai area, New Zealand: dating the transition between volcanism in the Coromandel Arc and the Taupo Volcanic Zone. *New Zealand Journal of Geology and Geophysics* 48: 459–469.

Briggs, R. M., Lowe, D. J., Esler, W. R., Smith, R. T., Henry, M. A. C., Wehrmann, H., Manning, D. A., 2006. *Geology of the Maketu area, Bay of Plenty, North Island, New Zealand – Sheet V14 1:50 000.* Department of Earth Sciences, University of Waikato, Occasional Report 26. 44 pp + map.

BSI, 1999. *Methods of test for Soils for Civil Engineering Purposes – Part 8: Shear Strength Tests (Effective Stress).* BS1377-8:1990 Incorporating Amendment No 1.

Churchman, G. J., 2015. The identification and nomenclature of halloysite (a historical perspective). In: Pasbakhsh, P. and Churchman, G. J. (eds), *Natural Mineral Nanotubes: Properties and Applications.* Apple Academic Press, Waretown, NJ, pp. 51-92.

Churchman, G. J. and Lowe, D. J., 2012. Alteration, formation, and occurrence of minerals in soils. In: Huang, P. M., Li, Y. and Sumner, M. E. (eds) *Handbook of Soil Sciences. 2nd edition. Vol. 1: Properties and Processes.* CRC Press (Taylor & Francis), Boca Raton, FL, pp. 20.1-20.72.

Churchman, G. J., Pasbakhsh, P., Lowe, D. J. and Theng, B. K. G., 2015. Water, pH and iron: are these the keys to selecting halloysites for applications as nanotubes? *Euroclay 2015*, Edinburgh, UK: quadrennial meeting of the European Clay Groups Association (ECGA) jointly with annual meeting of The Clay Minerals Society (CMS) and in association with the International Natural Zeolite Association (INZA) and the Geological Society (Edinburgh University, Appleton Tower, 5–10 July) (abstract in press).

Churchman, G. J., Pontifex, I. R. and McClure, S. G., 2010. Factors influencing the formation and characteristics of halloysites or kaolinites in granitic and tuffaceous saprolites in Hong Kong. *Clays and Clay Minerals* 58: 220–237.

Churchman, G. J., Whitton, J. S., Claridge G. G. C. and Theng B. K. G., 1984. Intercalation method using formamide for differentiating halloysite from kaolinite. *Clay Minerals* 19: 161–175.

Cunningham, M. J., 2013. *Sensitive rhyolitic pyroclastic deposits in the Tauranga region: mineralogy, geomechanics and microstructure of peak and remoulded states.* MSc thesis, University of Waikato, New Zealand.

Dahlgren, R. A., Saigusa, M. and Ugolini. F. C. 2004. The nature, properties, and management of volcanic soils. *Advances in Agronomy* 82: 113–182.

Danišík, M., Shane, P. A. R., Schmitt, A. K., Hogg, A. G., Santos, G. M., Storm, S., Evans, N. J., Fifield, L. K. and

Lindsay, J. M., 2012. Re-anchoring the late Pleistocene tephrochronology of New Zealand based on concordant radiocarbon ages and combined $^{238}U/^{230}Th$ disequilibrium and (U-Th)/He zircon ages. *Earth and Planetary Science Letters* 349-350: 240-250.

Dixon, J. B., 1989. Kaolin and serpentine group minerals. In: Dixon, J. B. and Weeds, S. B. (eds) *Minerals in Soil Environments*. Soil Science Society of America Book Series 1, pp. 467-525.

Eilertsen, R. S., Hansen, L., Bargel, T. H., Solberg, I.-L., 2008. Clay slides in the Målselv valley, northern Norway: Characteristics, occurrence, and triggering mechanisms. *Geomorphology*, 93: 548–562.

Fieldes, M., 1955. Clay mineralogy of New Zealand soils. Part II: allophane and related mineral colloids. *New Zealand Journal of Science and Technology* 37: 336–350.

Geertsma, M. and Torrance, J. K., 2005. Quick clay from the Mink Creek landslide near Terrace, British Columbia: Geotechnical properties, mineralogy, and geochemistry. *Canadian Geotechnical Journal* 42: 907–918.

Gislason, S.R. and Oelkers, E. H., 2003. Mechanism, rates, and consequences of basaltic glass dissolution: II. An experimental study of the dissolution rates of basaltic glass as a function of pH and temperature. *Geochimica et Cosmochimica Acta* 67: 3817–3832.

Gulliver, C. P. and Houghton, B. F., 1980. *Omokoroa Point land stability investigation*. Report prepared by Tonkin & Taylor for Tauranga County Council (New Zealand). 54 pp.

Hatrick, A. V., 1982. *Report of Committee to Inquire into the Failure of the Ruahihi Canal*. Ministry of Works and Development, New Zealand.

Head, K. H., 1986. *Manual of Soil Laboratory Testing. Volume 3: Effective Stress Tests*. Pentech Press, London.

Head, K. H., 1992. *Manual of Soil Laboratory Testing. Volume 1: Soil Classification of Compaction Tests*. Pentech Press, London.

He, P., Ohtsubo, M., Higashi, T. and Kanayama, M., 2015. Sensitivity of Salt-leached Clay Sediments in the Ariake Bay Area, Japan. *Marine Georesources and Geotechnology* 33: 429–436.

Hiradate, S. and Wada, S.-I., 2005. Weathering processes of volcanic glass to allophane determined by ^{27}Al and ^{29}Si solid-state NMR. *Clays and Clay Minerals* 53: 401–408.

Hodder, A. P. W., Green, B. E. and Lowe, D.J., 1990. A two-stage model for the formation of clay minerals from tephra-derived volcanic glass. *Clay Minerals* 25: 313–327.

Hodder, A. P. W., Naish, T. R. and Lowe, D. J., 1996. Towards an understanding of thermodynamic and kinetic controls on the formation of clay minerals from volcanic glass under various environmental conditions. In: Pandalai, S. G. (ed) *Recent Research Developments in Chemical Geology*. Research Signpost, Trivandrum, India, pp. 1–11.

Itami, K. and Fujitani, H., 2005. Charge characteristics and related dispersion/flocculation behaviour of soil colloids as the cause of turbidity. *Colloids and Surfaces A: Physicochemical and Engineering Aspects* 265: 55–63.

Jacquet, D., 1990. Sensitivity to remoulding of some volcanic ash soils in New Zealand. *Engineering Geology* 28 (1–2): 1–25.

Jorat, M. E., Kreiter, S., Mörz, T., Moon, V. G. and de Lange, W. P. 2014a. Utilizing cone penetration tests for landslide evaluation. *Submarine Mass Movements and Their Consequences, Advances in Natural and Technological Hazards Research* 34: 55–71.

Jorat, M.E., Mörz, T., Schunn, W., Kreiter, S., Moon, V. and de Lange, W. 2014b. Geotechnical Offshore Seabed

Tool (GOST): CPTu measurements and operations in New Zealand. In: *Proceeding, 3nd International Symposium on Cone Penetration Testing*, Las Vegas, Nevada, pp. 217–223.

Jorat, M. E., Mörz, T., Schunn, W., Kreiter, S., Moon, V. and de Lange, W., 2014c. Geotechnical Offshore Seabed Tool (GOST): a new cone penetrometer. In: *Proceeding, 3nd International Symposium on Cone Penetration Testing*, Las Vegas, Nevada, pp. 207–215.

Jorat, M. E., Mörz, T., Moon, V., Kreiter, S. and de Lange, W. P. 2015. Utilizing piezovibrocone in marine soils. *Geomechanics and Engineering* (in press).

Joussein, E., Petit, S., Churchman, G. J., Theng, B., Righi, D. and Delvaux, B., 2005. Halloysite clay minerals – a review. *Clay Minerals* 40: 383–426.

Keam, M. J., 2008. *Engineering geology and mass movement on the Omokoroa Peninsula, Bay of Plenty, New Zealand*. MSc thesis, University of Auckland, New Zealand.

Keeling, J. L., 2015. The mineralogy, geology and occurrences of halloysite. In: Pasbakhsh, P. and Churchman, G. J. (eds) *Natural Mineral Nanotubes: Properties and Applications*. Apple Academic Press, Waretown, NJ, pp. 95–138.

Kirkman, J. H., 1981. Morphology and structure of halloysite in New Zealand tephras. *Clays and Clay Minerals* 29: 1–9.

Lal, R. (ed), 2006. *Encyclopedia of Soil Science, 2nd edition*. CRC Press. 1600 pp.

Lowe, D. J., 1986. Controls on the rates of weathering and clay mineral genesis in airfall tephras: a review and New Zealand case study. In: Colman, S. M. and Dethier, D. P. (eds) *Rates of Chemical Weathering of Rocks and Minerals*. Academic Press, Orlando, FL, pp. 265–330.

Lowe, D. J., 1995. Teaching clays: from ashes to allophane. In: Churchman, G. J., Fitzpatrick, R.W. and Eggleton, R. A. (eds) *Clays: Controlling the Environment*. Proceedings, 10th International Clay Conference, Adelaide, Australia (1993). CSIRO Publishing, Melbourne, pp. 19–23.

Lowe, D. J., 2011. Tephrochronology and its application: a review. *Quaternary Geochronology* 6: 107–153.

Lowe, D. J. and Alloway, B. V., 2015. Tephrochronology: a volcanic correlational and dating tool. In: Finney, S. and Oleinik, A. (eds) *Encyclopaedia of Stratigraphy*. Springer, New York (in press).

Lowe, D. J. and Palmer, D. J., 2005. Andisols of New Zealand and Australia. *Journal of Integrated Field Science* 2: 39–65.

Lowe, D. J., Tippett, J. M., Kamp, P. J. J., Liddell, I. J., Briggs, R. M. and Horrocks, J. L., 2001. Ages on weathered Plio-Pleistocene tephra sequences, western North Island, New Zealand. *Les Dossiers de l'Archéo-Logis* 1: 45–60.

Lundström, K., Larsson, R. and Dahlin, T., 2009. Mapping of quick clay formations using geotechnical and geophysical methods. *Landslides* 6: 1–15.

McDaniel, P. A., Lowe, D. J., Arnalds, O. and Ping, C.-L., 2012. Andisols. In: Huang, P. M., Li, Y. and Sumner, M. E. (eds) *Handbook of Soil Sciences. 2nd edition. Vol. 1: Properties and Processes*. CRC Press (Taylor & Francis), Boca Raton, FL, pp. 33.29–33.48.

Moon, V. G., Cunningham, M. J., Wyatt, J. B., Lowe, D. J., Mörz, T. and Jorat, M.E., 2013. Landslides in sensitive soils, Tauranga, New Zealand. *19th New Zealand Geotechnical Society Geotechnical Symposium: "Hanging by a thread – lifelines, infrastructure, and natural disasters"*, Queenstown, Vol. 38 (1), pp. 537–544.

Moon, V. G., de Lange, W. P., Garae, C. P., Mörz, T., Jorat, M. E. and Kreiter, S., 2015. Monitoring the landslide at Bramley Drive, Tauranga, New Zealand. *ANZ 2015 Changing the Face of the Earth, Geomechanics and*

Human Influence, 12th Australia New Zealand Conference on Geomechanics, pp. 737–744.

Moore, R., 1991. The chemical and mineralogical controls upon the residual strength of pure and natural clays. *Géotechnique* 41 (1): 35–47.

Neall, V. E., 2006. Volcanic soils. In:Verheye, W. (ed) *Land Use and Land Cover, Encyclopaedia of Life Support Systems (EOLSS)*. EOLSS Publishers with UNESCO, Oxford, U.K., pp. 1–24. (http://www.eolss.net)

New Zealand Geotechnical Society, 2001. *Guideline for hand held shear vane test*. New Zealand Geotechnical Society.

New Zealand Geotechnical Society, 2005. *Guidelines for the field classification and description of soil and rock for engineering purposes*. New Zealand Geotechnical Society.

Orense, R. P., Pender, M. J. and O'Sullivan, A. S., 2012. *Liquefaction Characteristics of Pumice Sand*. EQC Project 10/589. 123 pp. (Available at http://www.eqc.govt.nz/sites/public_files/375-liquefaction- pumice-sands.pdf)

Papoulis, D., Tsolis-Katagas, P. and Katagas, C., 2004. Progressive stages in the formation of kaolin minerals of different morphologies in the weathering of plagioclase. *Clays and Clay Minerals* 52: 275-286.

Parfitt, R. L., 1990. Allophane in New Zealand – a review. *Australian Journal of Soil Research* 28: 343–360.

Parfitt, R. L., 2009. Allophane and imogolite: role in soil biogeochemical processes. *Clay Minerals* 44: 135–155.

Parfitt, R. L., Russell, M. and Orbell, G. E., 1983. Weathering sequence of soils from volcanic ash involving allophane and halloysite, New Zealand. *Geoderma* 29: 41–57.

Parfitt, R. L., M. Saigusa, M., and Cowie, J. D., 1984. Allophane and halloysite formation in a volcanic ash bed under differing moisture conditions. *Soil Science* 138: 360–364.

Pullar, W. A., Birrell, K. S. and Heine, J. E., 1973. Age and distribution of late Quaternary pyroclastic and associated cover deposits of central North Island, New Zealand [with accompanying explanatory notes]. *New Zealand Soil Survey Report* 2. 32 pp. + 4 maps.

Robertson, I. D. M. and Eggleton, R. A., 1991. Weathering of granitic muscovite to kaolinite and halloysite and of plagioclase-derived kaolinite to halloysite. *Clays and Clay Minerals* 39: 113–126.

Singleton, P. L., McLeod, M. and Percival, H. J., 1989. Allophane and halloysite content and soil solution silicon in soils from rhyolitic volcanic material, New Zealand. *Australian Journal of Soil Research* 27: 67–77.

Smalley, L. J., Ross, C. W. and Whitton, J. S., 1980. Clays from New Zealand support the inactive particle theory of soil sensitivity. *Nature* 288: 576–577.

Stevens, K. F. and Vucetich, C. G., 1985. Weathering of Upper Quaternary tephras in New Zealand. 2. Clay minerals and their climatic interpretation. *Chemical Geology* 53: 237–247.

Theng, B. K. G. and Yuan, G., 2008. Nanoparticles in the soil environment. *Elements* 4: 395–399.

Torrance, J. K., 1992. Discussion on sensitivity to remoulding of some volcanic ash soils in New Zealand, by D. Jacquet. *Engineering Geology* 32: 101–105.

Wesley, L. 1973. Some basic engineering properties of halloysite and allophane clays in Java, Indonesia. *Géotechnique* 23 (4): 471–494.

Wesley, L., 1977. Shear strength properties of halloysite and allophane clays in Java, Indonesia. *Géotechnique* 27 (2): 125–136.

Wesley, L., 2001. Consolidation behaviour of allophane clays. *Géotechnique* 51 (10): 901–904.

Wesley, L., 2007. Slope behaviour in Otumoetai, Tauranga. New Zealand. *Geomechanics News* 74: 63–75.

Wesley, L., 2009. Behaviour and geotechnical properties of residual soils and allophane clays. *Obras y Proyectos* 6: 5–10.

Wesley, L. D., 2010. Volcanic soils. Chapter 9. In: *Geotechnical Engineering in Residual Soils*. Wiley, Hoboken, NJ, pp. 189–222.

Wilson, C. J. N., Gravley, D. M., Leonard, G. S. and Rowland, J. V., 2009. Volcanism in the central Taupo Volcanic Zone, New Zealand: tempo, styles and controls. In: Thordarson, T., Self, S., Larsen, G., Rowland, S. K. and Hoskuldsson, A. (eds), Studies in Volcanology: the Legacy of George Walker. *Special Publications of IAVCEI (Geological Society, London)* 2: 225–247.

Wolff-Boenisch, D., Gislason, S. R., Oelkers, E. H. and C.V. Putnis, C. V., 2004. The dissolution rates of natural glasses as a function of their composition at pH 4 and 10.6, and temperatures from 25 to 74°C. *Geochimica et Cosmochimica Acta* 68: 4843–4858.

Wyatt, J. B., 2009. *Sensitivity and clay mineralogy of weathered tephra-derived soil materials in the Tauranga region*. MSc thesis, University of Waikato, New Zealand.

Wyatt, J., Lowe, D. J., Moon, V. G. and Churchman, G. J., 2010. Discovery of halloysite books in a ~270,000 year-old buried tephra deposit in northern New Zealand. In: Churchman, G.. J., Keeling, J. L. and Self, P. G. (eds) *Proceedings, 21st Biennial Australian Clay Minerals Society Conference*, QUT, Brisbane (7–8 August), pp. 39–42. (Available at http://www.smectech.com.au/ACMS/ACMS_Conferences/ACMS21/ACMS21.html)

Volcanic Rocks and Soils – Rotonda et al. (eds)
© 2016 Taylor & Francis Group, London, ISBN 978-1-138-02886-9

Mechanical behavior of volcanic rocks

Áurea Perucho
Laboratorio de Geotecnia, CEDEX. Madrid, Spain

ABSTRACT: Some relevant aspects related to the mechanical behavior of volcanic rocks are considered, mainly based on laboratory tests performed on rocks from the Canary Islands. Ranges of variation of the most relevant geotechnical parameters from the point of view of mechanical behavior and some correlations are given for different types of rocks. More attention is paid to the analysis of pyroclastic rocks, as their mechanical behavior is more peculiar. An empirical yield criterion for low density pyroclasts is adjusted to test results.

1 INTRODUCTION

Two main wide groups of volcanic rocks can be distinguished:

1. Rocks from lava flows of basalts, trachytes or phonolites. They will be referred to as 'highly cohesive and dense' rocks.
2. Pyroclastic rocks, which may originate from pyroclastic falls, surges or flows.

The two groups exhibit a very different mechanical behavior. The first group is formed by materials with much higher densities and strength, though highly dependent on their weathering degree. On the other hand, pyroclastic rocks are formed by fragmented materials with very different grain sizes and textures and with very high porosities and low densities. Commonly, these ones are easily weathered rocks, with very low strength and high deformability, with the exception of the group of ignimbrites (welded and non-welded), which corresponds to hard rocks and has been studied included in the first group.

These two types of rocks frequently appear mixed together in alternative and interlocked layers (Figure 2), this fact greatly affecting the global rock mass mechanical behavior.

Studies on geotechnical behavior of different volcanic rocks have been carried out in different countries (e.g.: Spain: Uriel & Bravo 1971, Uriel & Serrano 1973 and 1976, Serrano 1976, Uriel 1976, Serrano et al. 2002a & b, 2007 and 2010; Lomoschitz Mora-Figueroa 1996; Peiró Pastor 1997; González de Vallejo et al. 2006, 2007 and 2008; Rodríguez-Losada et al. 2007 and 2009; Conde 2013; Hernández-Gutiérrez 2014); Italy: Pellegrino 1970, Aversa et al. 1993 Aversa & Evangelista 1998, Evangelista & Aversa 1994, Cecconi & Viggiani 1998, 2001 and 2006, Cecconi 1998, Tommasi & Ribacchi 1998, Rotonda et al. 2002, Cecconi et al. 2010, Tommasi et al. 2015); Japan: Adachi et al. 1981; New Zealand: Moon 1993).

Several studies on volcanic rocks from the Canary Islands (Figure 1) have been performed at CEDEX's Laboratorio de Geotecnia in the last years. Some papers related to the mechanical behavior of low density pyroclasts were published in the international congresses on volcanic rocks held in Madeira (Serrano et al. 2002a &b), Azores (Serrano et al. 2007) and Canary Islands (Serrano et al. 2010). New results from the most recent study carried out (CEDEX 2013), associated to a phD thesis (Conde, 2013), are presented by the authors in several papers to this Congress.

In this paper some relevant aspects related to the mechanical behavior of these two types of volcanic rocks are considered, mainly based on laboratory tests performed on rock samples from the Canary Islands. Furthermore, some published geotechnical data of volcanic rocks from other sites are included in graphs and correlations.

Aspects related with the mechanical behavior of low density pyroclasts will be analyzed with more extension as their mechanical behavior is more peculiar and problematic.

2 MECHANICAL BEHAVIOR OF 'HIGHLY COHESIVE AND DENSE' VOLCANIC ROCKS

2.1 Introduction

Rocks from lava flows usually have high strength properties, but the mechanical behavior of rock massifs is usually strongly affected by their sets of discontinuities, like other rock masses. Moreover, these types of rocks often alternate with weaker scoriaceous or pyroclastic layers, cavities or lava tubes, forming highly anisotropic and discontinuous rock masses (Figure 2). Usually, the main problems are related to their spatial heterogeneity, both in vertical and horizontal directions, which is the cause of collapses and instability problems. These relevant aspects will not be

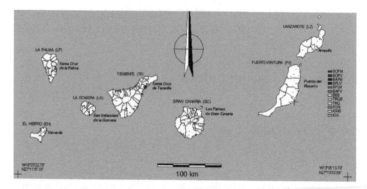

Figure 1. Canary Islands. Location of samples tested from the group of highly cohesive and dense rocks.

Figure 2. Levels of basalts, pyroclasts and scoria. Photos from La Palma.

considered in this paper, but just the mechanical behavior of the rock matrix. González de Vallejo et al. (2007) provide data of typical values for discontinuities in these materials as well as RMR and Q values of basaltic lava flows. Muñiz Menéndez & González-Gallego (2010) point out the difficulty of applying classification schemes to volcanic rocks. Barton (2010) indicates some suggestions to estimate Q values for columnar basalts.

2.2 Mechanical behavior: strength and deformability

Rodríguez-Losada et al. (2007) published a new classification of the Canarian volcanic rocks defining twelve lithotypes based on lithologic, textural and voids criteria. It is intended to be a useful and simple classification in groups with similar geo-mechanical behavior. The defined lithotypes are summarized in Table 1.

Later on, Hernández-Gutiérrez (2014) reduced the number of lithotypes to only ten, eliminating these two: BES, which was very little represented and studied in Canary islands, and TRQB, whose mechanical behavior is very similar to BAFM lithotype, making it possible to analyzed them together.

Pictures of the ten lithotypes are shown in Figure 3.

An extensive study was performed on these Canarian types of rocks, based on the results of test on 369 specimens, with the main objective of defining

Table 1. Defined lithotypes for "highly cohesive and dense" Canarian volcanic rocks (Rodríguez-Losada et al. 2007).

Lithotype	Description
BAFM	Massive aphanitic basalts
BAFV	Vesicular aphanitic basalts
BES	Scoriaceous basalts
BOPM	Olivine pyroxene massive basalts
BOPV	Olivine pyroxene vesicular basalts
BPLM	Plagioclase massive basalts
BPLV	Plagioclase vesicular basalts
FON	Phonolites
IGNS	Non welded ignimbrites*
IGS	Welded ignimbrites*
TRQ	Trachytes
TRQB	Trachybasalts

*Ignimbrites have been studied in this group despite being pyroclastic rocks

their geotechnical properties. A detailed description of all the works carried out for sampling the materials, the origin of each sample and all the test results may be found in Hernández-Gutiérrez (2014).

The geotechnical tests performed on specimens in the laboratory were mainly: determination of specific gravity, dry and wet specific weight, open and total porosity, water absorption, velocity of elastic waves, point load, tensile strength (Brazilian test), uniaxial compressive strength and triaxial strength with confining pressures up to 10 MPa. Measures of hardness

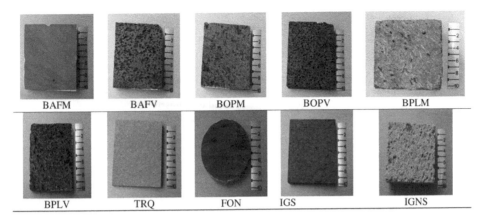

Figure 3. Some pictures of the lithotypes defined in Table 1 (from Hernández-Gutiérrez 2014).

Table 2. Summary of geotechnical data obtained for Canarian volcanic rocks (mean values (standard deviation) (number of tests)).

Lithotype	γ_d (kN/m³)	nt	na	vu (m/s)	UCS (MPa)	UTS (MPa)	E (MPa)
BAFM	27.3	5.5	3.6	4752	104.4	50.3	64600
	(1.0) (7)	(−) (1)	(0.8) (7)	(975) (21)	(54.8) (24)	(21.3) (13)	(−) (1)
BAFV	20.8	29.8	11.0	3824	31.3	21.7	16923
	(4.4) (8)	(13.9) (3)	(10.1) (8)	(980) (21)	(16.0) (22)	(11.4) (14)	(3270) (2)
BES	20.6	34.4	19.8	2964	31.2	22.8	4400
	(3.3) (4)	(−) (1)	(11.2) (4)	(912) (10)	(34.2) (11)	(16.5) (8)	(566) (2)
BOPM	28.0	9.6	4.1	5040	114.5	47.9	30720
	(1.4) (40)	(5.1) (15)	(2.3) (40)	(843) (109)	(59.8) (114)	(16.3) (86)	(5599) (4)
BOPV	23.4	31.1	8.2	4435	47.7	28.2	15550
	(3.2) (5)	(13.3) (2)	(7.0) (5)	(901) (31)	(35.7) (31)	(16.1) (23)	(5728) (2)
BPLM	24.0	15.5	4.8	4071	60.9	25.0	–
	(1.5) (3)	(1.1) (2)	(1.1) (3)	(739) (7)	(27.9) (7)	(11.7) (5)	–
BPLV	23.7	–	3.3	3052	36.1	22.1	24550
	(0.6) (2)		(3.1) (2)	(713) (7)	(14.8) (7)	(4.2) (4)	(−) (1)
FON	24.7	–	3.0	4858	118.9	44.8	47597
	(2.7) (9)		(1.4) (9)	(910) (24)	(76.8) (27)	(17.2) (21)	(15278) (2)
IGNS	16.4	44.4	26.6	2592	16.5	22.0	8265
	(4.6) (5)	(−) (1)	(10.8) (5)	(897) (15)	(19.5) (159	(18.1) (9)	(6152) (2)
IGS	21.5	9.2	13.3	3649	48.2	33.5	50154
	(1.7) (14)	(-) (1)	(5.6) (13)	(738) (48)	(29.1) (51)	(14.4) (38)	(−) (1)
TRQ	24.1	8.9	4.6	4485	95.5	42.4	32861
	(2.2) (15)	(4.4) (5)	(2.3) (15)	(887) (31)	(62.8) (34)	(17.7) (21)	(17835) (2)
TRQB	24.8	14.3	6.3	4513	75.4	46.8	51125
	(2.9) (10)	(10.5) (5)	(4.5) (10)	(904) (14)	(61.1) (15)	(17.4) (11)	(21390) (2)
Whole	24.7	14.7	7.3	4377	81.1	39.3	29316
	(3.8) (122)	(11.4) (37)	(7.4) (121)	(1107) (338)	(61.6) (358)	(18.7) (253)	(28) (21)

(γ_d: dry spec. weight; nt: total porosity; na: open porosity; vu: ultrasonic velocity; UCS: uniaxial compressive strength; UTS: uniaxial tensile strength (Brazilian test); E: Young modulus).

with Schmidt hammer were also taken on blocks. Apart from these physical and mechanical tests a petrographic study with thin sections and a geochemical analysis of oxides, carbonates, sulfates and halides, were also performed.

Some of these data have been previously published (Rodriguez-Losada et al. 2007 and 2009). A summary of the most relevant results from the mechanical point of view is provided in Table 2. The boxplots included in Table 4 show the ranges of variation of the different parameters for each lithotype. These boxes represent the interquartile range (50% of the data) and the marked line represents the median. The histograms of these data are shown in Table 5.

Many correlations were obtained with these results. Only a few of them related to the mechanical behavior will be shown in the next section.

González de Vallejo et al. (2007) provide the following data (Table 3) for intact basaltic rock, fully in agreement with the values given in the previous tables.

More than 100 triaxial tests were performed in a Hoek cell with confining pressures up to 10 MPa.

Table 3. Data from basalts and ignimbrites (González de Vallejo et al. 2007).

	γ_d (kN/m³)	UCS (MPa)
Basalt (most frequent)	15–31 (23–28)	25–160 (40–80)
Vesicular basalt	15–23	<40
Massive basalt	>28	>80
Ignimbrites	13–20 sometimes >20	15–70 <5 if weathered

However, it was not possible to obtain representative parameters for the lithotypes, as the results showed a great dispersion, mainly due to the different densities for specimens in the same lithotypes. Nevertheless, a clear relationship between the rock strength and the specific weight of the rock is observed if all tests results are put together, regardless of the lithotype, as shown in Figure 4.

2.3 Correlations

Quite a few correlations were obtained, the most interesting of them, related to the mechanical behavior of these volcanic rocks, being shown in the following figures. Figure 5 shows the relation between the mean value of the point load index ($I_{s(50)}$) and uniaxial compressive strength (UCS) for each lithotype. Considering all the samples, not only the mean values of the lithotypes, a relation of $UCS = 15 * I_{s(50)}$ is obtained; and if only basalts are considered then it results $UCS = 14 * I_{s(50)}$. This last is the same than the one obtained by Mesquita Soares et al. (2002) for the basalts from the volcanic complex of Lisbon.

However, this relation is not constant but increases with the strength of the rock, as Figure 6 shows.

Kahraman (2014) shows a wide variety of relations provided by different authors and for different types of rocks. Most of them are linear relations of the type $UCS = 12$ to $24 * Is$, and a few ones are non-linear

Table 4. Ranges of variation of the different parameters (boxes with interquartile range (50%) and median).

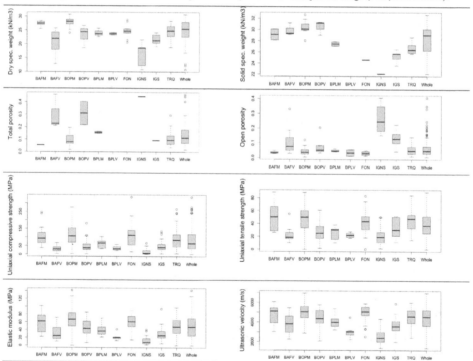

Table 5. Histograms of the different parameters shown in Table 4.

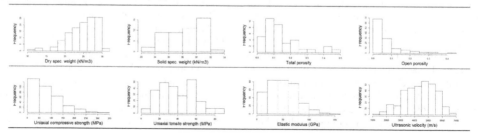

(exponential or power expressions). He studies this relation for pyroclastic rocks and concludes that it is probably non-linear and that further research is needed to check if that relation is also non-linear for all types of soft rocks. Probably, it is non-linear for all types of rocks, not only soft ones.

In other respects, Rodríguez-Losada et al. (2007) recommend increasing an 18% the compression strength deduced from the Schmidt hammer to estimate the uniaxial compressive strength: $UCS = 1.18 * UCS_{Schmidt}$.

Figure 7 shows the relation between the uniaxial compressive strength and the uniaxial tensile strength for the mean values of each lithotype. If all the samples are considered, the relation $UCS = 2.2 * UTS$ is obtained; and if only basalts are considered it results $UCS = 2.4 * UTS$.

In Figure 8 the modulus ratios obtained for these samples are shown, ranging from 250 to 1250.

In Figure 9 and in Figure 10 the relationship between the dry specific weight of these rocks and the uniaxial compressive strength and Young modulus, respectively, can be observed.

3 MECHANICAL BEHAVIOR OF LOW DENSITY PYROCLASTS

3.1 Introduction

Pyroclastic rocks are usually very little affected by discontinuities, conversely to the lava flow masses. Therefore, it is mainly the strength of the matrix rock that determines the mechanical behavior of these rock masses.

Low density pyroclasts have a peculiar mechanical behavior. At low pressures they behave like a rock, with high elasticity modulus and therefore low deformations, whereas for pressures higher than a threshold value their structure is broken and then their deformability increases greatly and they behave more like soils. This phenomenon is referred to as mechanical collapse, and the materials that suffer this process are known as mechanically collapsible rocks.

Despite all the studies carried out since 1970, as the ones mentioned before, not many data referring to the mechanical behavior of these materials can be found in literature. Actually, it is difficult to obtain reliable results from strength tests due to several factors:

– On the one hand, it is difficult to obtain proper specimens as these materials often have jointed irregular and angular fragments, which frequently

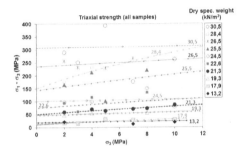

Figure 4. Results of triaxial tests for these volcanic rocks.

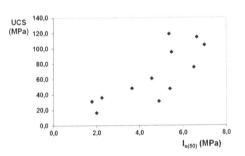

Figure 5. Relation between Point load index, I_s, and UCS (data from Hernández-Gutiérrez 2014).

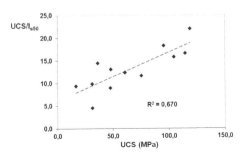

Figure 6. Relationship between the UCS/Is ratio and UCS (Hernández-Gutiérrez 2014).

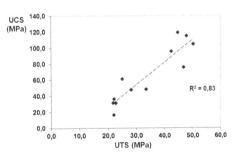

Figure 7. Relationship between the UCS/UTS ratio and UCS (data from Hernández-Gutiérrez 2014).

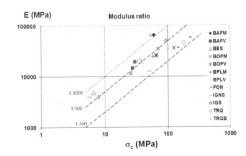

Figure 8. Modulus ratio for samples from Canary Islands.

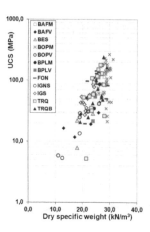

Figure 9. Uniaxial compressive strength versus dry specific weight for rocks from Canary Islands (volcanic flows).

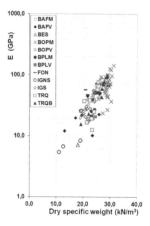

Figure 10. Young modulus versus dry specific weight for rocks from Canary Islands (volcanic flows).

Figure 11. Sample of pyroclast immersed in a colored resin to allow for a better observing of the macrostructure.

break when cutting the samples. It is not easy to obtain good quality specimens, which often show rough walls. Particularly, for poorly cemented or welded pyroclasts, the weaker they are the more difficult it becomes to obtain proper specimens and reliable test results. As a consequence, there is an important dispersion of test results, making it necessary to perform a large number of tests to be able to extract convincing conclusions and correlations.

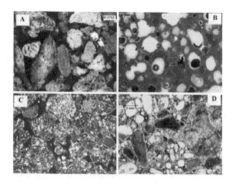

Figure 12. Microphotographs of several pyroclasts with the four types of macroporous structure defined (A- Reticular, B-Vacuolar, C-Mixed and D-Matrix) (CEDEX 2007, Santana el al. 2008).

– On the other hand, due to the roughness of the samples walls the membranes needed for triaxial testing are frequently broken, particularly when high confining pressures are applied, becoming difficult to obtain test results of isotropic collapse pressures or values of strength at high confining pressures for many of these materials.

In addition to that, these rocks are usually quite heterogeneous and specimens often have appreciable differences in densities that crucially influence their strength and deformability, even when coming from a same block.

In the following sections the mechanical behavior observed in a large number of samples is described, as well as the adjustment of a strength criterion proposed by Serrano (2012) to define the yield surface of these materials. Previously, some relevant aspects related to the macroporosity will be commented, as well as other issues related to manufactured samples ('ideal pyroclast'), the specimen preparation and the classification used to designate the tested samples.

3.1.1 Macroporosity

It is well known that the structure has a great influence on the strength of pyroclasts, as many authors point out (e.g.: Pellegrino 1970, Serrano 1976, 1997, Leorueil & Vaughan 1990, Aversa & Evangelista 1998, Serrano et al. 2007, Conde 2013, Tommasi et al. 2015). In the studies carried out at CEDEX since 2002 the macroporosity of the specimens has been carefully observed through a microscope, in some cases after immersing the samples in colored resins in order to better visualize the structure of the macropores of different types of pyroclasts (Figure 11).

Four main types of porosity were defined (Figure 12):

– Reticular porosity, with particles cemented at their points of contact and surrounded by macropores; it was observed in non-altered pumice samples.
– Vacuolar porosity, when a more or less homogeneous mass with a glassy appearance presents a

Figure 13. Left: Pyroclasts with reticular porosity (pumice);
Right: Pyroclasts with vacuolar porosity (scoria).

large number of pseudo-spherical vacuoles inside;
it was found in scoria.

– Mixed porosity, sharing both characteristics of reticular porosity between particles and visible vacuolar porosity inside particles; it was observed in non-altered lapilli samples.

– Matrix porosity, characterized by the presence of a fine grain material filling the macropores that surround the particles; it was observed in altered samples of either lapilli or pumice type. Moreover, it could be defined as the type of porosity of ashes too, because, due to their small particle size, they do not present macropores.

3.1.2 'Ideal pyroclast'

Uriel & Bravo (1970) had found a collapsible behavior in porous concrete that was later on adjusted reasonably well to a theoretical energetic model defined for low density pyroclasts by Serrano (1976, also in Serrano et al. 2002).

With this precedent and also due to the great difficulty to obtain reliable results from strength tests and to the great heterogeneity of the samples, an attempt was made to manufacture at geotechnical laboratory of CEDEX an artificial material with similar structures and strengths to those of real pyroclasts, but homogeneous and easily testable, i.e. an 'ideal pyroclast', that could be helpful in finding the shape of a strength criterion for these type of rocks.

Two structures were simulated, defined by reticular and vacuolar porosity, respectively (Figure 13). Many different combinations were tried, using mainly cement, bentonite, arlite particles and small spherical particles of porexpan in different proportions. Bentonite was added to reduce the strength of cement. Reticular porosity was simulated by a combination of the four mentioned components (Figure 14 and Figure 16) whereas vacuolar porosity was simulated by combining cement and bentonite with porexpan particles (Figure 15 and Figure 16). Arlite particles were used to simulate the low density pumice particles and in both cases porexpan spherical particles were added to simulate the macroporosity of the pyroclasts. Many trials were done until similar values of the uniaxial and isotropic strengths of real pyroclasts were achieved. The procedure is described in detail in CEDEX (2013) and Conde (2013).

However, despite all the efforts applied in manufacturing homogeneous samples, the dispersion of results from the strength test was not smaller than for real

Figure 14. 'Ideal pyroclast' with reticular structure.

Figure 15. 'Ideal pyroclast' with vacuolar structure.

Figure 16. 'Ideal pyroclasts' with reticular (left) and vacuolar (right) structure. Voids (simulated by porexpan spheres) are painted in green color and particles (arlite rounded pieces) in yellow.

Figure 17. Pumice specimen being hand trimmed.

pyroclasts but of the same order of magnitude. Nevertheless the study permitted to define the best ways to perform triaxial tests on these materials with such irregular walls avoiding the breakage of membranes.

Figure 19. Impervious membranes (left) and rubber type oilcloth (right) used for the isotropic and triaxial compression tests.

Table 6. Defined lithotypes in the classification of Hernández-Gutiérrez & Rodríguez-Losada (CEDEX 2007; Consejería de Obras Públicas y Transportes del Gobierno de Canarias 2011). Particle sizes: Lapilli (2–64 mm), Scoria (>64 mm), Ashes (<2 mm) and Pumice (>2 mm).

BASALTIC	LAPILLI (LP)	LOOSE (S)	LPS
PIROCLASTS		WELDED (T)	LPT
	SCORIA (ES)	LOOSE (S)	ESS
		WELDED (T)	EST
	BASALTIC	LOOSE (S)	CBS
	ASHES (CB)	WELDED (T)	CBT
SALIC	PUMICE (PZ)	LOOSE (S)	PZS
PYROCLASTS		WELDED (T)	PZT
	SALIC ASHES	LOOSE (S)	CSS
	(CS)	WELDED (T)	CST

Figure 18. Some specimens partly or totally covered with plaster to enhance parallelism of the bases and to smooth the irregularities in the wall.

3.1.3 Specimen preparation of real pyroclasts

After trying different methods of trimming cylindrical samples it was decided that the best way to obtain good quality specimens from the blocks was to freeze them to −20°C beforehand, particularly with the pyroclasts with lower strength and welding, which frequently ended up broken during carving. Furthermore, the weakest samples, mainly of pumice type, had to be carefully hand trimmed (Figure 17). Cecconi (1998) and Cecconi & Viggiani (1998) study the effect of freezing on a pyroclastic soft rock by measuring compression wave velocities and conclude that after freezing the velocity reduction is only around 6%.

On the other hand, in order to enhance the parallelism of the bases of specimens and in some cases also to smooth the walls a plaster was used (Figure 18). Finally, the most common solution adopted for softening the irregularities in the walls consisted in covering laterally the specimens with a strong oilcloth type rubber, over three rubber normal impervious membranes (Figure 19). Such a strong protection may produce a small increase in the rigidity of the samples, but not a relevant change in the test results.

3.1.4 Classification of pyroclasts

Hernández-Gutiérrez & Rodríguez-Losada defined in 2007 the geotechnical classification shown in 3.2 as a result of an extensive study carried out by the Infrastructures Department of the Canary government (CEDEX 2007; Consejería de Obras Públicas y Transportes del Gobierno de Canarias 2011). It is a qualitative classification that defines the different types of low density materials from the point of view of magma composition, particle size and degree of welding, regardless of the genesis of the material. The pyroclasts referred in the following sections will be classified in the main five groups indicated in the second column of this table, without indication of the qualitative degree of welding specified in it. So, the pyroclasts will be named as lapilli, pumice, scoria and basaltic or salic ashes.

Conde (2013) and Conde et al. (2015) propose a new classification derived from this one, with the main contribution of defining new lithotypes for the five groups of pyroclasts to distinguish the ones that have matrix type porosity, as defined before. These materials probably correspond to the ones classified as tuffs by other authors (González de Vallejo et al. 2007). Their main feature and difference from the other lithotypes of the classification is that they have a fine material filling the macropores, which could come from the alteration of the particles but could also have a different origin (deposits from fluids, etc.). As it will be shown in the stress-strain curves, the presence of a matrix in the macropores influences crucially the mechanical behavior of low density pyroclasts.

Apart from the mentioned materials there could be added the volcanic agglomerates and breccia, formed by fragments of different sizes and not included in these classifications.

Although it is a difficult task with these very complex volcanic materials, it should be very convenient to establish a unique and universal classification for them so that published test results could be more easily compared.

Figure 20. Some specimens before being tested.

3.2 Mechanical behavior: strength and deformability

3.2.1 Influential factors

The geomechanical behavior of low density pyroclasts mainly depends on the following factors (Serrano et al. 2002):

1. Overall compaction.
2. Particle welding or cementing.
3. Imbrication of the particles.
4. Intrinsic strength of the particles.
5. Weathering.

The first three factors define the structure and are usually associated with each other: in general, the more imbricated and welded the particles are, the greater is their degree of compaction, but there are cases with a very low density and a strong bonding between the particles, or a low density due to presence of vesicles. On the other hand, the rock could be very compact, but at the same time it might be very friable if the particles were not firmly welded.

The overall compaction is the most important factor and it is vital when estimating the strength of the rock mass. The strong importance of the specific weight of the rock – sometimes even more than the lithotype-shown for the highly cohesive and dense rocks, is also true in the case of low density pyroclasts, as will be shown in the next sections.

The intrinsic strength of the particles is very influential when high confining pressures are applied. When confining pressures are low the strength of the bonding between particles is usually more influential as the breaking is mainly produced at these contacts.

3.2.2 Experimental mechanical characterization

Around 250 specimens were subjected to uniaxial, triaxial and isotropic compression tests. The five types of pyroclasts indicated before were tested: lapilli, pumice, scoria and basaltic or salic ashes. All of them came from pyroclastic falls, with the possible exemption of the ashes, which origin is not certain, as they could also come from pyroclastic flows.

Tests were performed on oven-dried specimens and the axial strains were measured in all cases. In isotropic compression tests volumetric strains were also measured through the volume of water expelled from the cell during the tests. The magnitude of measured volumetric strains may not be very accurate, but the curves of volumetric strain helped to locate yield points in the stress-strain curves, in the cases where they were not very clear. Triaxial and isotropic tests were performed

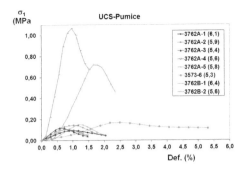

Figure 21. Some examples of uniaxial compression curves of lithified pumice. (In the legend, together with specimen number, the dry specific weight in kN/m^3 is reflected). Samples tested at CEDEX.

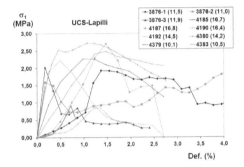

Figure 22. Some examples of uniaxial compression curves of welded lapill. (In the legend, together with specimen number, the dry specific weight in kN/m^3 is reflected). Samples tested at CEDEX.

in a triaxial cell in which cell pressures up to 3.5 MPa can be applied.

All specimens were approximately 5 cm in diameter (7 in some cases for uniaxial compression) and 10–13 cm in height, except a few ones that were shorter (Figure 20). Those ones were tested on uniaxial compression and a correction was applied in the results, according to ISRM suggestions.

Yield stresses were determined for each sample on the bases of the results of all these compression tests looking for a strength criterion that could define the shape of the yield surface.

Apart from these strength tests, other identification tests were performed, mainly the following ones: specific weight and gravity, water content, carbonates, sulfates and organic content and ray-X fluorescence and diffraction.

A summary of the main results is shown in Table 7.

Tests results show the influence of the structure of the rock. Some representative examples of the main results related to the mechanical behavior of these pyroclasts are shown in the curves of Figure 21 to Figure 31.

In Figure 21 to Figure 23 examples of the uniaxial compression curves are shown for pumice, lapilli and ashes, respectively. In the uniaxial compressive failure the strength of the bonding of the particles is crucial,

Table 7. Summary of geotechnical properties of Canarian pyroclasts tested at CEDEX (mean value (standard deviation) (number of tests)).

Lithotype	γ_d (kN/m³)	γ_s (kN/m³)	n	UCS (MPa)	Def. UCS (%)	p_c (MPa)	Def. p_c (%)	Et$_{50}$ (MPa)
Lapilli	10.6	29.2	0.64	1.4	0.90	1.3	0.80	100.2
	(2.9)	(1.7)	(0.10)	(1.6)	(0.67)	(0.6)	(0.41)	(69.4)
	(58)	(47)	(48)	(39)	(39)	(19)	(19)	(19)
Altered lapilli	12.8	28.3	0.55	1.3	1.23	–	–	77.9
(with matrix)	(2.4)	(1.3)	(0.07)	(0.9)	(0.40)	–	–	(53.6)
	(14)	(14)	(14)	(14)	(14)	–	–	(9)
Pumice	5.2	23.7	0.78	0.3	1.1	0.2	0.57	11.1
	(0.6)	(0.8)	(0.03)	(0.3)	(0.5)	(0.09)	(0.32)	(5.2)
	(28)	(28)	(28)	(11)	(11)	(17)	(17)	(2)
Altered pumice	9.7	24.6	0.61	1.1	1.0	–	–	148.3
(with matrix)	(1.1)	(1.0)	(0.05)	(0.7)	(0.3)	–	–	(45.1)
	(10)	(10)	(10)	(10)	(10)	–	–	(7)
Ashes	13.1	26.7	0.50	2.9	0.85	–	–	171.0
	(4.6)	(1.9)	(0.16)	(3.7)	(0.27)	–	–	(110.7)
	(13)	(10)	(10)	(13)	(13)	–	–	(5)
Scoria	10.0	28.8	0.59	1.6	0.74	1.4	0.90	311.2
	(2.2)	(0.7)	(0.01)	(1.1)	(0.46)	(–)	(–)	(–)
	(6)	(3)	(3)	(5)	(5)	(1)	(1)	(1)

γ_d: dry specific weight; γ_s: solid specific weight; n: porosity; UCS: uniaxial compressive strength; Def. UCS: deformation at uniaxial failure; p_c: isotropic collapse load; Def. p_c: deformation at isotropic failure; Et$_{50}$: tangent deformation modulus at 50 % of uniaxial compressive strength.

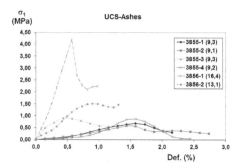

Figure 23. Some examples of compression curves of lithified ashes in uniaxial compression. (In the legend, together with specimen number, the dry specific weight in kN/m³ is reflected). Samples tested at CEDEX.

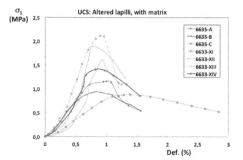

Figure 24. Some examples of uniaxial compression curves of altered lapilli (matrix-type porosity). Samples tested at CEDEX.

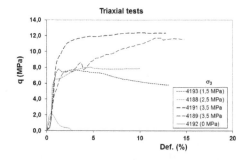

Figure 25. Some examples of compression curves in triaxial compression (lapilli). Confining pressures (in brackets) lower than yield pressures. Samples tested at CEDEX.

as the failure planes observed pass through them. In Figure 21 it is observed that the uniaxial compressive failure of pumice gives soft failure curves and reflects a linear elastic behavior for these pyroclasts before failure. Conversely, the stress-strain curves of uniaxial compression in lapilli show much more irregular shapes and sometimes a non-linear behavior before failure. For altered lapilli, with matrix-type porosity, curves tend to smooth (Figure 24). Pola et al. (2010) study the relationship between porosity and physical mechanical properties in weathered volcanic rocks and also find smoothed curves for higher alteration degrees.

Figure 23 shows uniaxial compression curves in pyroclastic ashes, with a soft shape in general.

Despite the strong importance of the strength of the bonding between particles, in all cases a strong

influence of the density of the samples in the uniaxial compressive strength is observed. In the case of lapilli this strength is mainly due to welding whereas in the case of pumice and ashes it is more due to lithification,

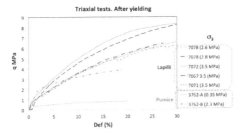

Figure 26. Some examples of compression curves in triaxial compression. Confining pressures (in brackets) higher than yield pressures. Samples tested at CEDEX.

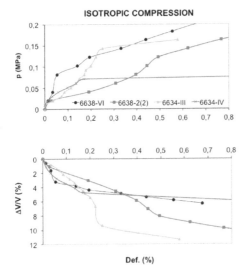

Figure 27. Some examples of isotropic compression and volumetric strains curves (values of volumetric strains not accurate). Some pumice samples tested at CEDEX.

as due to their low density when they deposit they have already cooled, at least to a certain point.

In Figure 25 some representative examples of triaxial tests results are shown for samples tested with confining pressures lower than isotropic yield pressure. The behavior is approximately linear elastic up to pressures close to failure ones and brittleness decreases as the confining pressure rises.

A different behavior is observed on samples tested with confining pressures higher than isotropic yield pressure as shown in Figure 26. A strain hardening elasto-plastic behavior is observed and failure is reached after large axial strains (up to 30% and higher sometimes).

In Figure 27 an example of volumetric strain curves is shown. Although, due to the low precision of the measure, the values are not accurate, they reflect the increase in volumetric strains produced when yield occurs.

In Figure 28 to Figure 31 some representative curves of the stress-strain behavior of these materials in isotropic compression are shown. All the curves are

represented in a semi-logarithmic scale (curves at the left side) and in a natural form (at the right side).

Mesri and Vardhanabhuti (2009) study the compression of granular materials and conclude that most of the existing data on primary compression of granular soils can be summarized in three types of behavior named A, B and C. As pyroclasts are formed by granular fragments the same behavior can be assumed. In Figure 28 and Figure 29 some typical results from lapilli and pumice samples are shown respectively and in both cases a behavior similar to A-type described by those authors is mainly observed: there is a first stage in which the deformation modulus is slightly increasing or constant, followed by a second stage in which the modulus decreases. In some cases a third stage is observed, in which an increase of modulus is again observed. The increasing modulus in the first stage can be attributed to the closing of the structure due to the approaching of particles. In Figure 32 and in Figure 33 some pictures from microscope of pumice and lapilli specimens are shown respectively, where, apart from macropores, considerable gaps between particles can be observed that will tend to close when the specimens are subjected to compression stresses.

Isotropic yield pressures are higher for lapilli, most of them in a range of 1.09–1.54 MPa (95% confidence interval), and lower for pumice samples, most of them in a range of 0.18–0.26 MPa (95% confidence interval).

Tommasi et al. (2015) also found an A-type behavior for a poorly cemented pyroclast from a flow deposit, locally known as pozzolana.

According to Mesri and Vardhanabhuti (2009), in the first stage of the materials with type A behavior some small particle movements are produced enhancing interparticle locking and some level I and II particle damage may occur (abrasion or grinding of particle surface asperities and breaking or crushing of particle edges or corners, respectively). In the second stage level III particle damage is produced (fracturing, splitting or shattering of particles) and there is a collapse of the load-bearing aggregate framework. When major particle fracturing and splitting is complete the third stage can be observed, during which there is an increase in the modulus due to the improved particle packing.

A different behavior is observed in altered lapilli and pumice samples, in which the macropores are filled with a fine grain material, having therefore matrix-type porosity. In Figure 30 some typical results of these samples when isotropically compressed are shown in a semi-logarithmic form (left graph) and in a natural scale (right graph). A behavior of type B according to Mesri and Vardhanabhuti (2009) is observed, with a first stage in which the deformation modulus gradually increases followed by a second stage in which it remains constant. There could be a third stage in which the modulus would increase again, but this has not been observed in the performed tests either because they were stopped before reaching that stage or because it does not occur. In the first stage levels I and II

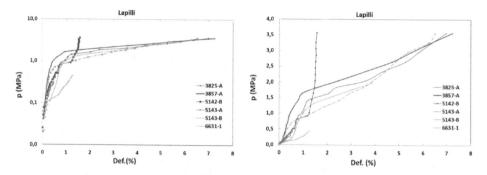

Figure 28. Compression curves of welded lapilli in isotropic tests in semi-logarithmic (left) and natural scale (right) (p':
mean isotropic pressure; def.: axial deformation). Samples tested at CEDEX.

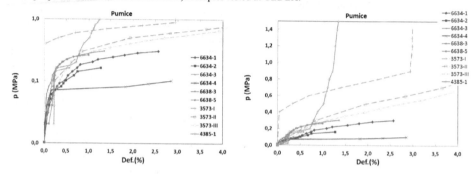

Figure 29. Compression curves of welded pumice in isotropic tests in semi-logarithmic (left) and natural scale (right) (p':
mean isotropic pressure; def.: axial deformation). Samples tested at CEDEX.

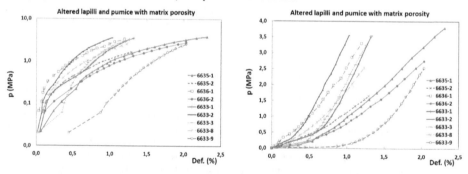

Figure 30. Compression curves of altered pyroclasts with matrix porosity in isotropic tests in semi-logarithmic (left) and
natural scale (right) (p': mean isotropic pressure; def.: axial deformation). Samples tested at CEDEX.

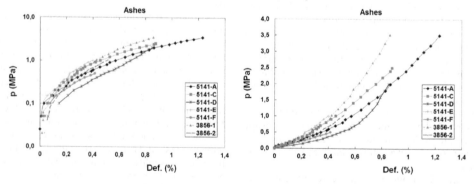

Figure 31. Compression curves of ashes in isotropic tests in semi-logarithmic (left) and natural scale (right) (p': mean
isotropic pressure; def.: axial deformation).

Figure 32. Microscope pictures of pumice specimens. Macropores and gaps between particles are observed.

Figure 33. Microscope pictures of lapilli specimens. Macropores and gaps between particles are observed.

of particle damage are probably caused, while in the second stage level III particle damage occur but it is balanced by the improved packing, so no collapse happens and the modulus remains constant. From a civil engineering point of view these altered pyroclasts with a matrix-type porosity are much less dangerous, as they do not suffer from mechanical collapse like the lapilli and pumice pyroclasts shown in Figure 28 and Figure 29, or if they did it would occur at much higher pressures and deformations.

In Figure 31 typical compression curves of ashes in isotropic compression tests are shown in semi-logarithmic (left curve) and natural scale (right curve) and a type B behavior similar to the altered pyroclasts is observed, so the same comments can be made for these types of pyroclasts, where no sudden collapse is expected.

According to Conde (2013) it seems that in isotropic compression the collapse is more gradual in pumice samples and brusquer in lapilli samples. Figure 28 and Figure 29 seem to corroborate this as a general tendency, although some pumice specimens exhibit a brusque collapse too.

Figure 34 shows the ratio between deformation modulus before and after isotropic collapse in different specimens. In general, there is a greater variation of the modulus in lapilli samples with a ratio varying between 3 to 12, while in pumice specimens this ratio varied from less than 2 to 4.

3.2.3 Yield criteria

3.2.3.1 Behavior at low confining pressures

For low confining pressures the Hoek & Brown (1980) triaxial failure criterion can be adjusted to these materials (Figure 35). However, as shown in Figure 36, for low density pyroclasts even at low confining pressures it can be observed that the strength criterion tends to close in the right part of the figure, for higher pressures as will be seen later on.

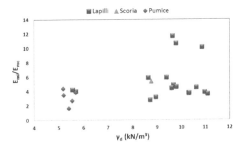

Figure 34. Ratio between deformation modulus before (E_{mi}) and (E_{mc}) after collapse in isotropic compression (Conde 2013).

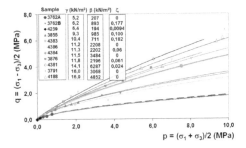

Figure 35. Example of adjusting Hoek & Brown failure criterion to low density pyroclasts for low confining pressures. Note the great influence of specific weight (γ) on the strength (β and ζ are the Serrano & Olalla strength parameters) (Serrano et al, 2007).

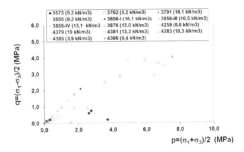

Figure 36. Tests results for low density pyroclasts (CEDEX, 2007).

As in highly cohesive and dense volcanic rocks, the strength of these rocks is also greatly influenced by density.

3.2.3.2 Mechanical collapse

Some authors such as Serrano (1976), Adachi et al (1981), Aversa & Evangelista (1998) and Serrano et al (2002, 2007, 2010) have developed different equations trying to obtain the yield surface for pyroclastic materials.

Mainly two types of models have been proposed for pyroclastic rocks in order to reproduce mechanical collapse:

– Theoretical models, from which two types of models can be distinguished: structural models (e.g. Uriel &

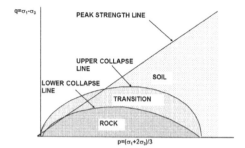

Figure 37. Behavior domains (Serrano, 1976).

Bravo 1970; Uriel & Serrano 1973) and energetic models (e.g. Serrano 1976, Aversa & Evangelista 1998, Serrano et al. 2002);
– Empirical models (e.g. Serrano et al. 2015a).

Besides, Del Olmo & Serrano (2010) tried to model the collapse of macroporous materials in a discrete way.

In the following sections a few remarkable features of some of these models are emphasized.

a. Energetic models for mechanical collapse
According to Serrano (1976), the behavior pattern shown in Figure 37 could be established in the effective stress space (q, p) defined as in Cambridge convention:

$$p = \frac{\sigma_1 + 2\sigma_3}{3} \; ; q = \sigma_1 - \sigma_3$$

There is a domain containing the origin in which the material behaves elastically in a normal load process. For stresses within this domain the pyroclast behaves as a rock. The boundary for the elastic domain is what is referred to as the lower collapse line. When the stresses reach this line, the structure of the material starts being affected, and when they reach the upper collapse line all the structure is totally destroyed. The failure occurs within this transition zone defined between the lower and the upper collapse lines. Once the structure has been totally destroyed the material behaves as a soil, being able to have one peak strength and another residual strength. The behavior will either be stable or unstable depending on the point reaching the upper collapse line.

For homogeneous materials the lower and upper collapse lines are the same and the transition zone disappears. These lines, which are in fact no more than the boundaries for the behavior domains, depend upon the stress path.

Four potential stress paths are considered and the stress-strain laws produced depend on these stress paths. There are four possibilities (Figure 38):

a) Isotropic compression tests (curve 1). Some examples of these types of curves were shown in Figure 28 for lapilli and in Figure 29 for pumice.
b) Tests with a high consolidation pressure (curve 2). In Figure 39 an example of this type of curve from lapilli is shown (red curve).

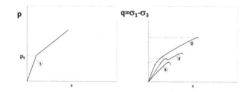

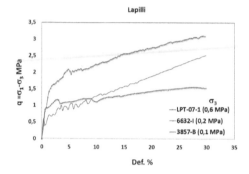

Figure 38. Stress-strength laws indicated in Figure 40.

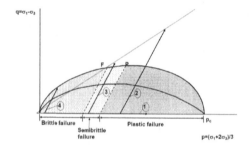

Figure 39. Some examples of curves of type 2 and 3 indicated in Figure 38 (intermediate and high confining pressures).

Figure 40. Types of failure (Serrano et al, 2002).

c) Tests with an intermediate consolidation pressure (curve 3). In Figure 39 some examples of these type of curves for lapilli are shown.
d) Tests with a low consolidation pressure (curve 4). These types of curves were shown in Figure 21 to Figure 24.

The zones corresponding to these different paths can be seen in Figure 40, with regard to the type of failure behavior.

The mathematical model developed by Serrano (1976; Serrano et al. 2002)) assumes the energy consumption law proposed by Roscoe and Burland (1968) for soft clays but slightly modified by a factor, $\lambda(\eta)$, depending on the uniformity and anisotropy of the material in relation to the stress path imposed. The proposed law for macroporous rocks is:

$$F* = \lambda(\eta)\sqrt{\psi^2 + M^2}$$

and taking into account Drucker's local stability theory (1959) expressed in a more general way by Wong and

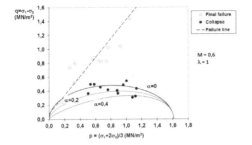

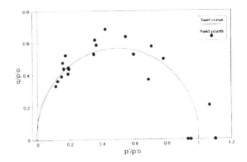

Figure 41. Adjustment of energetic model to pyroclastic samples (Serrano 1976; Serrano et al. 2002).

Figure 42. Proposed yield curve and yield stressed measured in a volcanic tuff (Aversa & Evangelista 1998).

Michel (1975) the differential equation of the collapse lines is:

$$p\frac{d\eta}{dp} + \psi + \eta - \alpha = 0$$

where η is the obliquity of the stress, $\eta = q/p$, ψ is the plastic dilatancy ratio, $\psi = dv^p/d\gamma^p$ (being v^p and γ^p the plastic volumetric and shear strains, respectively), M is a frictional parameter and α a parameter being 0 for associated plastic flow and a different value for non-associated plastic flows. Wong and Michel (1975) found a constant value of $\alpha \approx 0.25$ for sensitive cemented clays and Serrano (1976) obtained values between 0 and 0.4 for volcanic agglomerates if it is assumed $\lambda = 1$ and $M = 0.6$. According to Serrano, the assumption of $\lambda = 1$ simply derives from the lack of knowledge of this value, that should be investigated for each material.

Aversa & Evangelista (1998) obtain a good adjustment of test results on a volcanic tuff to a yield curve derived from Modified Cam-Clay as shown in Figure 42.

b. Empirical models: Unified failure model (Serrano (2012)

After the experimental study performed at CEDEX in the last years, Serrano (2012) proposes a unified failure criterion as follows (Figure 43):

$$\varphi(q) = Mf(p_0)$$

with:

$$\varphi(q) = q^{(k+1)} + 2kq$$

where:

$$f(p_0) = p_0\left(1 - \frac{p_0}{p_{c0}}\right)^\lambda$$

– k is a coefficient indicating the strength law at low pressures: if $k = 0$ the strength criterion is that of a parabolic collapsible type, representing an evolution of the Mohr-Coulomb criterion (Figure 44); if $k = 1$ the strength criterion is that of an elliptic collapsible type, representing an evolution of Hoek and Brown criterion (Figure 45);

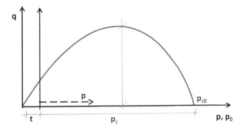

Figure 43. Unified strength criterion in Cambridge variables (Serrano, 2012).

– λ is a coefficient that satisfies $0 < \lambda < 1$;
– q and p are variables defined according to Cambridge convention, the deviatoric and the mean stress, respectively, made dimensionless by dividing them by a strength modulus β, i.e.:

$$q = \frac{\sigma_1 - \sigma_3}{\beta}$$

$$p = \frac{\sigma_1 + 2\sigma_3}{3\beta}$$

– p_0 is the dimensionless variable p referred to a translated q axis so that:

$$p_0 = \frac{\sigma_1 + 2\sigma_3}{3\beta} + \frac{t}{\beta}$$

being t the isotropic tensile strength of the material;
– p_{c0} is the dimensionless isotropic collapse pressure referred to the translated q axis:

$$p_{c0} = \frac{p_c + t}{\beta}$$

Both criteria can be expressed in simple forms as follow:

$$q = Mp_0\left(1 - \frac{p_0}{p_{c0}}\right)^\lambda$$

Parabolic collapsible criterion

$$q^2 + 2q = 6p_0\left(1 - \frac{p_0}{p_{c0}}\right)^\lambda$$

37

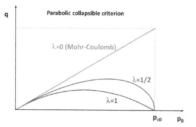

Figure 44. Parabolic collapsible criterion (Serrano, 2012).

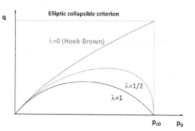

Figure 45. Elliptic collapsible criterion (Serrano, 2012).

Elliptic collapsible criterion.

In case of elliptic criterion, coefficient M must be equal to 6 to coincide with the Hoek & Brown criterion for low pressures (Figure 35). This criterion depends on four parameters: two explicit ones, λ and p_{c0}, and two implicit ones, β and t.

In the case of parabolic criterion if the strength modulus used to formulate it in dimensionless form is $\beta = p_{c0}$, it simplifies as:

$$q = Mp_0(1 - p_0)^\lambda$$

This criterion depends on four parameters: two explicit ones, M and λ, and two implicit ones, p_{c0} and t.

Both models were adjusted to the extent database obtained from all the samples tested at CEDEX and a few from published data (CEDEX 2013, Conde 2013) fitting well the test results. Nevertheless, in order to determine the parameters of the tested samples, the parabolic collapsible model was selected, as, being both models equally adequate, this one was the simplest. From Figure 46 to Figure 49 some adjustments to different types of pyroclasts are shown. Figure 46 shows a very good fit for welded pumice and lapilli samples. Again, a remarkable influence of the density of materials is observed. Figure 47 shows the adjustment to altered pumice and lapilli samples. In these cases there is a good adjustment to the test results although the isotropic collapsible loads or values closed to them have not been obtained in the samples tested, so the right part of the curves has not been checked. Figure 48 shows a good adjustment to the test results in basaltic and salic cemented ashes. Finally, Figure 49 shows a very good adjustment to published data corresponding to different Italian pyroclastic flows.

More details of the fitting to the parabolic collapsible model can be found in Serrano et al. 2015a.

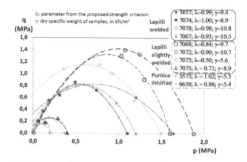

Figure 46. Adjustment of Serrano's parabolic collapsible criterion to welded pumice and lapilli samples (modified from CEDEX 2013 and Conde 2013).

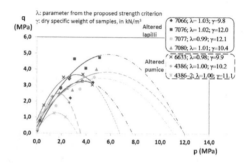

Figure 47. Adjustment of Serrano's parabolic criterion to altered pumice and lapilli samples (modified from CEDEX 2013 and Conde 2013).

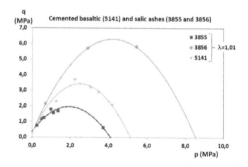

Figure 48. Adjustment of Serrano's parabolic criterion to pyroclastic cemented ashes samples (modified from CEDEX 2013 and Conde 2013).

As it can be observed, there is a similarity in the curves from the energetic model and the empirical elliptic collapsible model. The authors are working now in the way of obtaining the parameters for both models from the empirical results and the relation between them, looking for the theoretical basis that may lie under the empirical model.

From this criterion a relationship between the isotropic collapse pressure and the uniaxial compressive strength has been deduced as well as a depth of collapse in the ground due to the internal stresses overcoming the failure criterion of these materials. Both studies are included in Serrano et al. 2015b.

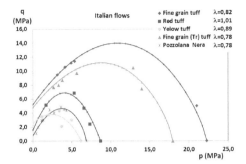

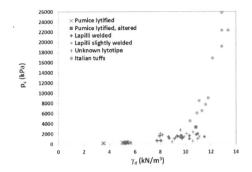

Figure 49. Adjustment of Serrano's parabolic criterion to published data from Italian tuffs (data from Aversa et al. 1993, Evangelista et al. 1998, Aversa & Evangelista 1998, Tommasi & Ribacchi 1998; Cecconi & Viggiani 1998 and 2001).

Figure 51. Results using tests of Canary Islands pyroclasts with a larger database of results (Serrano et al. 2015c).

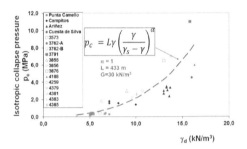

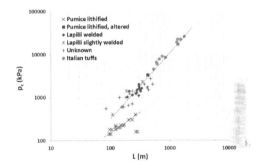

Figure 50. Results using tests of Canary Islands pyroclasts (Serrano et al 2010).

Figure 52. Structural parameter L versus the isotropic collapse ressure, p_c, for the studied materials (Serrano et al. 2015c).

c. Isotropic collapse pressure as a function of the macroporosity

Serrano et al. (2010) propose a single equation for the isotropic collapse pressure of macroporous rocks, p_c, theoretically deduced for pyroclasts with reticular and vacuolar porosity (Figure 50):

$$p_c = L\gamma \left(\frac{\gamma}{\gamma_s - \gamma}\right)^\alpha \qquad (1)$$

where L and α are parameters to be empirically adjusted, L is a structural parameter with length dimensions, γ_s is the specific gravity and γ the unit weight of the rock.

Terms from equation (1) can be rearranged so that the following expression is obtained for the adopted value of $\alpha = 1$:

$$p_c = L\gamma_s \frac{(1-n)^2}{n} \qquad (2)$$

where n is the porosity of the material.

Serrano et al. (2015c) adjust this equation to a large database and obtain the results shown in Figure 51 and in Figure 52. Two different trends are observed in this last figure, for the pumice lithified materials, with lower densities, and for the rest of the pyroclasts. If parameter L were known for each type of material the isotropic collapse pressure could be easily estimated, just by determining its specific gravity and porosity.

An empirical adjustment has been done and ranges of parameter L, although very wide, have been estimated. More details are given in Serrano et al. (2015c).

3.3 Empirical values and correlations

3.3.1 Ranges of values

The boxplots included in Table 8 show the ranges of variation of the different parameters for each lithotype. The boxes show the interquartile range 50% of the data) and the marked line represents the median of the data. The histograms of these data are shown in Table 9.

3.3.2 Correlations

Some correlations are obtained from the database created with test results. Some of them were published before (Serrano et al. 2007) and are now enlarged with new data.

Figure 53 shows the relationship between dry specific weight and uniaxial compressive strength for different lithotypes and for other samples of unknown lithotype, all of them tested at CEDEX. The following figures show a different trend for pyroclastic falls and pyroclastic flows between uniaxial or isotropic compressive strength and dry specific weight or porosity (Figure 54 to Figure 57). Finally, in Figure 58 the values for the modulus ratio obtained for different

39

Table 8. Ranges of variation of the different parameters (boxes with interquartile range (50%) and median).

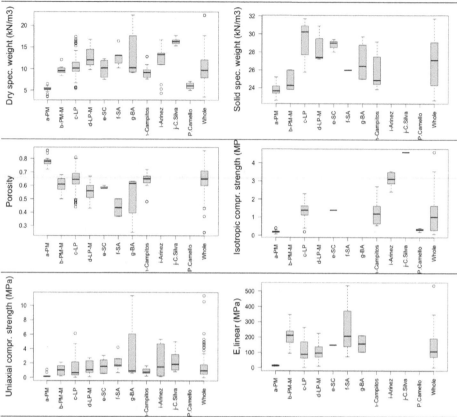

Lithotypes: PM: Pumice; PM-M: Altered pumice, with matrix (tuff); LP: Lapilli; LP-M: Altered lapilli, with matrix (tuff); SC: Scoria; SA: Salic ashes; BA: Basaltic ashes; Campitos: origin (Tenerife), unknown lithotype; Arinez: origin (Tenerife), unknown lithotype; C. Silva: origin (Tenerife), unknown lithotype; P. Camello: origin (Tenerife), unknown lithotype.

Table 9. Histograms of the different parameters shown in Table 7.

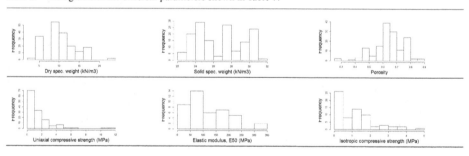

pyroclasts are shown for the whole group. A mean value of $E_{linear} = 160 * \sigma_c$ is obtained with a good correlation coefficient.

3.3.3 Some other remarkable features

In civil engineering and in construction, in general, the main problems related to low density pyroclasts from the point of view of their mechanical behavior are a low bearing capacity and a high deformability as well as collapsibility, particularly when it can occur brusquely. One way to avoid foundation ground collapse is to induce it before the construction of a structure, using heavyweight compactors (Uriel, 1976).

In addition to the features already analyzed in previous sections, there are some other important aspects that should be considered in relation to the mechanical behavior of low density pyroclast. Some of them are mentioned next:

– A rheological behavior of the pyroclasts is very often observed. For instance, in a reservoir constructed in El Hierro a deferred settlement due to creep around 33% of the instantaneous settlement was

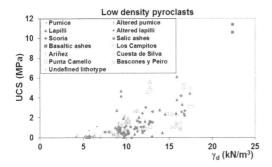

Figure 53. Dry specific weight (γ_d) vs. uniaxial compressive strength (σ_c). Low density pyroclasts (149 data, samples tested at CEDEX).

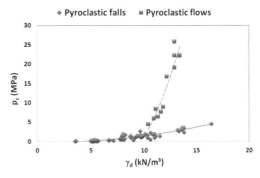

Figure 56. Dry specific weight (γ_d) vs. isotropic collapse load (p_c). Low density pyroclasts from pyroclastic falls and flows (from Conde 2013).

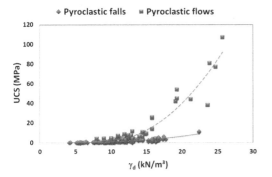

Figure 54. Dry specific weight (γ_d) vs. uniaxial compressive strength (UCS). Low density pyroclasts from pyroclastic falls and flows (from Conde 2013).

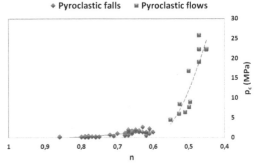

Figure 57. Porosity (n) vs. isotropic collapse load (p_c). Low density pyroclasts from pyroclastic falls and flows (from Conde 2013).

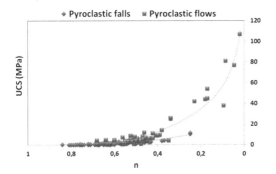

Figure 55. Porosity (n) vs. uniaxial compressive strength (UCS). Low density pyroclasts from pyroclastic falls and falls (from Conde 2013).

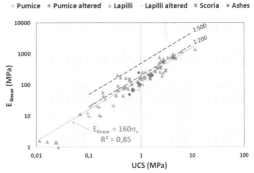

Figure 58. Modulus ratio for low density pyroclasts.

measured (Martín-Gómez et al. 2010); Fe Marqués & Martínez Zarco (2010) perform a back-analysis of a tunnel section excavated in basaltic pyroclasts based in measures of convergences and they observe that more than four months are needed for almost 100% of the final deformation to be produced, and still small long term flow deformations remain. Evangelista & Aversa (1994) published some test results of creep tests on tuffs.

– Influence of the water on the strength and deformability: Vásárheli (2002) studies this aspect on volcanic tuffs (andesite, basalt and rhyolite) with uniaxial compressive strengths ranging approximately from 2.5 to 60 MPa and finds a strength reduction about 27% and a modulus reduction around 20% for saturated samples. González de Vallejo et al. (2007) indicate a 30% reduction of strength for volcanic tuffs.

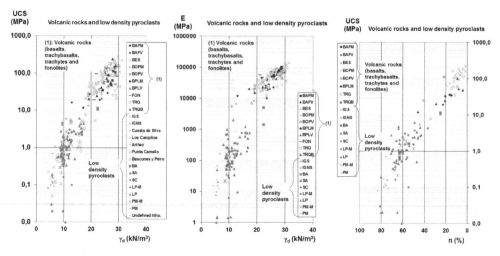

Figure 59. Relationship in semilog. scale between: a) dry specific weight (γ_d) and uniaxial compressive strength (UCS); b) dry density (γ_d) and Elastic modulus (E); c) dry specific weight (γ_d) and porosity (n), for rocks from Canary Islands proceeding from volcanic lava flows and from pyroclastic falls.

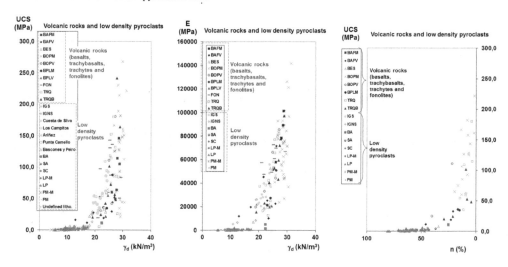

Figure 60. Relationship in natural scale between: a) dry specific weight (γ_d) and uniaxial compressive strength (UCS); b) dry density (γ_d) and Elastic modulus (E); c) dry specific weight (γ_d) and porosity (n), for rocks from Canary Islands proceeding from volcanic lava flows and from pyroclastic falls.

4 SOME CORRELATIONS WITH ALL VOLCANIC ROCKS STUDIED AT CEDEX

General trends shown in Figure 59 (semi-logarithmic scale) and in Figure 60 (natural scale) are obtained for all the volcanic rocks studied at CEDEX corresponding to the relation between the dry specific weight or the porosity and the uniaxial compressive strength and the elastic modulus, respectively.

5 FINAL REMARKS

In volcanic areas two main groups of volcanic rocks can be distinguished: rocks from lava flows and pyroclastic rocks. These two types of rocks have very different mechanical behaviors.

Some relevant aspects related to the mechanical behavior of these two types of volcanic rocks have been considered, mainly based on laboratory tests performed on rocks from the Canary Islands. Ranges of variation of the most relevant geotechnical parameters and some correlations are given for both types of rocks.

More detail has been paid to the analysis of low density pyroclasts, as their mechanical behavior is more peculiar and problematic. In civil engineering works and in construction, in general, the main problems with low density pyroclasts from the point of view of their mechanical behavior are the low bearing capacity and high deformability as well as the collapsibility, particularly when it can occur brusquely.

As pointed out by many researchers, the structure of pyroclasts has a strong influence on their mechanical behavior. The macroporosity of the samples tested at CEDEX was carefully analyzed and four different types of macroporosity were defined: reticular, vacuolar, mixed and matrix porosity. It was found that pyroclasts with a matrix-type porosity are much less dangerous as they do not suffer from mechanical collapse or if they did, it would be at much higher pressures and deformations than the others. This matrix-type of porosity is characterized by the presence of a fine grain material filling the macropores that surround the particles and was observed in altered samples of either lapilli or pumice type, and also in ashes.

Stress-strain curves from uniaxial, triaxial and isotropic tests on a large number of specimens tested at CEDEX were analyzed in detail studying the type of mechanical behavior and yield stresses, looking for a failure criterion that could be experimentally adjusted and could have a similar shape to theoretical models.

Serrano (2012) defined an empirical model that fits very well into test results of all samples from lapilli, pumice and ashes tested at CEDEX and into other published data of pyroclastic flows.

Although it is a difficult task with these complex volcanic materials, it should be very convenient to establish a unique and universal classification for them so that published test results could be more easily compared.

ACKNOWLEDGEMENTS

The author is very grateful to professor Serrano for sharing his wide knowledge and brilliant ideas on rock mechanics and particularly on volcanic rocks. The author would also like to thank the other experts with whom she has collaborated: professors Olalla and Rodríguez-Losada and particularly to Dr. Hernández-Gutiérrez, who not only performed an outstanding phD but also was responsible for selecting and sending all the samples tested at CEDEX Geotechnical Laboratory, and Dr. Conde, who performed a brilliant phD study. Also to Clemente Arias and José Toledo, who did a great job carefully cutting and preparing all the specimens tested at CEDEX.

REFERENCES

Adachi, T., Ogawa, T. & Hayashi, M. Mechanical properties of soft rock and rock mass. *Proc. 10th ICSMFE 1, 527–530* (1981).
Aversa, S., Evangelista, A., Leroueil, S. & Picarelli, L. Some aspects of the mechanical behaviour of "structured" soils and soft rocks. *International symp, on Geotechnical engineering of hard soils and rocks,* Athens, 1, 359–366 (1993).
Evangelista, A. and Aversa, S. Experimental evidence of non-lineal and creep behavior of pyroclastic rocks. *Viscoplastic behavior of geomaterials*, Ed, Springer (1994).

Aversa, S. & Evangelista, A. The mechanical mehaviour of a pyroclastic rock: failure strength and "destruction" effects. *Rock mechanics and Rock engineering*, 31, 25–42 (1998).
Barton, N. Low stress and high stress phenomena in basalt flows. *3rd Int. Workshop on Rock Mechanics and Geo-Engineering in Volcanic Environments*. Tenerife (2010).
Cecconi, M. & Viggiani, G. Physical and structural properties of a pyroclastic soft rock. *The Geotechnics of Hard Soils-Soft Rocks. 2do International Symposium on Hard Soils-Soft Rocks/Naples/Italy*, 1, 85–91 (1998).
Cecconi, M. Sample preparation of a problematic pyroclastic rock. *Problematic Soils*, Yanagisawa, Moroto & Mitachi (1998).
Cecconi, M. and Viggiani, G.M. Structural features and mechanical behavior of a pyroclastic weak rock. *International Journal for Numerical and Analytical Methods in Geomechanics, 25, 1525–1527.* (2001).
Cecconi, M. & Viggiani, G. Pyroclastic flow deposits from the colli albani. *Characterization and Engineering Properties of Natural Soils*, Taylor & Francis (2006).
Cecconi, M. Scarapazzi, M. & Viggiani, G. On the geology and the geotechnical properties of pyroclastic flow deposits of the colli albani. *Bulletin of Engineering Geology and the Environment*, 69, 185–206 (2010).
CEDEX. *Informe geotécnico sobre el terreno de cimentación de la presa de los campitos*. Laboratorio del transporte y mecánica del suelo". Internal report, 1972.
CEDEX. Caracterización geotécnica de los piroclastos canarios débilmente cementados. *Final Report* (April 2007).
CEDEX. Estudio del comportamiento geomecánico de los pi-roclastos canarios de baja densidad para su aplicación en obras de carreteras. *Final Report* (April 2013).
Drucker, D.C. 1959. A definition of stable inelastic materials. Trans. *A.S.M.E. Jour. Appl. Mech.* 26:1.
Conde M. *Caracterización geotécnica de materiales volcánicos de baja densidad*. PhD Thesis. Universidad Politécnica de Madrid (2013).
Del Olmo, D. & Serrano, A. Modeling of the collapse of a macroporous material. *3rd International Workshop on Rock Mechanics and Geo-Engineering in Volcanic Environments*. Tenerife (2010).
Evangelista, A. & Aversa, S. Experimental evidence of non-lineal and creep behavior of pyroclastic rocks. *Viscoplastic behavior of geomaterials,* Ed, Springer (1994).
Evangelista, A., Aversa, S., Pescatore, T.S. & Pinto, F. Soft rocks in southern Italy and the role of volcanic tuffs in the urbanization of Naples. The Geotechnics of Hard Soils-Soft Rocks. 2do *International Symposium on Hard Soils-Soft Rocks/Naples/Italy*, 3, 1243–1267 (1998).
Fe Marqués, M. & Martínez Zarco, R. Geotechnical parameters of basaltic pyroclastics in La Palma Island, based on convergences measured in a tunnel. *3rd International Workshop on Rock Mechanics and Geo-Engineering in Volcanic Environments*. Tenerife (2010).
Gobierno de Canarias. *Guía para la planificación y realización de estudios geotécnicos para la edificación en la comunidad autónoma de Canarias*. GETCAN 011 (2011).
González de Vallejo, L.I., Hijazo, T., Ferrer, M., & Seisdedos, J. Caracterización Geomecánica De Los Materiales Volcánicos De Tenerife, *ed. M. A. R. G. Nº 8*, Madrid: Instituto geológico y minero de España (2006).
González de Vallejo, L.I., Hijazo, T., Ferrer, M. & Seisdedos, J. Geomechanical characterization of volcanic materials in Tenerife. *ISRM International Workshop on Volcanic Rocks*. Ponta Delgada, Azores. (2007).

Gonzalez de Vallejo, L. I., Hijazo, T. & Ferrer, M. Engineering Geological Properties of the Volcanic Rocks and Soils of the Canary Islands. *Soils and Rocks*, Sao Paulo 31 (1): 3–13, (2008).

Hernández-Gutiérrez, L-E. Caracterización geomecánica de las rocas volcánicas de las islas Canarias. PhD Thesis. Universidad de La Laguna, Tenerife (2014).

Hoek, E. & Brown T. Empirical strength criterion for rock masses. *J. Geotech. Eng. Div., ASCE 106 (GT9)*. (1980).

Kahraman, S. The determination of uniaxial compressive strength from point load strength for pyroclastic rocks. *Engineering Geology* 170 (2014).

Lerouiel, S. & Vaughan, P.R. The general and congruent effects of structure in natural solis and weak rocks. *Geotechnique* 40, No 3, 467–488 (1990).

Lomoschitz Mora-Figueroa, A. Caracterización geotécnica del terreno con ejemplos de Gran Canaria y Tenerife. *Universidad de Las Palmas de Gran Canari*,(1996).

Martín-Gómez M.R., Fernández-Baniela, F., Arribas-Pérez de Obamos, J.J. & Soriano, A. Deformational behavior of pyroclastic rocks beneath the upper reservoir of the hydro-wind plant at El Hierro. *3rd International Workshop on Rock Mechanics and Geo-Engineering in Volcanic Environments*. Tenerife (2010).

Mesquita Soares, S., Dinis da Gama, C. & Reis e Sousa, M. Geomechanical properties of basalts from the volcanic complex of Lisbon-Some interesting correlations. *Workshop on Volcanic Rocks*, Funchal, Madeira. Eurock (2002).

Moon, V. G. Geotechnical characteristics of ignimbrite: a soft pyroclastic rock type. *Eng. Geology* 35(1–2): 33–48 (1993).

Muñiz Menéndez, M. & González-Gallego, J. Rock mass classification schemes in volcanic rocks. 3rd International *Workshop on Rock Mechanics and Geo-Engineering in Volcanic Environments*. Tenerife (2010).

Peiró Pastor, R. Caracterización geotécnica de los materiales volcánicos del Archipiélago Canario. *Tierra y Tecnología* 16 y 17 (1997).

Pellegrino, A. Mechanical behaviour of soft rocks under high stresses. *ISRM, II Congreso Internacional de Mecánica de Rocas*, Tomo II, Belgrado, R,3,25 (1970).

Pola, A., Crosta, G.B., Castellanza, R., Agliardi, F., Fusi, N., Barberini, V. Norini, G. & Villa, A. Relationship between porosity and physical mechanical properties in weathered volcanic rocks. *3rd Int. Workshop on Rock Mechanics and Geo-Engin. in Volcanic Environments*. Tenerife (2010).

Rodríguez-Losada, J.A., Hernández-Gutiérrez, L.E., Olalla, C., Perucho, A., Serrano, A. & Potro, R. D. The volcanic rocks of the Canary Islands. Geotechnical properties. *ISRM International Workshop on Volcanic Rocks*. Ponta Delgada, Azores (2007).

Rodríguez-Losada, J.A., Hernández-Gutiérrez, L.E., Olalla, C., Perucho, A., Serrano, A. & Eff-Darwich, A. Geomechanical parameters of intact rocks and rock masses from the Canary Islands: Implications on their flank stability. *Journal of Volcanology and Geothermal Research* 182 (2009).

Rotonda, T., Tommasi, P. & Ribacchi, R. Physical and mechanical characterization of the soft pyroclastic rocks forming the Orvieto cliff. *Workshop on Volcanic Rocks*. Funchal, Madeira. Eurock (2002).

Santana, M., de Santiago, C., Perucho, A. & Serrano, A. Relación entre características químico-mineralógicas y propiedades geotécnicas de piroclastos canarios. *VII Congreso Geológico de España*. Geo-Temas 10 (2008).

Serrano, A. Aglomerados volcánicos en las Islas Canarias. *Memoria del Simposio Nacional de Rocas Blandas*. Tomo II. pp. 47–53. Madrid (1976).

Serrano, A. Mecánica de rocas I: Descripción de las rocas; II Propiedades de las rocas. *Publicaciones de la E.T.S. de I.C.C. y P*. Madrid (1997).

Serrano, A., Olalla, C. & Perucho, Á. Mechanical collapsible rocks. *Workshop on Volcanic Rocks*, Funchal, Madeira. Eurock (2002)a.

Serrano, A, Olalla, C, & Perucho, Á. Evaluation of non-linear strength laws for volcanic agglomerates. *Workshop on Volcanic Rocks*, Funchal, Madeira. Eurock (2002)b.

Serrano, A., Olalla, C., Perucho, A. & Hernández, L. Strength and deformability of low density pyroclasts. *ISRM Int. Workshop on Volcanic Rocks*. P. Delgada, Azores (2007).

Serrano, A., Perucho, Á. & Conde, M. Isotropic collapse load as a function of the macroporosity of volcanic pyroclasts. *3rd International Workshop on Rock Mechanics and Geo-Engineering in Volcanic Environments*. Tenerife (2010).

Serrano, A. Unified failure criterion for collapsible materials. *Personal communication*. (2012).

Serrano, A., Perucho, Á. & Conde, M. Failure criterion for low density pyroclasts. *Workshop on volcanic rocks and soils*. Ischia (2015)a.

Serrano, A., Perucho, Á. & Conde, M. Relationship between the isotropic collapse pressure and the uniaxial compressive strength, and depth of collapse, both derived from a new failure criterion for low density pyroclasts. *Workshop on volcanic rocks and soils*. Ischia (2015)b.

Serrano, A., Perucho, Á. & Conde, M. Correlation between the isotropic collapse pressure and the unit weight for low density pyroclasts. *Workshop on volcanic rocks and soils*. Ischia (2015)c.

Tommasi, P. & Ribacchi, R. Mechanical behaviour of the orvieto tuff. The Geotechnics of Hard Soils-Soft Rocks. *2do International Symposium on Hard Soils-Soft Rocks*/Naples/Italy, 2: 901–909 (1998).

Tommasi , P., Verrucci, L. & Rotonda, T. Mechanical properties of a weak pyroclastic rock and their relationship with microstructure. *Can. Geotech. J*. 52: 1–13 (2015).

Uriel, S. & Bravo, B. La rotura frágil y plástica en un aglomerado volcánico de las palmas de gran canaria. *I Congreso hispano-luso-americano de geología económica*, Sesión 5. (1971).

Uriel, S. & Serrano, A. Geotechnical properties of two collapsible volcanic soils of low bulk density at the site of two dams in canary island (Spain). *8th Congress I.S.S.M.F.E*. Vol. I: 257–264. Moscú (1973).

Uriel, S. Comportamiento de suelos colapsibles. *Mem. Simp. Nac. de Rocas Blandas*. Tomo 2. Madrid (1976).

Uriel, S. & Serrano, A. Propiedades geotécnicas de algunos aglomerados volcánicos en las islas canarias. *Mem. Simp. Nacional de Rocas Blandas*. Tomo I, A-10. Madrid (1976).

Wong, P.K.K. & Mitchell, R.J. Yielding and plastic flow of sensitive clay. *Geotechnique* 25. No 4. (1975).

Volcanic Rocks and Soils – Rotonda et al. (eds)
© *2016 Taylor & Francis Group, London, ISBN 978-1-138-02886-9*

From micro to macro: An investigation of the geomechanical behaviour of pumice sand

R.P. Orense & M.J. Pender
University of Auckland, Auckland, New Zealand

ABSTRACT: Pumice deposits, which are found in several areas of the North Island, NZ, are lightweight, highly crushable and compressible, making them problematic from an engineering and construction viewpoint. In this paper, the results of a research programme undertaken to investigate the properties of pumice sand at grain-size level in order to understand its macro-response are presented. SEM and X-ray CT images were acquired to visualize and quantify the particle shape characteristics and distribution of voids within the grains. Single particle crushing tests were performed to characterize the particle strength. Geotechnical laboratory tests, including bender element tests, dynamic deformation tests and monotonic/cyclic undrained triaxial tests, were conducted on pumice sand specimens and the results were compared with those obtained from hard-grained sands. The results indicate that due to the complex surface shape and low crushing strength of pumice, interlocking effect appeared to be significant, resulting in higher mobilised peak frictional angle and higher liquefaction resistance when compared with hard-grained sands. These results confirmed that the micro-level characteristics of the pumice particles have significant effect on their macro-level response.

1 INTRODUCTION

The recent Canterbury earthquake sequence of 2010–2011 caused significant damage to many parts of Christchurch, New Zealand mainly due to widespread liquefaction and amplified ground motion (e.g., Cubrinovski and Orense, 2010; Orense et al. 2011 and 2012). Consequently, the engineering community, as well as local and regional councils, has focused attention on understanding the geotechnical characteristics and seismic behaviour of various local soils for the purpose of designing earthquake-resistant structures.

A cursory review of the current state of research on soil liquefaction showed that nearly all the work on this topic has been directed towards understanding the properties of hard-grained (quartz) sands; very little research has been done on the dynamic characteristics of volcanically-derived sands. However, it is known that New Zealand's active geologic past has resulted in widespread deposits of volcanic soils throughout the country. The M 6.3 1987 Edgecumbe earthquake showed localized patches of liquefaction of sands of volcanic origin across the Rangitaiki Plains.

Pumice deposits are found in several areas of the North Island. They originated from a series of volcanic eruptions centred in the Taupo and Rotorua regions, called the "Taupo Volcanic Zone" (see Figure 1). The pumice material has been distributed initially by the explosive power of the eruptions and associated airborne transport; this has been followed by erosion and river transport. Presently, pumice deposits exist mainly as deep sand layers in river valleys and flood plains, but are also found as coarse gravel deposits in hilly areas. Although they do not cover wide areas, their concentration in river valleys and flood plains means they tend to coincide with areas of considerable human activity and development. Thus, they are frequently encountered in engineering projects and their evaluation is a matter of considerable geotechnical interest.

Because it is lightweight, highly crushable and compressible, pumice is problematic from an engineering and construction viewpoint. Although significant studies have been carried out on many crushable soils, such as carbonate sands and calcareous soils, pumice is a unique material – the particles crush easily under fingernail pressure. It is possibly the most delicate of the suite of crushable soils found at various locations around the world. Because pumice sand occurs widely in many areas of the country and is important in infrastructure construction, and yet not readily characterized in situ, better understanding of its micro-mechanical properties is important.

Previous research at the University of Auckland showed that the q_c values obtained from cone penetration tests (CPT) on pumice sand were only marginally influenced by the density of the material, as shown in Figure 2. The reason for this behaviour is possibly because the stresses imposed by the penetrometer are so severe that particle breakage forms a new material and that the properties of this are nearly independent of the initial state of the sand. It is also noted that pumice sand CPT resistance shows very gentle increase with confining stress as compared to normal (i.e. hard-grained) sands (Wesley et al. 1999). Thus

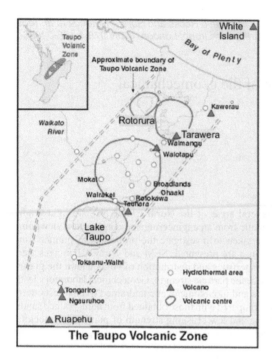

The Taupo Volcanic Zone

Figure 1.   The Taupo Volcanic Zone (TVZ) in the North Island of New Zealand (from www.explorevolcanoes.com).

cone resistance (kPa)

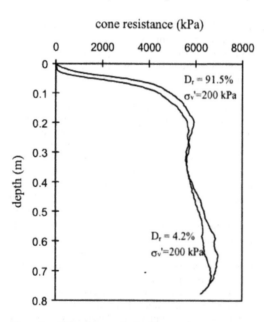

Figure 2.   CPT resistance obtained from calibration chamber testing of loose and dense pumice sand (Wesley et al., 1999).

conventional relationships between the q_c value, relative density, and confining stress are not valid for these soils. Therefore, alternative relationships specifically for pumice sands need to be developed.

In working with pumice, it is important to distinguish between surface and internal voids within

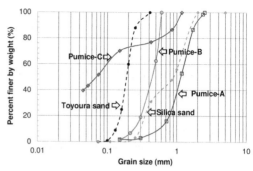

Figure 3.   Particle size distribution of samples used in the tests.

particles and the conventional voids defined as the spaces between particles. For crushable soils like pumice, such distinction is necessary because when particles are crushed, the release of internal voids would contribute to the void space between the particles.

Noting that the micro-level characteristics of the sand particles have significant effect on their macro-level response, a comprehensive research programme was undertaken not only to characterise pumice at the particle level but also to investigate its geomechanical properties. This paper presents some of the major observations made regarding the relation between mechanical properties of New Zealand pumice at grain-size level and its geomechanical behavior. Other papers (Kikkawa et al. 2013a; Pender et al. 2014) showed that the properties of New Zealand and Japanese pumice sands have similarities.

2   SOILS SAMPLES USED

The material used in the testing programme was the commercially-available pumice sand. This is not a natural deposit but was derived by processing sand from the Waikato River, located near Mercer. The particles were centrifugally separated from the other river sand particles so that the samples consisted essentially of pumice grains. This commercially-available material has been used extensively in the Geomechanics Laboratory of the University of Auckland (e.g. Marks et al., 1998; Wesley et al., 1999; Wesley, 2001; Pender, 2006; Pender et al., 2006; Kikkawa et al., 2009 and 2011). In this research, pumice sands with three different grading curves were used: Pumice-A sand (0.075–2.5 mm); Pumice-B sand (0.15–0.60 mm); and Pumice-C sand, which was prepared by crushing Pumice-A sand to obtain a target fines content, $Fc \sim 50\%$. The grain size distribution curves of the soils used are shown in Figure 3, while the index properties based on NZ Standard (1986) are summarised in Table 1. For comparison purposes, Toyoura sand and Silica sand, both hard-grained sands, were also used.

A comparison of the void ratio characteristics of pumice sands with those of typical geomaterials is

Table 1.

Table 1. Properties of soils used.

Material	Specific Gravity	Maximum void ratio	Minimum void ratio
Pumice-A sand	1.95	2.584	1.760
Pumice-B sand	2.10	2.640	1.420
Pumice-C sand	2.38	2.091	1.125
Toyoura sand	2.64	0.974	0.613
Silica sand	2.65	0.850	0.524

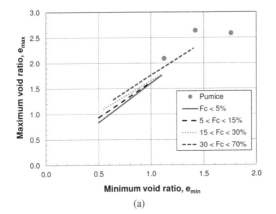

(a)

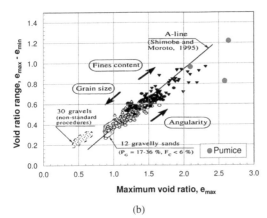

(b)

Figure 4. Comparison of void ratios of pumice with natural sands: (a) e_{max} vs. e_{min} relation; and (b) e_{max} vs. $e_{max}-e_{min}$ relation (modified from Cubrinovski and Ishihara, 2002).

shown in Figure 4, where it is observed that pumice more or less fits the trend for most natural (hard-grained) sands, although it is characterized with higher values of e_{max} and e_{min}.

3 CHARACTERISATION AT GRAIN-SIZE LEVEL

In the first stage of the study, the properties of pumice sands at grain-size (micro-)level were investigated through particle shape characterisation and particle crushing tests. This was done through scanning

electron microscope (SEM) and computed tomography (CT) imaging, as well as single particle crushing tests.

3.1 SEM imaging

In order to observe the grain structure of pumice in detailed but qualitative way, a series of micro-graphs of the original particles (Pumice A sand) with different sizes and of the crushed particles (Pumice C sand) was obtained by scanning electron microscope (SEM), and the images are shown in Figure 5. The images give a clear qualitative indication that as the particle size decreases the shape and surface texture tend to be less uniform and more angular. Especially when the particles are crushed, the surface is more jagged and irregular, and this could lead to more interlocking potential under shear stress application.

3.2 Micro X-ray CT scanning and image analysis

In addition to SEM imaging, we scanned many particles of Pumice A sand individually using Skyscan 1172 high-resolution micro-CT scanning machine. For each particle, the scanned images consisted of many box-shaped cells (called *voxels*), with each voxel having a gray scale value between 0 and 255, which correspond to black and white colours, respectively. With the adopted scheme of binarising each CT images, the solid portion in the particle was delineated from the air/water portion. Overall, 30 particles were scanned, with size ranging from 1.18–2.36 mm. Example images are shown in Figures 6 where the solid portion of the particle is shown as white colour and the black region is the void. Figure 6(a) shows the 3D pumice particle reconstructed using all longitudinal binarised images. Figure 6(b) indicates the locations of the cross-sectional planes while Figures 6(c), 6(d) and 6(e) are the images in the y-z, z-x and x-y cross-sectional planes, respectively.

It can be seen from the figure that almost all the voids open to the exterior of the pumice particle, so that there are more surface voids than internal voids. Using an imaging analysis algorithm, we attempted to quantify the bulk volume of the particle, the volume of solids and the volume of internal voids. The calculation details are provided in Kikkawa et al. (2013b; 2013c). Based on the analysis of 23 particles of varying sizes, the internal void ratio, defined as the ratio of the volume of internal voids to the solid volume of the particle, ranges from 10^{-5} and 10^{-2}, with an average value of 6×10^{-3} It was noted that a pumice particle has roughly the same proportion of internal void volume to the total particle volume, regardless of size (even for particles $<75\,\mu m$). Such internal void ratio range appears to be surprisingly small, especially when comparing with cross-sectional images in 2-dimension (see Figure 6). When the 3-dimensional image of the particle is constructed, as shown in Figure 6(a), the said range appears to be correct because in reality, the apparent internal voids in the cross-section

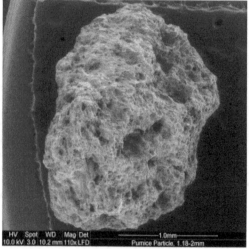

(a) Original > 2mm

(b) Original 0.6 – 1.18mm

(c) original 0.075 – 0.15mm

(d) crushed particles

Figure 5.   Scanning electron microscope (SEM) images of pumice sands with different sizes.

of the particle are actually connected to the outside surface in 3-dimension; thus, interpretations based on 2-D and 3-D may be very different.

Next, using a very sensitive mass balance with a resolution of 10^{-5} g, measurement of the particle mass was conducted with the aim of quantifying the solid density of the pumice particle. Figure 7 shows the variation of the solid density of pumice with particle size, based on image analysis. For comparison purposes, the densities measured by Wesley (2001) using the New Zealand standard procedure and a "direct" procedure without vacuum extraction are included. From the figure, the computed values are generally higher than those from the laboratory-obtained values. We believe that CT scanning could measure the solid density of pumice particle more accurately than the standard procedure because it could measure the

volume of solid particle with due consideration to the volume of surface and internal voids within a particle.

Based on CT-scanned images of the 30 particles, the average solid density calculated was 2.2 g/cm^3, which is almost the same as the density of a particle with ~75 μm size as measured by Wesley (2001) using the standard method. Similar range of solid densities (2.330–2.413 g/cm^3) was reported by de Lange et al. (1991) on the pumiceous sediments they sampled in the vicinity of Tauranga Harbour, New Zealand, which is also located within the Taupo Volcanic Zone.

It should be acknowledged that the possibility exists that the scanning resolution adopted in this study, 4 μm and 8 μm, may not be sufficient to accurately measure the solid density, especially for finer particles. Higher resolution may lead to higher density because more details of the microstructure would be available

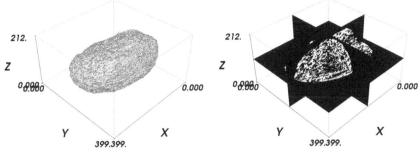

(a) Three-dimensional reconstructed image   (b) locations of cross-sectional planes

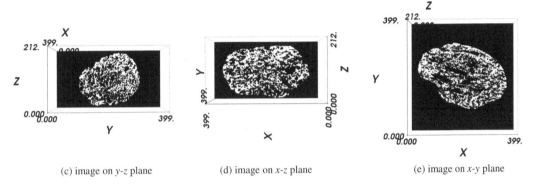

(c) image on *y-z* plane   (d) image on *x-z* plane   (e) image on *x-y* plane

Figure 6. CT images of a pumice particle, showing the cross-sectional images and the reconstructed particle through image analysis (Kikkawa et al. 2013b).

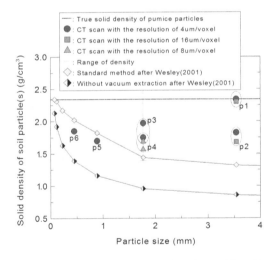

Figure 7. Variation of solid density of pumice particles with particle size (Kikkawa et al. 2013c).

particularly the volume of internal voids of smaller dimension than the resolution used to date.

3.3 *Particle shapes*

Next, we characterise the shape of the pumice particles by analysing the horizontal cross-sectional images of several particles and then quantifying the particle

shape through roundness coefficient, R_c, aspect ratio, A_r, and angular coefficient, A_c. These indices were calculated as follows.

$$R_c = \frac{L^2}{4\pi A} \tag{1}$$

$$A_r = \frac{b}{a} \tag{2}$$

$$A_c = \left| R_c - \frac{1 + (A_r)^2}{2(A_r)} \right| \tag{3}$$

In the above equations, L is the perimeter measured on plan image of particle in its most stable position, A is the cross-sectional area of the same image, and a and b are widths of the particle along the minor and major axes, respectively. Note that R_c is equal to unity for a perfectly circular particle cross-section and becomes >1 with increasing particle roughness.

Comparing the relationship between roundness coefficient and aspect ratio of pumice and other sands, pumice plots together with other crushable soils in Japan per the classification scheme proposed by Kato (2002), i.e. crushable soils are characterised with $A_r \geq 1.5$ and $1.4 \leq R_c < 2.0$. However, as illustrated in Figure 8, which shows the relationship between angular coefficient and aspect ratio for different types of sands, pumice particles have higher angular coefficient than the other crushable soils.

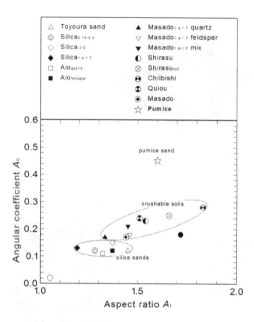

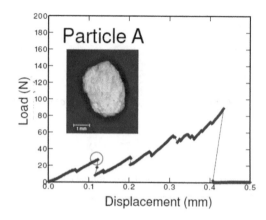

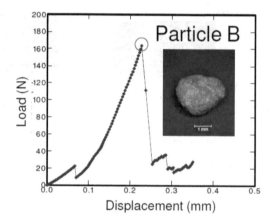

Figure 8. Relation between angular coefficient and aspect ratio (after Kato 2002).

Therefore, based on imaging and particle shape characterisation, it is recognised that pumice particles are not only porous but also very angular as well. Such complex surface shape is the reason for the large void range of pumice when compared to other natural sands.

3.4 *Single particle crushing tests*

Single particle crushing tests were conducted on pumice particles. In the test, a particle was placed in stable direction on the bottom bearing plate and the top plate was lowered at constant speed (0.1 mm/min) to crush it. During the test, axial load and displacement were measured and recorded with a computer. Details of the tests are described in Orense et al. (2013).

In this study, particle crushing tests were performed on 60 Pumice A sand particles with diameter between 0.6–2.5 mm. The samples of pumice grains appear to be of two types: yellow-coloured particle (referred as particle A; and light brown-coloured particle (referred as particle B). Representative results for Particle A and B are shown in Figure 9 in terms of load vs displacement relations. It can be seen that for both particles, a "saw tooth" pattern is observed, with the load increasing and decreasing with displacement as crushing of fragments/corners occurs. For the purpose of the analysis, the first peak load, F_c (denoted by red circle in the figure) is considered as the fracturing of significant portions of the particle and indicates the crushing strength, σ_f, of the particle which can then be calculated by dividing the first peak load, F_c, by the square of the initial height of the particle, i.e.,

$$\sigma_f = \frac{F_c}{h_0{}^2} \tag{4}$$

Figure 9. Typical force-displacement relations obtained for different types of pumice particles in single particle crushing test.

The definition of crushing strength adopted here is dependent on how the particles were crushed. The first peak load is considered as the crushing of the particle core (i.e. main portion of the particle). In addition to monitoring the load-displacement relation, visual inspection of the crushing process was employed. In Figure 9, significant splitting of the core of Particle A occurred at around 0.12 mm of plate movement, the succeeding readings represent the crushing of the already broken particles. Thus, only the portion up to say 0.15 mm should be considered. For Particle B, on the other hand, the first peak in the plot at 0.06 mm plate movement represents crushing of corners, while the particle core was powdered at 0.21 mm; hence, the crushing strength was taken at the higher load level. Note that while the crushing strength of the hard-grained sand is depicted by splitting of the particle into 2–3 blocks at the maximum load, the crushing of soft-grained pumice is characterized by gradual breakage followed by a large drop in load when the particle core is crushed.

Figure 10 plots the relation between single particle crushing strength and initial height of the pumice specimen. It can be seen that there is a general trend of

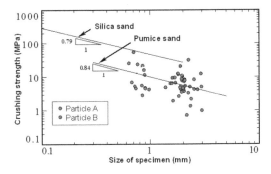

Figure 10. Variation of crushing strength with particle size (modified from Orense et al. (2013).

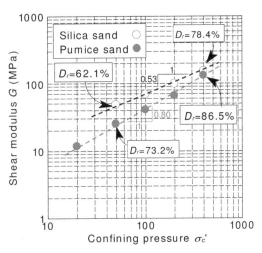

Figure 11. Results of BE tests showing variation of initial shear modulus with effective confining pressure for both pumice sand and Silica sand.

decreasing strength with increase in particle size For comparison purposes, the trend for Silica sand is also indicated in the figure, based on tests conducted by Nakata et al. (2001). It can be seen that the trends are similar, but the particle crushing strength of pumice is one order of magnitude less, showing the highly crushable nature of pumice.

4 INITIAL SHEAR MODULUS

The initial shear modulus of the soil, G_{max}, is an important design parameter. For this purpose, bender element (BE) tests were performed on reconstituted Pumice A specimens with target density $D_r = 80\%$. The dimensions of the specimen were 50 mm in diameter and 100 mm high. The specimens were saturated in similar manner as the triaxial specimens (as discussed below) after which they were isotropically consolidated at a prescribed effective confining pressure, σ_c', which ranged from 20–400 kPa.

The BE system consisted of WF1943 function synthesizer generator and a VC6723 digital oscilloscope. Sine waves were applied with frequency $f = 2.5, 5, 10, 15, 20, 25$ and 30 Hz. The transmitted and received signals were downloaded directly to a computer for post-analysis. The travel distance of the wave was calculated between tip to tip of bender element while the peak to peak arrival time was chosen to determine the arrival time, t_s.

The shear wave velocities, V_s, of reconstituted pumice specimens were obtained using a single specimen consolidated at different levels of σ_c'. Figure 11 plots the variation of G_{max} of pumice as a function of σ_c'. For comparison purposes, the trend for Silica sand of slightly lower relative density is also presented in the figure. The values were calculated from the shear wave velocity using the relation

$$G_{max} \rho_{bulk} (V_s)^2 \qquad (5)$$

where ρ_{bulk} is the bulk density of the specimen. The effect of σ_c' is obvious, with a higher increase in the G_{max} of pumice sand for a prescribed change in confining pressure when compared to Silica sand. From

the relation $G_{max} \propto (\sigma_c')^m$, the best-fit lines indicated in the figure show $m = 0.80$ for pumice, which is higher than $m = 0.53$ for Silica sand; this is a manifestation of the crushability of pumice sands under isotropic consolidation pressure.

5 DYNAMIC DEFORMATION PROPERTIES

The strain-dependent shear modulus and damping characteristics of soils are very important input parameters in performing seismic ground response analyses. To examine this, a hollow cylindrical torsional shear apparatus was employed, with the specimens prepared by air pluviation method. The hollow cylindrical specimen, with an inner diameter $d_i = 60$ mm, outer diameter $d_o = 100$ mm and height, $h = 100$ mm, was enclosed laterally by two flexible membranes and vertically by rigid top and bottom caps. After ensuring high degree of saturation, the specimen was then isotropically consolidated to the target mean effective stress, p'. Two types of pumice sands (Pumice A and Pumice C sands) and two levels of confining pressure ($p' = 20$ and 100 kPa) were used.

Undrained cyclic torsional shear tests were carried out as per JGS 0543-2000 (JGS 2000; Tatsuoka et al. 2001). All the specimens were subjected to sinusoidal undrained cyclic loading at a frequency of 0.1 Hz and 11 cycles were applied in each loading stage. At the end of each loading stage, the excess pore-water pressure was dissipated by opening the drainage valves. In the calculation of shear stresses and shear strains, the volume of specimen immediately before the start of each loading stage was employed. The dynamic properties were calculated using the tenth cycle data.

It is customary to represent the variation in shear modulus at any shear strain level by normalizing it with the initial shear modulus at a strain level equal

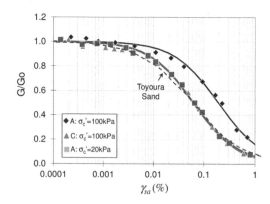

Figure 12. Normalized shear modulus versus single amplitude shear strain for Pumice A and C sands. The curve for Toyoura sand was obtained by Iwasaki et al. (1978).

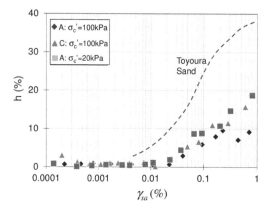

Figure 13. Damping ratio versus single amplitude shear strain relation for Pumice A and C sands. The curve for Toyoura sand was obtained by Tatsuoka and Iwasaki (1978).

to 0.0001%. This facilitates a comparison of the relationship of soils under different conditions. Figure 12 shows the strain-dependent normalized shear modulus G/G_0 for the three cases. It is evident from the plots that Pumice A sand (with no fines content) shows elastic behaviour for a wider range of shear strains when compared to Pumice C sand (with higher F_c). This is in agreement with the behaviour of Shirasu (Hyodo 2006). Similarly, the rate of reduction in shear modulus of pumice with shear strain decreases with increase in confining pressure, consistent with the observation of Kokusho (1980) who conducted cyclic triaxial tests on Toyoura sand for confining pressures ranging from 20–300 kPa. Also shown in the figure is the shear modulus ratio of Toyoura sand ($p' = 100$ kPa) as reported by Iwasaki et al. (1978) using hollow torsional shear apparatus and considering different sample preparation techniques. It is observed that the trend line for Toyoura sand is closer to Pumice C sand, although the reduction rate is a bit higher.

The relation between damping ratio, h, and single amplitude shear strain, γ_{sa}, is illustrated in Figure 13. It is observed that at strain level less than 0.2%, the

damping ratio of pumice is not influenced by the level of confining pressure and grading characteristics (or fines content); however, beyond 0.2%, damping ratio seems to increase with increase in fines content and with decrease in confining pressure. These trends are consistent with those observed for Shirasu (Hyodo 2006) and Toyoura sand (Kokusho 1980), respectively. Toyoura sand, on the other hand, has higher damping ratio at shear strain near failure, almost twice that of pumice, as shown by the plot obtained from the test results of Tatsuoka and Iwasaki (1978) using hollow torsional shear apparatus ($p' = 100$ kPa) with samples prepared through different methods. This can be attributed to the crushability of pumice, where the structure of the specimen becomes more stable with continuous shearing, resulting in a more deformation-resistant response and therefore lower damping at large strain level.

6 UNDRAINED TRIAXIAL TESTS

Undrained monotonic and cyclic tests were conducted on reconstituted pumice sand specimens. Because of the presence of surface and interior voids, it was not easy to completely saturate the pumice sand. Instead, saturated specimens were made using de-aired pumice sands, i.e., sands were first boiled in water to remove the entrapped air. To prepare the test specimens, the sand was water-pluviated into a two-part split mould which was then gently tapped until the target relative density, D_r, was achieved. Next, the specimens were saturated with appropriate back pressure and then isotropically consolidated at the target σ_c'. B-values > 0.95 were obtained for all specimens. The cyclic triaxial test specimens were 75 mm in diameter and 150 mm high, while for the monotonic tests, the specimen dimension was 38 mm in diameter and 80–85 mm high.

The undrained monotonic triaxial tests were performed using a screw-jack loading machine. Undrained monotonic loading was applied using axial compression at a rate of 0.015 mm/min. The target strain was 30% to observe the specimen behaviour at large deformation state. Some specimens did not reach this strain level due to a number of factors, rendering the data unreliable. For the undrained cyclic testing, a sinusoidal cyclic axial load was applied by a servo-hydraulic-powered loading frame at a frequency of 0.1 Hz. In all tests, the axial load, cell pressure, pore pressure, volume change and axial displacement were all monitored electronically and these data were recorded via a data acquisition system onto a computer for later analysis.

6.1 Undrained monotonic test results

The results of the consolidated undrained monotonic triaxial tests on reconstituted pumice sands are expressed in terms of the effective stress path (deviator stress vs. mean effective stress) and deviator

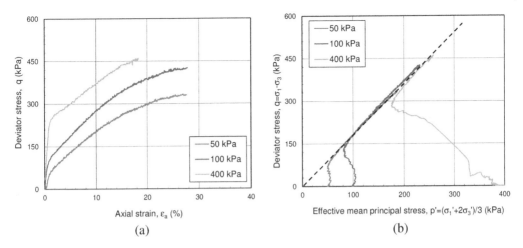

Figure 14. CU test results for loose Pumice A sands: (a) stress-strain relation; and (b) effective stress paths.

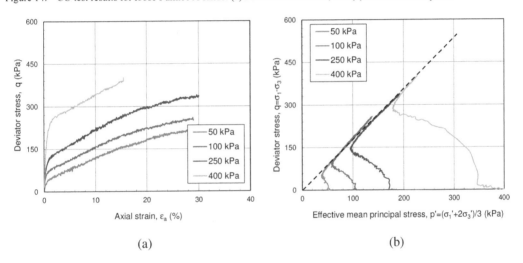

Figure 15. CU test results for medium dense Pumice A sands: (a) stress-strain relation; and (b) effective stress paths.

stress-axial strain relation. Taking σ_1' and σ_3' as the maximum and minimum effective principal stresses the triaxial specimen is subjected to, then the deviator stress, $q = \sigma_1 - \sigma_3$ and the mean effective stress: $p' = (\sigma_1' + 2\sigma_3')/3$. The axial strain is denoted as ε_a.

6.1.1 Effect of density and confining pressure

It is well known that changes in density and confining pressure affect the undrained response of natural sand. The effects were investigated for Pumice A sand under different conditions. The pumice samples were reconstituted at three different initial states: loose ($e = 2.20$–2.35, $D_r = 26$–32%), medium dense ($e = 1.97$–2.00, $D_r = 50$–54%), and dense ($e = 1.63$–1.68, $D_r = 79$–85%) states. Effective confining pressures, σ_c', ranging from 50 kPa to 1600 kPa were applied to the specimens.

The results for loose samples of Pumice A sand are presented in Figure 14 for σ_c' up to 400 kPa. The specimens showed strain hardening response at this range of pressure, with the deviator stress q increasing with the

increase in ε_a. Moreover, q increases with the confining pressure and at large strain level, the plots are more or less parallel to each other. From the effective stress paths, the specimen under lower confining pressure was less contractive than those under higher confining pressure. The stress-strain relations show a stiffer response at small strain level, followed by development of large strains and greater dilatancy when the phase transformation state (from contractive to dilative) is reached. Compared to natural sands where the stress-strain curves generally merge at large strain range (i.e., steady state of deformation), the curves for pumice sand do not converge, at least within the strain level investigated.

The results for two other series of tests, one on medium dense ($D_r = 50$–54%) and another on dense ($D_r = 79$–85%) states are presented in Figures 15 and 16, respectively. Looking at the influence of initial confining pressure on the stress-strain relation and pore water pressure response, similar tendencies are observed for dense and loose pumice specimens. Thus,

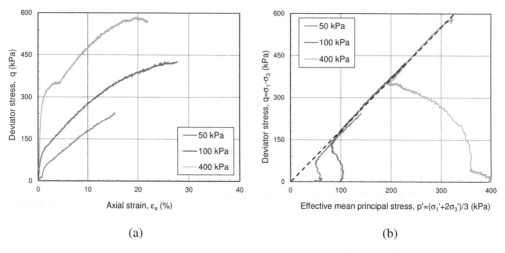

Figure 16. CU test results for dense Pumice A sands: (a) stress-strain relation; and (b) effective stress paths.

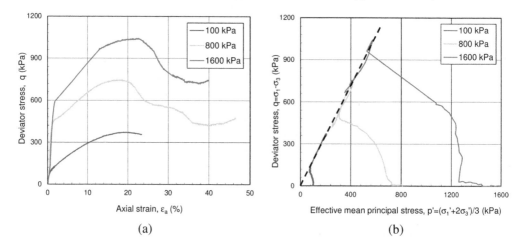

Figure 17. CU test results for dense Pumice B sands: (a) stress-strain relation; and (b) effective stress paths.

relative density is not a good parameter to differentiate the response of pumice sands, consistent with the observation made by Wesley et al. (1999).

CU tests were also conducted at very high confining pressure; however, problems were encountered in some cases because of the limitation of the apparatus. Nevertheless, a clear trend was observed in the tests which were deemed successful. Figure 17 shows the monotonic undrained test results for dense Pumice B sands subjected to $\sigma_c' = 400$, 800 and 1600 kPa. Whereas the results for 400 kPa showed strain hardening behaviour, those at higher pressures manifested strain softening response, especially at large strain level. This behaviour is similar to those observed for Toyoura sand (Ishihara 1996). However, it can be seen from the figure that even at strain levels as large as 40%, the stress-strain curves did not converge to the steady state. Also noticeable is the stiff response of pumice at small strain range, followed by large deformation after the phase transformation was reached.

6.1.2 Effect of soil gradation

Next, the monotonic undrained response of Pumice A and Pumice B sand specimens are compared in Figure 18 at $\sigma_c' = 100$ kPa and 400 kPa. While the stress-strain curves are more or less similar, the development of excess pore water pressure appears to be faster for Pumice B sands. Similar general tendencies were also observed in the other test comparisons. Thus, the finer-grained Pumice B sand appears to be more contractive when compared to Pumice A sand.

For the cases examined in Figures 14–16, the effective stress paths for the specimens of a particular density become asymptotic to the failure line, indicated by the dashed lines in the figures. The slope of the line, M_f, can be correlated to the angle of inter-particle friction, ϕ_f, using the following equation:

$$M_f = \left(\frac{q}{p'}\right) = \frac{6\sin\phi_f}{3 - \sin\phi_f} \tag{6}$$

For all the densities considered, the values of ϕ_f were calculated for both Pumice A and B sands and

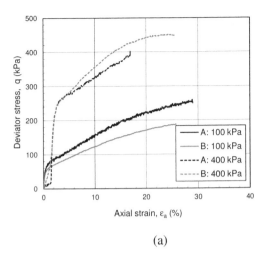

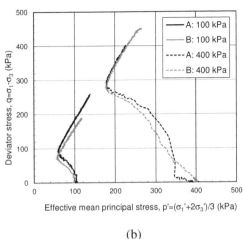

(a)   (b)

Figure 18. Comparison of test results between dense Pumice A and Pumice B sands: (a) stress-strain relation; and (b) effective stress paths.

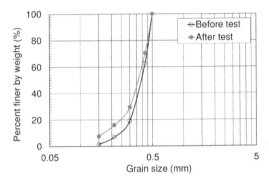

Figure 19. Grain size distribution curves of dense Pumice B specimen before and after test ($\sigma'_c = 400$ kPa).

the results show that, regardless of the relative density, ϕ_f for each type of pumice sand appears to be constant, with values of 42° and 44° for Pumice B and A, respectively. Thus, Pumice A has higher ϕ_f and therefore has a slightly higher shear resistance than Pumice B. For comparison purposes, typical values for loose natural sands are in the order of $\phi_f = 30°$. The higher ϕ_f of pumice may be attributed to their very angular shape as well as to its crushable nature, resulting in higher interlocking potential.

6.1.3 Investigation of particle crushing

To investigate whether particle breakage occurred during monotonic shearing, sieve analyses were conducted after most of the tests. A typical comparison of the grain size distribution before and after the test for dense Pumice B sand at $\sigma'_c = 400$ kPa is shown in Figure 19. For the level of shearing applied, considerable particle crushing occurred.

As pointed out earlier, pumice particles are very fragile not only because they are porous but also because they are angular in shape; it is the combination of these two factors that makes the particles highly

crushable. It is postulated that the axial strain measured during the test is the result of both load-induced compression of the specimen and particle crushing; as a result, the structure of pumice specimen becomes more stable with continuous shearing, accounting for the predominantly strain-hardening response observed in the tests. In addition, the changing particle size distribution during the course of shearing makes the pumice soil more resistant to deformation when compared to specimens consisting of hard-grained sands.

6.1.4 Steady state condition

Numerous studies on the undrained monotonic behaviour of sand have used the steady state concept to discuss the response. The steady state of deformation, also known as critical state, is defined as the state at which a sandy soil deforms under constant shear stress, constant effective stress and constant volume (Casagrande, 1976; Castro and Poulos, 1977). The strength and mean effective stresses which occur at the steady state of deformation change as the density of sand is varied, enabling a 'steady state line' to be defined in $e-q-p'$ space. The projection of this line in the $e-p'$ plane is often presented to discuss the response of sand at the steady state of deformation.

Looking at the monotonic test results for pumice, it is seen that the steady state condition was not reached in the tests. The deviator stress continues to increase even when large deformation was reached ($\varepsilon_a > 25\%$). The breakage of the particles does not allow for the pumice sand specimen to deform under constant shear stress, constant effective stress and constant volume. In fact, at large deformation, some of the specimens tend to deform irregularly. Thus, meaningful data at the end of the tests were not obtained in some tests.

Although limited in scope, the monotonic undrained tests on the reconstituted pumice sand specimens presented herein showed that the steady state was not reached in the tests. Because the particles are crushed as the deviator stress is applied, the soil structure

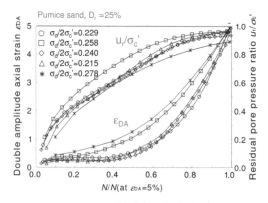

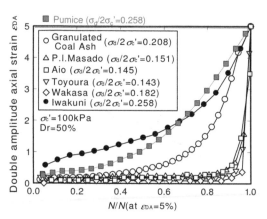

Figure 20. Double-amplitude axial strain ε_{DA} and residual pore pressure ratio u_r/σ_c' plots against the normalized number of cycles N/N (at $\varepsilon_{DA} = 5\%$) for pumice sand with $D_r = 50\%$.

Figure 21. Relationship between the normalized number of cycles N/N (at $\varepsilon_{DA} = 5\%$) and double-amplitude axial strain ε_{DA} for various geo-materials. Note that indicated values of $\sigma_d/2\sigma_c'$ are for $N = 20$ cycles (modified from Yoshimoto et al. 2014).

becomes more stable and resistance increases; as a result the condition of constant deformation under constant shear stress (and constant volume) was not achieved, at least within the strain range allowed by the triaxial apparatus used. More experiments are recommended to confirm this.

6.2 Undrained cyclic test results

As in monotonic tests, the results of the undrained cyclic triaxial tests are expressed in terms of cyclic deviator stress (σ_d) – axial strain (ε_a) relation and effective stress path (σ_d vs p'). Moreover, the liquefaction resistance curves are presented as the plot of the number of cycles, N, required for a specimen to reach either a double amplitude axial strain $\varepsilon_{DA} = 5\%$ or pore water pressure ratio, r_u, ($= u_r/\sigma_c'$) of 0.95, for a given cyclic shear stress ratio, CSR ($= \sigma_d/2\sigma_c'$).

6.2.1 Cyclic shear behavior

Figure 20 shows the plots of double amplitude axial strain ε_{DA} and residual pore pressure ratio u_r/σ_c' against normalized number of cycles N/N(at $\varepsilon_{DA} = 5\%$) obtained from undrained cyclic shear tests on Pumice A sand specimens ($D_r = 25\%$) with $\sigma_c' = 100$ kPa under different CSR values. The curves for each residual pore pressure ratio, u_r/σ_c', show practically similar behaviour, while those corresponding to double amplitude axial strain, ε_{DA}, are of similar trend. During the start of cyclic loading, the rate of development of pore water pressure is much higher than that of axial strain, while in the latter part the trend is reversed.

It would be interesting to compare the cyclic shear behaviour of pumice with those of natural sands in terms of the development of double amplitude axial strain ε_{DA} and residual pore pressure ratio u_r/σ_c' with the number of cycles. For this purpose, the trends shown in Figure 20 are compared with the findings of Yoshimoto et al. (2014) on natural sands and on granulated coal ash (GCA). Note that all the specimens investigated by Yoshimoto et al. (2014) have relative

density, $D_r = 50\%$, which is denser than that of the pumice sand tested.

The relation between normalized number of cycles N/N (at $\varepsilon_{DA} = 5\%$) and double amplitude axial strain ε_{DA} is presented in Figure 21. Because the cyclic shear resistances greatly differ for each material, the results for $\varepsilon_{DA} = 5\%$ corresponding to 20 cycles are shown next to the sample name indicated in the explanatory notes in the figure. In the case of natural sands, axial strain did not occur at the early stage of cyclic loading; however, at strain level of about $\varepsilon_{DA} = 5\%$, it suddenly increased to almost the maximum values at N/N (at $\varepsilon_{DA} = 5\%$) of about 0.9 to 0.95. Conversely, the axial strain in GCA and Iwakuni clay increased almost constantly from the start of cyclic loading. It can be seen that the result corresponding to pumice sand shows intermediate behaviour between GCA and clay; compared to natural sands, however, pumice showed faster development of strain at the early stage of cyclic load application, possibly because of particle crushing.

Figure 22 shows the relation between normalized number of cycles N/N (at $\varepsilon_{DA} = 5\%$) and residual pore pressure ratio u_r/σ_c' for all the geo-materials considered. Natural sands and Iwakuni clay initially show almost similar behaviour of constantly increasing pore pressure ratio with the number of cycles from the start of cyclic loading. For sands, however, the pore pressures are triggered to increase suddenly at about N/N (at $\varepsilon_{DA} = 5\%$) = 0.8, and continue to increase until a value equal to the initial effective confining stress is reached; this is typical of liquefaction behaviour. In the case of pumice, the pore pressure development is large from the start of cyclic loading, similar to GCA. This may be due to the crushable nature of both materials, with particle crushing occuring from the start of cyclic loading.

6.2.2 Liquefaction resistance

In general, the attainment of the state of soil liquefaction in a saturated sandy specimen is defined in two

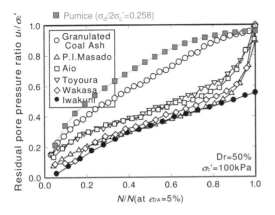

Figure 22. Relationship between the normalized number of cycles N/N (at $\varepsilon_{DA} = 5\%$) and residual pore pressure ratio u_r/σ'_c (modified from Yoshimoto et al. 2014).

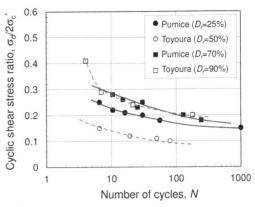

Figure 23. Liquefaction resistance curves for the reconstituted Pumice A sand specimens.

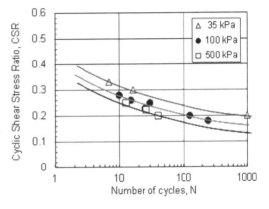

Figure 24. Comparison of liquefaction resistance curves for reconstituted dense pumice sands under different σ'_c.

ways: (a) development of high excess pore water pressure, usually 95% of initial confining pressure; or (2) development of high strain, typically expressed as 5% double amplitude axial strain for triaxial tests.

In order to specify the onset of liquefaction, the number of cycles needs to be specified for a given constant-amplitude uniform cyclic loading. The locus of points defining each state is called the liquefaction resistance (or cyclic strength) curve. In practice, the liquefaction resistance is specified in terms of the magnitude of cyclic stress ratio required to produce 5% double amplitude axial strain in 15–20 cycles of uniform load application (in view of the typical number of significant cycles present in many time histories of accelerations recorded during past earthquakes).

The following sections focus on the effect of various factors on the liquefaction resistance curve of pumice (defined in terms of $\varepsilon_{DA} = 5\%$).

6.2.2.1 Effect of relative density

Cyclic tests were performed on loose ($D_r = 25\%$) and dense specimens ($D_r = 70\%$) of Pumice A sands under $\sigma'_c = 100$ kPa, and the liquefaction resistance curves are shown in Figure 23. Also plotted in the figure are the curves for loose ($D_r = 50\%$) and dense ($D_r = 90\%$) Toyoura sand, as reported by Yamamoto et al. (2009). Comparing the curves for Toyoura sand (a hard-grained sand) and for reconstituted pumice sands, two things are clear: (1) loose specimens have gentler liquefaction resistance curves, while dense specimens have curves rising sharply as the number of cycles decreases; and (2) while the effect of relative density is very pronounced for Toyoura sand, the effect of relative density on pumice specimens appear to be not as remarkable. This observation is similar to that seen in monotonic undrained tests, as discussed earlier. Under the confining pressures considered, pumice undergoes considerable particle crushing when subjected to cyclic shear. As cyclic shearing and particle crushing occur, the soil structure is gradually stabilized, resulting in higher cyclic shear resistance. The cyclic shearing and the associated particle breakage

resulted in stable soil structure for both dense and loose cases, and therefore, the effect of density was not as remarkable when compared to the cyclic behaviour of Toyoura sand.

6.2.2.2 Effect of confining pressure

Next, the influence of initial effective confining pressure on the liquefaction resistance of reconstituted pumice sands was investigated. Dense Pumice A sand specimens (initial void ratio, $e_i = 1.90$–2.00) were subjected to three different levels of confining pressure, $\sigma'_c = 35$, 100 and 500 kPa under different levels of CSR. Figure 24 illustrates the confining pressure dependency of the liquefaction resistance of reconstituted pumice. The curves are almost parallel to each other, with the liquefaction resistance increasing as the confining pressure decreases, consistent with the observations made on natural sands (e.g., Rollins and Seed, 1988). The value of the correction factor for overburden stress, K_σ (i.e. CSR causing $\varepsilon_{DA} = 5\%$ in 15 cycles under any confining pressure normalized to the corresponding value of CSR at $\sigma'_c = 100$ kPa) is equal to 1.16 for $\sigma'_c = 35$ kPa and 0.88 for $\sigma'_c = 500$ kPa. These values appear to coincide with those reported

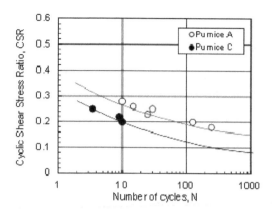

Figure 25. Comparison of liquefaction resistance curves for Pumice A and Pumice C sands.

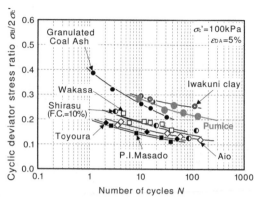

Figure 26. Comparison of liquefaction resistance curve of pumice ($D_r = 25\%$) corresponding to double-amplitude axial strain $\varepsilon_{DA} = 5\%$ and $\sigma'_c = 100$ kPa with the other geomaterials ($D_r = 50\%$) tested by Yoshimoto et al. (2014).

for reconstituted natural sands (e.g., Boulanger and Idriss, 2004).

6.2.2.3 Effect of gradation

Undrained cyclic tests were also performed on Pumice C sand specimens with initial void ratio range of $e = 1.32–1.33$ under $\sigma'_c = 100$ kPa. Attempts were made to form specimens similar to dense Pumice A specimens ($e = 1.9–2.0$) as reported in Figure 23; however, it was very difficult to make such specimens. In any case, the initial relative densities of Pumice C specimens were $D_r = 74–76\%$, whereas those of dense Pumice A specimens were $D_r = 68–74\%$.

The liquefaction resistance curves obtained for Pumice A and C specimens are shown in Figure 25. It is obvious that there is a reduction in liquefaction resistance as the fines content, F_c, increases. Based on previous experiments on soil mixtures, the liquefaction resistance of sands decreases with increase in the amount of non-plastic fines (e.g. Hyodo et al., 2006); hence, the trends presented herein are consistent with the results obtained for natural sands.

6.2.2.4 Comparison with other geomaterials

The comparison of cyclic shear resistance of pumice ($D_r = 25\%$) with those of other geomaterials at a higher density ($D_r = 50\%$) which were tested by Yoshimoto et al. (2014) is presented in Figure 26. Note that in general, the cyclic deviator stress ratio to cause failure after 15∼20 cycles is defined as liquefaction resistance. The cyclic resistance of pumice (with $D_r = 25\%$) is about 1.5∼2.5 times higher than those of natural sands (with $D_r = 50\%$), and almost similar to that of Iwakuni clay.

6.2.2.5 Development of particle crushing during cyclic loading

To elucidate further the development of particle crushing during a cyclic loading, a series of tests were performed such that cyclic shearing was terminated after a specified number of cycles, after which sieve analyses were performed. For these tests, virgin samples were used at each test. A confining pressure of $\sigma'_c = 100$ kPa was considered, with the initial void

ratio set at $e = 1.90–2.00$. For $CSR = 0.10$, the sieve analyses were carried out: (1) on the virgin samples; (2) after the end of consolidation stage; (4) after $N = 10$ cycles; (5) after $N = 100$ cycles; and (4) after $N = 1000$ cycles. On the other hand, for $CSR = 0.20$, sieving was done: (1) after $N = 10$ cycles; and (2) after $N = 83$ cycles where initial liquefaction (i.e. $r_u = 0.95$) occurred.

The grain size distribution of the specimen after each test was determined. Particle crushing occurred, but with the level of CSR and the number of cycles applied, it was difficult to use the grading curves to make reasonable comparison. Instead, a method of evaluating particle crushing originally proposed by Miura and Yamanouchi (1971) was used which involves the quantification of the surface area of the particles per unit volume. The specific surface area of the particles (in mm²/mm³) is calculated as:

$$S = \sum \frac{F}{100} \cdot \frac{4\pi(d_m/2)^2}{(4/3)\pi(d_m/2)^3 G_s \gamma_w} \cdot \gamma_d \quad (7)$$

where $d_m = (d_1 \times d_2)^{0.5}$, d_1 and d_2 are adjacent sieve sizes (e.g., 0.50 mm and 0.212 mm), F is the % by weight retained on the sieve, G_s is the specific gravity of the particles, γ_w is the unit weight of water and γ_d is the dry unit weight of the specimen.

Figure 27 shows the development of S for the different tests described above. Firstly, it was observed that consolidation at $\sigma'_c = 100$ kPa did not induce appreciable particle breakage to the pumice particles; however, the cyclic shearing did. Secondly, the degree of particle crushing increased with the amplitude of applied CSR. For the test with $CSR = 0.20$, the increase in surface area during the initial stage of cyclic loading was small; however, as the liquefaction stage was reached ($N = 83$), the surface area increased remarkably because large strains occurred with associated translation and rotation of particles causing the higher degree of crushing. For $CSR = 0.10$, the state of liquefaction did not occur even when $N = 1000$ cycles and

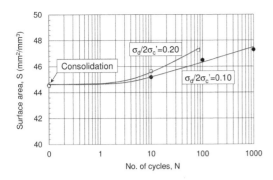

Figure 27. Relationships between specific surface area and number of cycles during cyclic undrained tests.

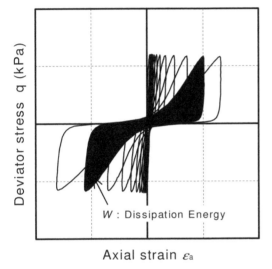

Figure 28. Schematic diagram of cumulative dissipated energy ΣW.

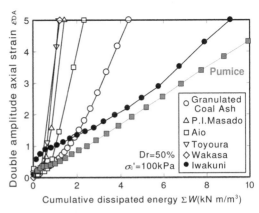

Figure 29. Relationship between cumulative dissipated energy ΣW and double-amplitude axial strain ε_{DA} for all materials tested.

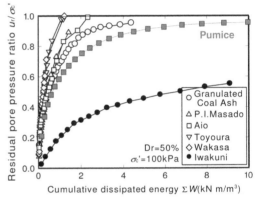

Figure 30. Relationship between cumulative dissipated energy ΣW and residual pore pressure ratio u_r/σ_c'.

the increase in particle breakage with the number of cycles was more or less gradual.

6.2.3 Shear work

The cumulative dissipated energy ΣW is indicated by the area under the cyclic stress-strain curve obtained from the relation between deviator stress and axial strain, as shown in Figure 28. Towhata and Ishihara (1985) used this concept to evaluate cyclic shear behaviour and liquefaction strength. Note that the cumulative dissipated energy ΣW is the energy consumed by the soil during plastic deformation until liquefaction. Thus, the energy absorption capacity is high when this value is large, and consequently, the liquefaction strength would also be high.

Figure 29 shows the plot of cumulative dissipated energy ΣW against double amplitude axial strain ε_{DA} for pumice ($D_r = 25\%$) and all the materials tested by Yoshimoto et al. (2014). Based on the figure, ΣW of pumice is about 5–6 times higher than those of natural sands, even higher than that of Iwakuni clay.

The same data is presented in Figure 30 in terms of the relation between cumulative dissipated energy ΣW and residual pore pressure ratio u_r/σ_c'. Sands are not resistant to liquefaction because they lack energy absorption capacity. In contrast, the residual pressure ratio in Iwakuni clay increased only up to about $u_r/\sigma_c' = 0.6$, corresponding to a double amplitude axial strain $\varepsilon_{DA} = 5\%$. It is also observed that natural sands and pumice show similar behaviour until $u_r/\sigma_c' = 0.6$; however, in the case of pumice, u_r/σ_c' does not reach 1.0 when $\varepsilon_{DA} = 5\%$. Pumice is therefore more resistant to liquefaction than natural sands.

7 SUMMARY OF MAJOR OBSERVATIONS

In terms of particle characterization at micro-level, the following are the major observations made:

• Both SEM and X-ray CT images showed that pumice particles are porous (with surface and internal voids) and have complex surface shape, which may lead to interlocking effects. As a result, pumice sands have larger range of void ratio in comparison with natural sands.

- Through CT images, the solid density of pumice particles was calculated to be 2.2 g/cm^3, which was almost the same as the value measured using the standard method for a particle with size $\sim$75 μm. In addition, imaging technique showed that the internal void ratio of pumice sand was in the range of 10^{-5} to 10^{-2}, which was much smaller than the surface void ratio. On average, the volume of the surface voids in a particle was greater than the sum of the volume of the internal voids and the volume of the solid in the particle.
- The single particle crushing strength of pumice is one order of magnitude less than that of natural hard-grained Silica sand.

In terms of the geotechnical response of pumice as observed from the laboratory tests, it was noted that the behaviour of this crushable sand is very different from natural hard-grained sands. More specifically:

- Because of its light weight, the initial shear modulus of pumice is lower than that of silica sand. However, the effect of confining pressure on the shear stiffness is more pronounced in pumice due to particle crushing at higher pressure.
- The reduction rate in shear modulus ratio of pumice with strain increased with fines content, similar to the trend observed on the Japanese volcanic soil Shirasu. Moreover, the modulus ratio increased with confining pressure.
- The damping ratio of pumice seemed to be unaffected by grading curve and confining pressure at strain level less than 0.2%. At strain level near failure, the damping ratio of pumice was about half that of Toyoura sand.
- Under monotonic undrained loading, reconstituted pumice specimens under loose and dense states showed similar response, indicating that relative density did not have significant effect on their behaviour.
- Within the range of effective confining pressures investigated, pumice specimens under monotonic shear showed contractive response followed by dilative behaviour, with the contractive response being more significant at high confining pressure. The stress-strain relations showed a stiffer response at small strain level, followed by development of large strains and greater dilatancy when the phase transformation state is reached.
- Pumice sands have higher friction angle at failure than hard-grained sands and steady state of deformation was not reached due to continuous particle breakage during the shear stress application.
- Although relative density had some noticeable effect on the liquefaction resistance of pumice, it was not as significant when compared to that observed for hard-grained sands.
- As the confining pressure and the fines content were increased, the liquefaction resistance of reconstituted pumice specimens decreased. These trends

were consistent with the observations made on hard-grained sands.

- During the initial stage of shearing, the degree of particle crushing was small; however, as the liquefaction stage was reached, particle crushing increased remarkably because large strains occurred with associated translation and rotation of particles.
- The pore pressure and deformation response of pumice sand showed intermediate behaviour between sand and clay. During cyclic triaxial testing, pumice sand underwent considerable particle crushing.
- The liquefaction resistance of pumice sand is about twice that of Toyoura sand. The cyclic deviator stress ratio of pumice sand is almost the same as that of Iwakuni clay.
- The cumulative dissipated energy for pumice sand is larger than that of natural sand, because some energy is spent as the particles undergo crushing. It is believed that this, along with the large friction angle, is the reason why the liquefaction resistance of pumice sand is higher than that of natural sand.

8 CONCLUDING REMARKS

To investigate the dynamic properties and liquefaction characteristics of pumice sands, attempts were made to understand the properties of pumice sands at grain-size level using SEM and X-ray CT imaging and single particle crushing tests. Next, bender element tests, dynamic deformation tests and monotonic/cyclic undrained tests were conducted on pumice sand specimens. In order to understand the response of this crushable material, comparisons were made with those obtained from hard-grained sands.

The results indicated that experimental trends observed for hard-grained sands were not generally applicable to pumice sands because of their complex surface shape and low crushing strength, increasing the interlocking potential between particles. Particle crushing appears to be more significant with application of shear stress, rather than by isotropic consolidation stress. As shearing and particle crushing occur, the soil structure of pumice is gradually stabilized, resulting in higher mobilised peak frictional angle and higher liquefaction resistance. Compared to hard-grained sands, pumice under shear stress has higher cumulative dissipated energy because the energy applied to the specimen is spent not only to induce sliding between particles but also to crush the particles.

These experimental results confirmed that the micro-level characteristics of the pumice particles have significant effect on their macro-level response.

ACKNOWLEDGMENTS

The authors would like to acknowledge the assistance of Dr N. Kikkawa, Dr. A. Tai, Messrs Y. Lu,

L. Liu, C. Chan and N. Taghipouran in performing and interpreting the laboratory tests presented herein. The particle crushing tests, dynamic deformation tests and bender element tests were performed with the assistance of Prof. M. Hyodo and Prof Y. Nakata while the first author was on research study leave at Yamaguchi University, Japan. A portion of the research funding was provided through the Earthquake Commission (EQC) Biennial Contestable Grant No. 10/589.

REFERENCES

Boulanger, R.W & Idriss, I.M. 2004. State normalization of penetration resistances and the effect of overburden stress on liquefaction resistance. *Proc., 11th International Conference on Soil Dynamics and Earthquake Engineering and 3rd International Conference on Earthquake Geotechnical Engineering*, Vol. 2, 484–491.

Casagrande, A. 1976. Liquefaction and cyclic deformation of sands: A critical review. *Harvard Soil Mechanics Series*, 88.

Castro, G. & Poulos, S. J. 1977. Factors affecting liquefaction and cyclic mobility. *J Geotech Eng Div, ASCE*, 103, 501–516.

Cubrinovski, M. & Orense. R. P. 2010. Case history: 2010 Darfield (New Zealand) Earthquake – Impacts of liquefaction and lateral spreading. *ISSMGE Bulletin*, 4 (3), 15–24.

Cubrinovski, M. & Ishihara, K. 2002. Maximum and minimum void ratio characteristics of sands. *Soils and Foundations*, 42, 65–78.

de Lange, W., Moon, V. & Healy, T. 1991. Problems with predicting the transport of pumiceous sediments in the coastal environment, *Coastal Sediments '91*, 990–996.

Hyodo, M., Orense, R., Ishikawa, S., Yamada, S., Kim, U.G. & Kim, J.G. 2006. Effects of fines content on cyclic shear characteristics of sand-clay mixtures. *Proceedings, Earthquake Geotechnical Engineering Workshop – Canterbury 2006*, Christchurch, 81–89.

Hyodo, T. 2006. Effects of fines on dynamic shear deformation characteristics of a volcanic soil "Shirasu". *Master Thesis*, Yamaguchi University (in Japanese).

Ishihara, K. 1996. *Soil Behaviour in Earthquake Geotechnics*, Oxford Science.

Iwasaki, T., Tatsuoka, F. & Takagi, Y. 1978. Shear moduli of sands under cyclic shear torsional loading. *Soils and Foundations*, 18 (1), 39–56.

Japanese Geotechnical Society 2000. *Soil Test Procedures and Commentaries*, First Revised Edition, Tokyo (in Japanese).

Kato, Y. 2002. Mechanical Properties of crushable materials based on single particle shape and strength, *PhD thesis*, Yamaguchi University, Japan (in Japanese).

Kikkawa, N., Pender, M.J., Orense, R.P. & Matsushita, E. 2009. Behaviour of pumice sand during hydrostatic and Ko compression. *Proc., 17th International Conference on Soil Mechanics and Geotechnical Engineering*, Alexandria (Egypt), Vol. 1, 812–815.

Kikkawa, N., Orense, R.P. & Pender, M.J. 2011. Mechanical behaviour of loose and heavily compacted pumice sand. *Proc., 14th Asian Regional Conference on Soil Mechanics and Geotechnical Engineering*, Paper 214, 6pp.

Kikkawa, N., Pender, M. J. & Orense, R. P. 2013a. Comparison of the geotechnical properties of pumice sand from Japan and New Zealand. *Proc. 18th International conference on Soil Mechanics and Geotechnical Engineering*, Paris, September. Vol. 1, 239–242.

Kikkawa, N., Pender, M.J. & Orense, R.P. 2013b. Micro-properties of pumice particles using Computed Tomography. *Proc. Experimental Micromechanics for Geomaterials – Joint Workshop of the ISSMGE TC101-TC105*.

Kikkawa, N., Orense, R.P. & Pender, M.J. 2013c. Observations on microstructure of pumice particles using computed tomography. *Canadian Geotechnical Journal*, 50(11): 1109–1117.

Kokusho, T. 1980. Cyclic triaxial test of dynamic soil properties for wide strain range. *Soils and Foundations*, 20 (2), 45–60.

Marks, S., Larkin, T.J. & Pender, M.J. 1998. The dynamic properties of a pumiceous sand. *Bulletin of the New Zealand Society for Earthquake Engineering*, 31 (2), 86–102.

Miura, N. & Yamanouchi, T. 1971. Drained shear characteristics of Toyoura sand under high confining stress, *Proc. Japan Society of Civil Engineers*, 260, 69–79 (in Japanese).

Nakata, Y., Kato, Y., Hyodo, M., Hyde, A.F.L. & Murata, H. 2001. One-dimensional compression behavior of uniformly graded sand related to single particle crushing strength. *Soils and Foundations*, 41 (2), 39–51.

New Zealand Standard 1986. *NZS 4402: 1986 – Methods of Testing Soils for Civil Engineering Purposes*. Part 2 Soil classification tests.

Orense, R.P., Kiyota, T., Yamada, S., Cubrinovski, M., Hosono, Y., Okamura, M. & Yasuda, S. 2011. Comparison of liquefaction features observed during the 2010 and 2011 Canterbury earthquakes. *Seismological Research Letters*, 82 (5), 905–918.

Orense, R.P., Pender, M.J. & Wotherspoon, L.M. 2012. Analysis of soil liquefaction during the recent Canterbury (New Zealand) earthquakes. *Geotechnical Engineering Journal of the SEAGS & AGSSEA*, 43 (2), 8–17.

Orense, R.P., Pender, M.J., Hyodo, M. & Nakata, Y. 2013. Micro-mechanical properties of crushable pumice sands. *Géotechnique Letters*, 3, Issue April–June, 67–71.

Pender, M. J. 2006. Stress relaxation and particle crushing effects during Ko compression of pumice sand. *Proc. International Symposium on Geomechanics and Geotechnics of Particulate Media*, Yamaguchi, Japan, Vol.1, 91–96.

Pender, M.J., Wesley, L.D., Larkin, T.J. & Pranjoto, S. 2006. Geotechnical properties of a pumice sand. *Soils and Foundations*, 46 (1), 69–81.

Pender, M.J., Orense, R.P., & Kikkawa, N. (2014). Japanese and New Zealand pumice sands: Comparison of particle shapes and surface void structures. *Geomechanics from Micro to Macro – Proceedings of the TC105 ISSMGE International Symposium on Geomechanics from Micro to Macro, IS-Cambridge 2014*, 2, 1111–1116.

Rollins, K.M. & Seed, H. B. 1988. Influence of buildings on potential liquefaction damage. *Journal of Geotechnical Engineering, ASCE*, 116, GT2, 165–185.

Tatsuoka, F. & Iwasaki, T. 1978. Hysteretic damping of sands under cyclic loading and its relation to shear modulus. *Soils and Foundations*, Vol. 18 (2), 25–40.

Tatsuoka, F., Shibuya, S. & Kuwano, R. 2001. *Advanced Laboratory Stress-Strain Testing of Geomaterials*, Balkema, Rotterdam, The Netherlands, 92–110.

Towhata, I. & Ishihara, K. 1985. Shear work and pore water pressure in undrained shear. *Soils and Foundations*, 25(3), 73–84.

Wesley, L. D., Meyer, V. M., Pronjoto, S., Pender, M. J., Larkin, T. J. & Duske, G. C. 1999. Engineering properties of pumice sand. *Proc. 8th Australia-NZ Conference on Geomechanics*, Hobart, Vol. 2, 901–908.

Wesley, L. D. 2001. "Determination of specific gravity and void ratio of pumice materials," *Geotechnical Testing Journal*, 24 (3), 418–422.

Yamamoto, Y., Hyodo, M. & Orense, R. 2009. Liquefaction resistance of sandy soils under partially drained condition. *Journal of Geotechnical and Geoenvironmental Engineering, ASCE*, 135(8), 1032–1043.

Yoshimoto, N., Orense, R., Hyodo, M. & Nakata, Y. 2014. Dynamic behaviour of granulated coal ash during earthquakes. *Journal of Geotechnical and Geoenvironmental Engineering, ASCE*, 140 (2), Paper 0413002, (11pp).

Volcanic Rocks and Soils – Rotonda et al. (eds)
© *2016 Taylor & Francis Group, London, ISBN 978-1-138-02886-9*

Climatic effects on pore-water pressure, deformation and stress mobilization of a vegetated volcanic soil slope in Hong Kong

C.W.W. Ng
Department of Civil and Environmental Engineering, Hong Kong University of Science and Technology, Hong Kong

A.K. Leung
Division of Civil Engineering, University of Dundee, Dundee, UK

ABSTRACT: Rainfall-induced landslide has been increasingly reported in the past 20 years due to increasing intense rainfall under the impact of global climate change. Previous research has investigated the hydrological behaviour of soil slopes extensively, including the responses of pore-water pressure, water content and groundwater table. Mechanical slope behaviour in relation to stress-deformation characteristics is rarely studied, especially for residual and volcanic soils that are typically found in the sub-tropical region of the world. In this keynote paper, some new findings from a comprehensive field monitoring work conducted in a volcanic soil slope in Hong Kong are summarised and reported. The field monitoring programme included two parts: Part I – *in-situ* measurements of soil hydraulic properties; and Part II – field monitoring of groundwater responses and slope movements due to climatic effects. *In-situ* measured stress-dependent soil-water retention curves (SDSWRCs) and permeability functions of unsaturated volcanic soils are used to interpret the slope hydrological responses through a three-dimensional anisotropic transient seepage analysis. Based on the improved understanding of the slope hydrology, some observed seasonal slope movements and the associated mobilisation of soil stress are explained under the framework of unsaturated soil mechanics.

1 INTRODUCTION

Rainfall-induced slope failure has been increasingly reported in the past 20 years (Sassa & Canuti, 2008), especially in sub-tropical regions like Hong Kong, Brazil and parts of Italy as well as tropical regions like Singapore and Malaysia in the world (Brand, 1984; Angeli et al., 1996; Tsaparas et al., 2003; Corominas et al., 2006; Tommasi et al., 2006; Ng & Menzies, 2007). This natural hazard has caused significant socio-economic losses. Under the impact of global climate change, it has been predicted that over the next 50 years, *"Annual precipitation is very likely to **increase** in most of northern Europe… In central Europe, precipitation is likely to **increase** in winter but decrease in summer… Extremes of daily precipitation are very likely to **increase** in northern Europe"* (IPCC, 2007). Such environmental change imposes increasing threat to slope instability. Previous field studies (Skempton, 1970; Leroueil, 2001; Smethurst et al., 2012), physical modelling (Take & Bolton, 2011) and numerical analyses (Potts et al., 1990; 1997) have shown that seasonal drying-wetting events promote progressive slope failure due to repeated mobilisation of dilatancy and softening of soil in successive wet seasons. Most of these studies concerned on the behaviour of clay slopes under temperature climates. Studies of the climatic effects of slope responses for other soil types,

such as volcanic and residual soils, under tropical and subtropical climates are limited.

Hydrological responses of volcanic and residual slopes have been widely studied in the field (Angeli et al., 1996; Lim et al., 1996; Li et al., 2005; Rahardjo et al., 2005; Cheuk et al., 2009; Tommasi et al., 2006; among others). Typically, these studies measured PWP, water content and groundwater level (GWT), which were then used for subsequent water balance and slope stability calculations. In contrast, mechanical slope behaviour (i.e., slope deformation and the associated stress mobilisation) received little attention only. A field study conducted by Ng et al. (2003) is an exception that measured stress mobilisation of an expansive clay slope during rainfall. The field monitoring results revealed that there was a significant increase in total horizontal stress (> three times higher than the total vertical stress) when the slope was subjected to a rainfall event with an average daily amount of 62 mm for seven days. The mobilised stress ratio was close to an effective passive stress ratio, indicating the possibility of passive failures of the clay. This field observation highlights the importance to study mechanical slope response including stress mobilisation. This information is crucial and relevant for engineers to carry out more reliable design for their slope retaining systems.

This keynote paper summarises some key findings of a comprehensive field monitoring work conducted

Figure 1. Overview of the hillslope and the study area.

in a volcanic slope situated in Hong Kong (Ng et al., 2011; Leung et al., 2011; Leung and Ng, 2013a, b; Leung and Ng, in press). The slope was heavily-instrumented with a wide range of sensors. These sensors monitored not only the hydrological responses (PWP, water content, GWT) but also the mechanical behaviour (slope displacement and stress mobilisation). The interpretation of data was facilitated by a three-dimensional, anisotropic transient seepage analysis. The improved understanding of the slope hydrological responses is then used to interpret observed seasonal slope movements and the associated mobilisation of soil stress.

2 A LANDSLIDE HAZARD IN HONG KONG

A slowly-moving landslide body was identified in a volcanic soil slope at Tung Chung, Lantau Island in Hong Kong (Fig. 1). The hillslope is in the vicinity of North Lantau Highway, which is the primary access to the Hong Kong International Airport. Based on detailed engineering field mapping (GEO, 2007), the active landslide body was identified to be approximately 45 m wide and 50 m long (see plan view in Fig. 2). Sub-parallel tension cracks were found at the elevations between +84 to +86 mPD (mPD

stands for metres above Principle Datum and it refers to an elevation 1.23 m above mean sea level; SMO, 1995). These cracks formed the main scarp features of the landslide body. Some lateral tension cracks were present along both the eastern and western flanks of the landslide body, extending north from the main scarp. At +64 mPD, some thrust features were encountered. Preliminary monitoring of the landslide body shows signs of movement with limited mobility (GEO, 2007).

Any slope failure from this site has been categorised to impose "high" risk and losses to the society (i.e., the highest level of awareness). This is because the failure mass could potentially block the North Lantau Highway, which is the critical transport corridor to the Hong Kong International Airport. On 7th June 2008, an extreme storm event (maximum one-hour rainfall of 133–140.5 mm with corresponding return period exceeding 240 years) resulted in a total of 38 landslides occurring on the hillside above the Highway (AECOM, 2012). Four of them developed into channelized debris flows, which transported about 540 m^3 of sediment downstream. Although no causalities were reported, the debris hit the Highway closing it for about 16 hours.

In order to better understand the behaviour of the landslide body, a comprehensive field study was implemented at the site. The study includes two parts: Part I: *in-situ* measurements of soil hydraulic

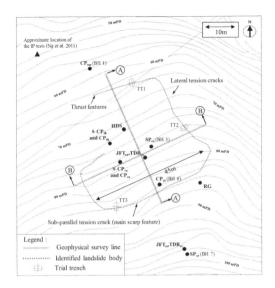

Figure 2. Plan view of the landslide body.

properties (Leung & Ng, 2010; Ng et al., 2011); and Part II: field monitoring of groundwater responses and slope movements due to climatic effects (Leung & Ng, 2011; 2013a, b; in press). It was aimed to use the measured soil hydraulic properties from Part I to interpret both the mechanical and hydrological responses of the volcanic slope obtained from Part II.

3 SITE CHARACTERISTICS

3.1 Overview of the research slope

As shown in Figure 1, the slope forms a blunt ridgeline located between a major stream channel to the northeast and a shallow topographic valley to the south and west. The main planar face of the blunt ridge forming the study area faces roughly north to northwest at an average slope gradient of 30° and overlooks the Tung Chung Eastern Interchange. The slope is densely vegetated. A detailed plant survey revealed that the most abundant species are a fern species, *Dicranopteris pedata,* and two woody species, *Rhodomyrtus tomentosa* and *Baeckea frutescens*. These species are common in Hong Kong and many parts of Asia including Malaysia, India and Vietnam (Corlett et al., 2000). The fern *Dicranopteris pedata* is found to be the most dominating and it covers almost the entire slope surface. In general, the height of vegetation was less than 3 m. The slope may be described as a typical short shrubland.

3.2 Climate of the site

Figures 3(a) to (d) show the daily climatic variation of the study area during the monitoring period from Mar 2008 to Dec 2009. The climatic data were monitored from a metrological station installed near the site. A tipping-bucket rain gauge was installed near the crown of the active landslide body to record rain

depth. As shown in Figure 3(a), the average air temperature increased (up to 30°C) during each wet summer and then decreased (down to 15°C) during each dry winter. When rainfall happened in both wet summers, the average RH was fairly constant at about 85% (see Fig. 3(b)). During the rainy period between May and July 2008, Hong Kong Observatory (HKO) issued black rainstorm signals (i.e., rainfall event with intensity higher than 70 mm/hr) for several times. On 7th June 2008, a signal was issued for more than 4.5 hours, which was the longest duration recorded since September 2001. During both the dry winters, air RH decreased and remained fairly constant at an average value of 70%. The measured wind speed depicted in Figure 3(c) showed large fluctuation. There has no observable trend and relationship with season. Nevertheless, it is worth noting that typhoon signal no. 8 (i.e., sustained wind speed of 63–117 km/hr and guts that exceed 180 km/hr), together with the black rainstorm signal, had been issued by the HKO on 7th June 2008. Variation of solar radiation with time is shown in Figure 3(d). The radiation in both summers fluctuated between 5–25 MJ/m²/day. It appears that in both years, the radiation reduced when dry winter arrived.

Based on the climate data, potential evapotranspiration (PET) can be estimated using the Peman-Monteith equation (Allen et al., 1998), as follows:

$$PET = \frac{0.408\Delta\left(R_n - G\right) + \gamma\dfrac{900}{T+273}u\left(e_s - e_a\right)}{\Delta + \gamma\left(1 + 0.34u\right)} K_c \tag{1}$$

where Δ is slope vapour pressure curve [kPa °C^{-1}]; R_n is net radiation intercepted by plant leaves [J m^{-2} day^{-1}] (taken to be 0.9 of measured solar radiation shown in Figure 3(d) by assuming albedo ratio of 0.1 for shrubs; Taha et al., 1988); G is soil heat flux density [J m^{-2} day^{-1}] (assumed to be negligible due to the relatively small magnitude when compared to R_n; Allen et al., 1998); γ is psychometric constant [kPa°C^{-1}]; T is air temperature [°C] (Fig. 3(a)); u is wind speed [m s^{-1}] (Fig. 3(c)); e_s is saturated vapour pressure [kPa]; e_a is actual vapour pressure [kPa] (Fig. 3(b)); and K_c is crop factor (taken to be 1.0 for general shrub species that were abundance in the site (Allen et al., 1998). PET refers to the amount of soil moisture extracted by vegetation under unrestricted condition when pore water is readily available for plant root-water uptake. The equation is derived based on energy balance of vegetated soil with due considerations of surface resistance (i.e., the resistance of vapour flow through stomata openings, total leaf area and soil surface) and aerodynamic resistance (i.e., the resistance from the vegetation involving friction from air flowing over leaf surfaces). The seasonal variation of estimated PET with time is shown in Figure 3(e). PET is relatively low during both wet summers, fluctuating between 0 to 5 mm/day. On the contrary, an increase in PET is found during the first two months of the dry winter 2008/2009. The peak PET was up to 10 mm/day. In the subsequent spring, PET decreased due to the substantial increase in air RH.

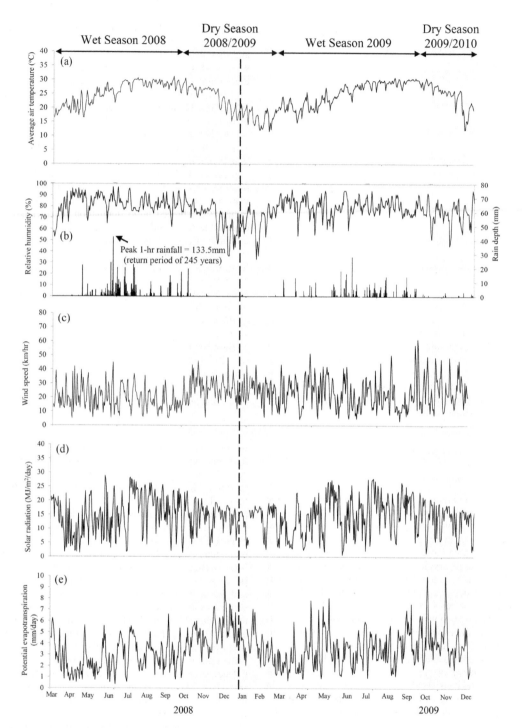

Figure 3. Site climate; (a) air temperature; (b) air relative humidity and rain depth; (c) wind speed; (d) solar radiation; and (e) calculated potential evapotranspiration.

3.3 *Slope geology*

Based on engineering field mapping and detailed ground investigation (including borehole exploration, trial pits and trial trenches excavation and geophysical resistivity survey), the site geology was identified.

Figures 4(a) and (b) show the ground profiles across section A-A and B-B, respectively (refer to Fig. 2). In the top 2 to 3 m, colluvial deposit, which consists of silty clay mixed with some cobbles of decomposed volcanic tuff, were encountered. Plant rootlets

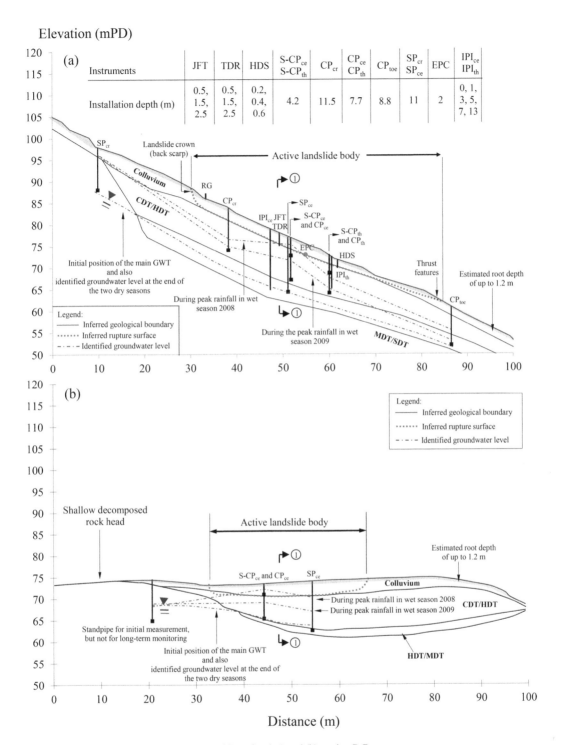

Figure 4. Idealised geological profiles across (a) section A-A and (b) section B-B.

with an average depth of 1.2 m were identified within this stratum. Below colluvium, a thick stratum of saprolites, namely completely decomposed volcanic tuff (CDT), was found to overly a 3–4 m thick stratum of highly decomposed volcanic tuff (HDT).

Borehole records suggest that there were several sets of *in-situ* relict joints in the CDT stratum. The dip angle ranged between 59 to 87°, while the dip direction varied from 010 to 316° (GEO, 2007). These joint sets were described as closely to medium spaced, smooth

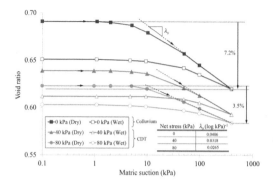

Figure 5. Laboratory measured volume change behaviour of unsaturated colluvium and CDT natural samples.

Table 1. Measured index properties of colluvium and CDT (Leung & Ng, in press).

Index properties	Colluvium	CDT
Compaction properties		
Maximum dry density (g/m³)	1.58	1.78
Optimum water content (%)	15.2	17.3
Particle-size distributions		
Gravel (≥2 mm, %)	18	2.5
Sand (63 μm–2 mm, %)	25.5	35
Silt (2 μm–63 μm, %)	39.5	42.5
Clay (≤2μm, %)	17	20
Atterberg limit		
Liquid limit (%)	41	34
Plastic limit (%)	17	20
Plasticity index (%)	24	14
Strength parameters		
Effective cohesion, c' (kPa)	0.3	7.4
Effective frictional angle, ϕ' (°)	35.2	33
Specific gravity	2.73	2.68
Unified Soil Classification System (USCS)	CL	CL

planar, iron stained and infilled with slightly silty sand. The information about the relict joints provide necessary basis for the modelling of permeability anisotropy in the anisotropic transient seepage analysis later.

At distance of 20 m, a substantial drop of the elevation of the underlying moderately to slightly decomposed volcanic tuff (MDT/SDT) was identified. Both SDT and MDT are classified as rock (GCO, 1988). Some rupture surfaces were found near the colluvium-CDT interface at 2–4 m depths. The inferred landslide body are shown in the figure. Before the field monitoring, the groundwater level (GWL) was identified at about 11 m depth and its profile nearly followed the rock head profile of MDT/SDT.

3.4 Soil properties

3.4.1 Index properties

Block samples at 0.5, 1, and 2 m depths were collected at the site for characterising the properties of colluvium and CDT in the laboratory. It is found that the *in-situ* water content (by mass) of colluvium was about 20% (slightly higher than the plastic limit of 17%), while that of CDT varied from 17% to 22% (the average of which is close to the plastic limit of 20%). The *in-situ* dry density of colluvium (1.5 g/m³) and CDT (1.6 g/m³) was about 95% and 90% of their maximum value obtained from the Standard Proctor Test, respectively. The gravel, sand, silt, and clay contents of colluvium are 0%–35%, 20%–25%, 30%–50% and 15%–25%, respectively, while the CDT has gravel, sand, silt, and clay contents of 2.5%, 35%, 42.5%, and 20%, respectively. According to the Unified Soil Classification System, both colluvium and CDT are classified as CL.

Shear strength parameters of saturated colluvium and CDT were determined through consolidated undrained (CU) triaxial tests (Leung & Ng, in press). The test results show that the effective cohesion, c', and the effective friction angle, ϕ', of colluvium are 0.3 kPa and 35.2°, respectively, while those of CDT are 7.4 kPa and 33°, respectively. Table 1 summarises the measured properties of colluvium and CDT.

3.4.2 Volume change behaviour

A series of laboratory tests were conducted to characterise shrink-swell behaviour of colluvium and CDT. A pressure-plate device (Ng & Pang, 2000), which adopts the axis-translation technique to control independently net stress and suction of a specimen, was used. Each natural (undisturbed) soil sample was trimmed into an oedometer ring and was then saturated. Two CDT samples were consolidated at a vertical net stress of 40 or 80 kPa, which represents approximately the *in-situ* overburden pressure at 2 and 4 m depths, respectively. No net stress was applied to a colluvium sample as it existed in shallow depths of the slope. Any effects of net stress on the shrink-swell behaviour of colluvium may be negligible. After consolidation, step increases in suction from 0 to 400 kPa were applied to each sample to undergo a drying path, while any applied net stress was maintained. At equilibrium, vertical displacement of each sample was recorded. When suction of 400 kPa was reached, similar test procedures were adopted for a wetting path. In this case, suction was controlled to reduce from 400 to 0 kPa in steps.

Figure 5 shows the measured variations of void ratio of each natural sample with suction. The observed different initial void ratio (at zero suction) was attributed to the different net stresses applied to each sample. The higher the applied stress, the lower the void ratio would be. For the soil sample subjected to zero net stress, plastic soil volume reduction is observed when suction reached a critical value of about 10 kPa, as indicated by the nonlinear decrease in void ratio (by 7.2%). Based on the elasto-plastic model framework such as Barcelona Basic Model (Alonso et al., 1990), the observed behaviour suggests that the soil stress state was likely to have reached and expanded the Suction Increase (SI) yield surface, as defined in the suction-net stress space. Similar plastic soil deformation is also found for the CDT samples loaded at higher net stresses of 40 and 80 kPa. By defining the gradient of the log-linear portion of the curve as soil shrinkability, λ_s, the λ_s for the soil at zero net stress is 0.0406 $(\log \text{kPa})^{-1}$. The values of λ_s of the two CDT samples were lower as CDT loaded to a higher stress level has higher stiffness to resist the plastic soil volume change due to suction increase (Wheeler et al., 2003). At 80 kPa of net stress, the decrease in void ratio of CDT was half of that of colluvium only.

Along the wetting path, only a slight increase in void ratio (i.e., elastic swelling) is observed for all specimens as suction reduced from 100 to 10 kPa. This means that only a portion of the plastic soil shrinkage during the previous drying was recovered by the swelling. When suction decreased further to values lower than 10 kPa, no swelling was observed.

4 INSTRUMENTATION PLAN

4.1 Part I: Instantaneous Profile (IP) tests

In order to interpret the responses of the volcanic slope, the in-situ hydraulic properties of colluvium and CDT, including stress-dependent soil water retention curves (SDSWRCs) and permeability function, were determined at the site (Fig. 2). SDSWRC is a measure of the water retention capability of an unsaturated soil for a given suction and net stress (Ng & Pang, 2000). The in-situ tests were based on a transient-state method, namely the Instantaneous Profile Method (IPM; Watson, 1966) and are therefore named instantaneous profile (IP) tests.

4.1.1 Theoretical background of IPM

The IPM considers one-dimensional (1D) water flow in an unsaturated soil column, where changes of vertical profiles of volumetric water content (VWC), θ_w, and hydraulic head, h, are monitored continuously. Consider two arbitrary vertical profiles of θ_w and h that are measured at elapsed time $t = t_1$ and t_2 and at depths z_A (Row A), z_B (Row B), z_C (Row C) and z_D (Row D). Each VWC profile may be extrapolated to the surface and the bottom of the soil column for determining water flow rate. Considering mass continuity, boundary outflow rate at the bottom of the soil column, $v_{ze,tave}$, can be determined by:

$$v_{ze,tave} = \frac{d}{dt} \int_{zB}^{ze} \theta_w \, z,t \cdot dz + v_{zB,tave} \quad (2)$$

where $\theta_w(z, t)$ is VWC profile as a function of depth z at specific time t; dt is time interval between two measurements (i.e., $t_2 - t_1$); $v_{zB,tave}$ is inflow rate at any depth z_B at average elapsed time $t_{ave} = (t_1 + t_2)/2$. After manipulation, the $v_{zB,tave}$ can be expressed as:

$$v_{zB,tave} = -\frac{\Delta V}{t_2 - t_1} + v_{ze,tave} \quad (3)$$

where ΔV is the area bounded by two profiles of $\theta_w(z, t)$ at $t = t_1$ and t_2. ΔV represents any water volume change between the depth under consideration (i.e., z_B) and one end of the soil (i.e., z_e). By estimating the slope of a hydraulic head profile (i.e., sum of PWP head and gravitational head), hydraulic head gradient, $i_{zB,tave}$, at any depth z_B at average elapsed time t_{ave} can be determined by:

$$
\begin{aligned}
i_{zB,tave} &= \frac{1}{2}\left(\frac{dh_{zB,t1}}{dz} + \frac{dh_{zB,t2}}{dz} \right) \\
&= \frac{1}{4}\left[\left(\frac{h_{zA,t1} - h_{zB,t1}}{z_A - z_B} + \frac{h_{zB,t1} - h_{zC,t1}}{z_B - z_C} \right) + \left(\frac{h_{zA,t2} - h_{zB,t2}}{z_A - z_B} + \frac{h_{zB,t2} - h_{zC,t2}}{z_B - z_C} \right) \right]
\end{aligned} \quad (4)
$$

where $h_{zi,tj}$ is hydraulic head at depth z_i ($i = A$, B, C and D) at elapsed time t_j ($j = 1, 2$).

By using the Darcy's law, permeability, $k_{zB,tave}$, at any depth z_B at average elapsed time t_{ave} can hence be calculated by dividing the water flow rate by the corresponding hydraulic head gradient:

$$k_{zB,tave} = -\frac{v_{zB,tave}}{i_{zB,tave}} \quad (5)$$

4.1.2 Test setup and arrangement of instruments

Based on the theoretical consideration of the IPM, IP tests were designed and conducted at the site. Figure 6 depicts a cross-section of the test setup. A flat test plot of 3.5 m × 3.5 m was formed near the toe of the slope. To achieve 1D vertical water flow assumption for the IPM, a 3 m deep and 1.2 m wide trench was excavated at the uphill side of the test plot to install a polythene sheeting. The sheet aimed to act as a cut-off to minimise any possible lateral groundwater flow and recharge from up-slope during testing. The trench was then backfilled. Subsequently, a circular steel test ring with a diameter of 3 m was installed at the ground surface and it was embedded 100 mm into the ground to retain water during each test.

In order to measure negative PWP, ten jet-fill tensiometers (JFTs) were installed at 0.36, 0.77, 0.95, 1.17, 1.54, 1.85, 2.13, 2.43, 2.6 and 2.99 m depths. Negative PWP is a gauge pressure measuring the amount of PWP below the atmospheric pressure. By assuming pore-air pressure to be atmospheric, negative PWP is equal to suction. In this paper, these two terms are used interchangeably. Four time-domain reflectometers (TDRs) were installed at 0.84, 1.85, 2.5 and 3.59 m depths to determine VWC profiles.

4.1.3 Test procedures

The test procedures were divided into four stages, consisting of two wetting-drying cycles. For each wetting cycle, a water head of 0.1 m was ponded on the ground surface inside the ring for four days. The water level was checked and refilled to the same elevation every 12 hours. Each wetting cycle was ended when the steady-state condition was reached. Subsequently, drying cycle was commenced by allowing the plot to dry under natural evaporation and internal drainage. Similarly, each drying cycle was stopped when all sensors showed steady-state readings. More details of the

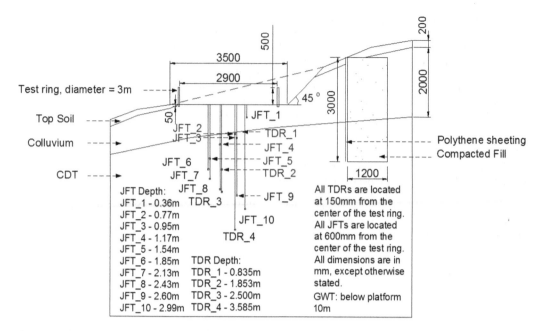

Figure 6. The IP test set up and the instrumentation plan.

test programme and procedures are given by Ng et al. (2011).

4.2 Part II: Monitoring seasonal slope performance

A wide range of different types of instruments was installed to monitor the climatic effects on the behaviour of the volcanic slope from March 2008 to December 2009. They include JFTs, heat dissipation matric water potential sensors (HDSs), TDRs, piezometers (CPs), standpipes (SP), earth pressure cells (EPCs) and inclinometers. The comprehensive set of instruments aims to quantify both the mechanical (i.e., stress-deformation characteristics) and hydrological (i.e., PWP and GWT) responses of the slope.

4.2.1 Monitoring PWP and VWC

As shown in Figure 4, three JFTs were installed at the central portion of the landslide body to measure PWP in colluvium at 0.5 and 1.5 m depths and CDT at 2.5 m depth. Due to the possibility of cavitation of water, the minimum PWP that can be recorded by each JFT is not less than −90 kPa. During the monitoring period, any accumulated air bubbles due to cavitation and air diffusion in each JFT were removed by pressing the jet-fill button.

In order to measure relatively high suction potentially induced by plant evapotranspiration (ET) in shallow depths, three HDSs were installed within the root zone at 0.2, 0.4 and 0.6 m depths. When the ceramic block of a HDS is in contact with soil, water exchange happens and any changes of water content of the ceramic block would result in a change of its thermal conductivity. A heating element inbuilt in the ceramic block releases a known amount of heat and

a portion of it would be dissipated, depending on the thermal conductivity of the ceramic. Any increase in temperature (ΔT) due to the heat stored in the ceramic block could then be correlated with suction via a calibration curve. Calibration method is reported in Fredlund and Wong (1989). Laboratory calibrations in this study revealed that the air-entry value (AEV) of the ceramic block of each HDS is 16 kPa. This means that when soil suction is lower than the AEV, changes of water content, and hence thermal conductivity, of the ceramic block become negligible. This casts the lower limit of suction measurement (i.e., 16 kPa) made by each HDS.

For monitoring VWC, three TDRs were installed 0.5 m away from the three JFTs at 0.5, 1.5 and 2.5 m depths. Each TDR was calibrated in the laboratory. The accuracy of each TDR is within 2% for VWC ranging from 10% to 40% (Leung & Ng, 2013a).

4.2.2 Monitoring groundwater level

In order to monitor any changes of GWL during the monitoring period, six CPs and a SP were installed. Each piezometer consists of a vibrating-wire type pressure transducer to record positive PWP. Any increase in positive PWP is equivalent to an increase in GWL with reference to the installation depth. Two piezometers, namely S-CP$_{ce}$ and S-CP$_{th}$, were installed at a shallower depth of 4 m near the colluvium-CDT interface. The other four, namely CP$_{cr}$, CP$_{ce}$, CP$_{th}$ and CP$_{toe}$, and the SP were installed in deeper depths (8 to 11 m) to monitor the responses of the main GWL at various locations of the slope.

4.2.3 Monitoring total horizontal stress

A pair of earth pressure cells (EPCs) was installed at 2 m depth in the central portion of the landslide body

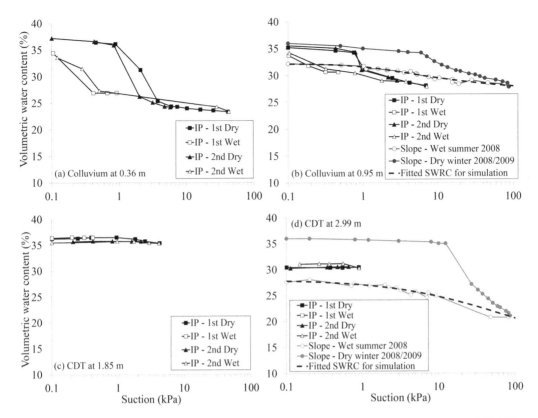

Figure 7. textitIn-siu SDSWRCs of colluvium at (a) 0.36 m and (b) 0.95 m depths and CDT at (c) 1.85 m and (d) 2.99 m depths.

to monitor any changes of total horizontal stress. The two EPCs were installed perpendicular to each other so that one of them measured total horizontal stress in the down-slope direction (σ_D), while the other measured the stress in the cross-slope direction (σ_C). At the installation location, a 1 m × 1 m (in plan) trial pit was excavated to a depth of 2.5 m. At 2 m depth, a slot with a size similar to the width of each EPC was cut and the EPC was inserted into the slot. Inevitably, these procedures caused stress release around the slot and hence, the initial total horizontal stress recorded by each EPC was reduced to zero. Thus, subsequent reading represents change of horizontal stress with reference to the initial zero total stress. To provide better contact between each EPC and the surrounding soil, the gap between them was filled with cement bentonite grout. This allows any tensile force to be transmitted between each cell and the soil (Brackley & Sanders, 1992).

4.2.4 Monitoring horizontal slope movement

In order to monitor the horizontal slope displacement, two in-place inclinometers (IPIs) were installed at the central portion (IPI$_{ce}$) and near the thrust features (IPI$_{th}$) of the landslide body. Four tilt accelerometers were mounted near the ground surface, at 1, 3 and 5 m depths along each IPI casing. As slope displaces, each accelerometer records tilt angle with respect to

the vertical casing. For the given spacing of accelerators and measured tilt angles, horizontal displacements at each depth can be determined. Note that the horizontal displacement measurements made before the storm event in June 2008 were referenced to 7 m depth, where zero displacement was assumed for both IPIs. After the storm event, the IPI$_{ce}$ recorded large slope displacements in the top 5 m (as discussed later), and therefore two extra tilt accelerometers were installed at 7 and 10 m depths on January 2009. Zero displacement was assumed at a 13 m depth thereafter.

5 SOIL HYDRAULC PROPERTIES

5.1 Water retention properties

Based on the results obtained from the IP tests, the measured VWC is related to the measured JFTs to investigate the soil water retention properties. Figure 7 shows the drying and wetting SDSWRCs of colluvium and CDT at various depths. For the colluvium at 0.36 and 0.95 m depths (Figs 7(a) and (b)), the AEV was about 2 kPa, beyond which VWC reduced substantially. At 0.36 m depth, the reduction of VWC was negligible when suction increased beyond 5.6 kPa in the first drying curve. Noticeable hysteretic loop is observed at both depths after each cycle of wetting-drying event. The loop size appears to decrease with an increase in depth. The observed trend may be because soil located in deeper depths is subjected to a higher confining pressure and lower degree of weathering (i.e., smaller average soil pore size). Stress effects on SDSWRCs of residual and volcanic soils have been investigated experimentally by Ng & Pang (2000), Ng & Leung (2012) and Ng et al. (2012). Zhou & Ng (2014) also examined the stress dependency of SDSWRCs through the consideration of stress-induced changes in soil pore-size and its distribution from a theoretical point of view. These studies consistently showed that soil subjected to a higher stress level would have a smaller AEV and smaller hysteresis loop size. For CDT at 1.85 and 2.99 m depths (Figs 7(c) and (d)), there were negligible changes in VWC and hysteresis loop. This might be because the suction range investigated in the IP tests was probably within the AEV of CDT.

Considering the fact that the ground at the site has been subjected to countless wetting and drying cycles in the past, it is not surprising to find that the SDSWRCs of both materials obtained from the first wetting-drying cycle were comparable to those from the second cycle (Ng & Pang, 2000). The measured SDSWRCs are referred to as scanning curves.

In-situ SDSWRCs obtained from the field monitoring (i.e., Part II) are shown in Figures 7(b) and (d) for comparison. The SDSWRCs of colluvium and CDT were obtained by relating the measured VWC with measured suction at 0.5 and 2.5 m depths, respectively (Leung & Ng, 2013a). For the colluvium (Fig. 7(b)), both the desorption and adsorption rates are similar to those found in the IP tests. However, the AEV obtained from the field monitoring (6 kPa) was much higher. Discrepancies are also found for the CDT (Fig. 7(d)). The observed different water retention properties between two sets of measurement are likely attributed to the heterogeneity nature of both soil types, as the measurements were made at two different locations of the site (Fig. 2).

5.2 *Permeability functions*

Based on the results of PWP and VWC obtained from the IP tests and Equations (2)–(5), permeability functions of colluvium and CDT were determined. Figure 8 shows the measured permeability functions at depths of 0.84 m (colluvium) and 1.85 m (CDT). A range of saturated permeability, k_s, of colluviums and decomposed volcanic (DV) soil recorded in the Mid-levels area in Hong Kong is also shown for comparison (GCO, 1982). It can be seen that the values of permeability of the colluvium fall within the range of k_s for suctions lower than 1 kPa. The permeability then reduced significantly by three orders of magnitude from 1×10^{-4} to 2×10^{-7} m/s as suction increased further from 1 to 4 kPa.

Compared to the colluvium, the CDT was less permeable as the values of permeability are lower by up to almost one order of magnitude. The permeability

functions of the CDT are less fluctuating than those of the colluvium. The values of permeability of CDT vary between 3×10^{-6} and 4×10^{-7} m/s. No significant drop of permeability is found in the CDT for suctions less than 5 kPa.

By considering the IPM, Ng & Leung (2012) developed a stress-controlled 1D soil column apparatus for measuring unsaturated permeability function of soils under the influences of both net normal stress and suction. Using the apparatus, CDT sampled from the site was re-compacted and loaded vertically at 40 kPa to achieve a dry density similar to the *in-situ* value. The laboratory measurement is compared with that obtained from the IP tests in Figure 8. It can be seen that the *in-situ* permeability function of the CDT is higher than the laboratory measurements by three orders of magnitude. However, the rate of change of permeability for suctions ranged from 0 to 5 kPa is found to be negligible in both the field and laboratory measurements. The observed higher permeability in the field may be attributed to preferential flow through cracks and fissures (which were not considered in the laboratory). Based on the laboratory test results, it can be seen that when suction increased from 0.1 to 80 kPa, the permeability dropped by almost two orders of magnitude and reached about 1×10^{-10} m/s at 80 kPa of suction.

6 SEASONAL HYDROLOGICAL RESPONSES

6.1 *Wet summers*

Figures 9(a) and (b) show the measured climatic responses of PWP (by both HDSs and JFTs) and VWC, respectively. The 1-hour rainfall intensity recorded at the site is depicted in each figure for reference. It should be noted that in Figure 9(a), the data recorded by HDSs during rainfall in both the wet summers 2008 and 2009 are not shown. This is to avoid confusion as each HDS was not able to measure low range of suction (i.e., <16 kPa; see Section 4.2.1) accurately. When rainfall happened in both the wet summers 2008 and 2009, rapid increases in PWP were recorded at all depths (0.5, 1.5 and 2.5 m). In general, PWPs at both 0.5 and 1.5 m depths varied between −5 and 10 kPa, while positive PWPs recorded at 2.5 m depth was as high as 15–25 kPa.

As PWPs increased during the wet summer 2008, VWCs at all depths increased correspondingly (see Fig. 9(b)). As compared to the responses observed in colluvium at 0.5 and 1.5 m depths, the measured changes of VWC in CDT at 2.5 m depth during most rainfall events were much larger, varying from 26% to 36% significantly. This may be because the CDT has relatively low water retention capability than colluvium when they were subjected to the same given change in PWP (compare SDSWRCs in Fig. 7). It can be identified in Figure 9 that when positive PWPs were recorded at 1.5 depth in colluvium and 2.5 m depth in CDT, peak VWC of 36% was recorded in both materials consistently. This peak value is found

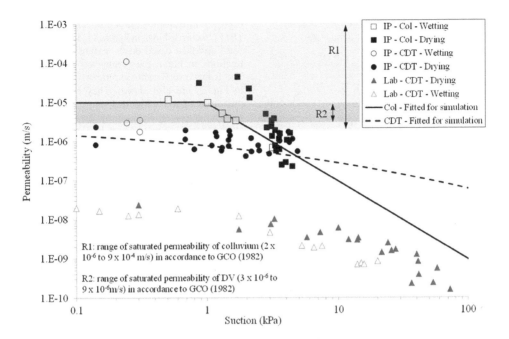

Figure 8. *In-situ* and laboratory measured permeability functions of colluvium and CDT.

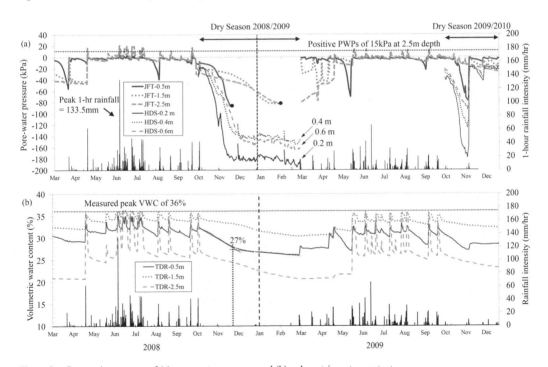

Figure 9. Seasonal responses of (a) pore-water pressure and (b) volumetric water content.

to be close to the saturated VWC as determined from the *in-situ* IP tests (Ng et al., 2011).

Figure 10 shows the distributions of PWP along depth (across section 1-1; refer to Fig. 4) during the largest rainfall event recorded during the two wet summers. Each PWP profile is obtained from the measurements made by the JFTs, S-CP$_{ce}$, CP$_{ce}$ and SP$_{ce}$. A hydrostatic line representing the initial elevation of the main GWT at 11 m depth is also depicted for reference. Initially before the start of the wet season 2008, PWP in colluvium was higher than the hydrostatic value, indicating a net downward

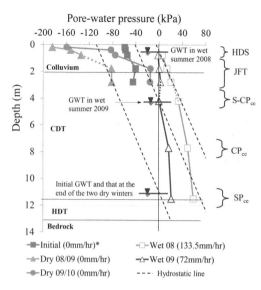

Figure 10. Seasonal distributions of PWP along depths.

infiltration. When the rainstorms with peak intensity of 133.5 mm/hr happened on 7th June 2008 (equivalent to return period of 245 years), positive PWPs were recorded at all depths. It can be observed that the PWP profile in the top 3 m of the ground was nearly hydrostatic. Significant building up of positive PWP of up to 60 kPa was recorded below 5 m depth. The measured PWP distribution was consistent with the observed GWT response. In Figure 4(a), monitoring data from S-CP$_{ce}$, CP$_{ce}$ and SP$_{ce}$ showed that during the storm, the GWT rose by about 6 m. A much substantial rise of about 10 m was recorded at the central portion of the landslide body (see also Fig. 4(b)). This explains the significant increase in positive PWPs in Figure 10.

When a peak rainfall intensity of 72 mm/hr happened in the 2009 summer, positive PWPs were recorded again, but the magnitude was smaller than that found in the previous summer. This is because the total rainfall recorded in 2009 summer was two times lower than that in the 2008 summer. This led to less significant rise of the GWT (Figs 4(a) and (b)) and increase in positive PWP in 2009 summer.

6.2 Groundwater flow mechanisms

In order to improve the understanding of the slope hydrology and to explain the observed local significant rise of the main GWT during the storm event on 7th June 2008, a three-dimensional (3D) anisotropic seepage analysis was carried out (Leung & Ng, 2013b). The FE software, SVFlux, was used. The analysis was based on the Richard's equation, which considers mass continuity and Darcy's law, to describe transient seepage in unsaturated soil.

6.2.1 Finite element mesh and boundary conditions

The FE mesh is shown in Figure 11. The ground profile was calibrated based on the extensive ground investigations, including the geophysical resistivity survey. The ground profile across sections A-A and B-B (refer to Fig. 2) are also shown. They are similar to the profiles depicted in Figure 4.

In order to simulate the initial main GWT identified at 11 m depth, a constant total head equal to the elevation of the CDT-HDT interface was specified on the right boundary and downstream boundary of the 3D mesh, while zero flux was applied on the left boundary and upstream boundary. The rock was assumed to be impermeable and zero flux was thus specified on the bottom boundary. The top surface boundary is presented in analysis procedures later.

6.2.2 Analysis plan

The numerical study considered the effects of (i) inclined relict joints identified in the CDT stratum and (ii) the shallow rock head along the cross-slope direction (Fig. 4(b)) on groundwater flow. Effects of inclined relict joint were modelled by considering permeability anisotropy in SVFlux.

The input parameters for each soil type, including SDSWRC and permeability function, are shown in Figures 7 and 8 (refer to the fitted lines according to the fitting equations suggested by van Genuchten (1980)). The permeability of CDT was considered to be anisotropy, by taking (i) the permeability in the horizontal direction 100 times higher than that in the vertical direction and (ii) a dip angle of 60°. These considerations were justified through the interpretation of borehole records discussed in Section 3.3. For both colluvium and HDT, the permeability functions remain isotropic.

6.2.3 Analysis procedures

Three stages of analyses were carried out. The first stage was to conduct a steady-state analysis by specifying a small rainfall intensity of 0.1 mm/day on the top surface boundary of the model. In the second stage, a transient analysis was performed by applying the same constant intensity of 0.1 mm/day. The analysis was stopped when the computed PWP profile and the elevation of the GWT were reasonably close to the field measurements recorded on 6th June 2008. In the last stage, the rainfall event shown in Figure 12 was applied on the top surface boundary for transient analysis.

6.2.4 Comparisons of measured and computed pore-water pressure distributions

Figure 13 compares the measured and computed PWP profiles across section 1-1 (refer to Fig. 4(b)). It can be seen that the initial PWP profile and the elevation of the main GWT were comparable between the measurements and simulations. Two PWP profiles are investigated at, namely t_1 (3 hours before the peak storm) and t_2 (at the peak storm; see Fig. 12 for definition). At t_1 when the slope was subjected to rainfall intensity of 80 mm/hr, a perched GWT is found above

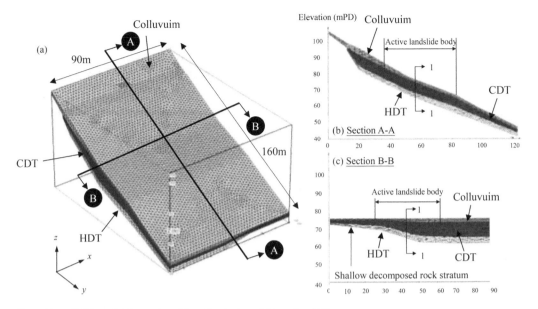

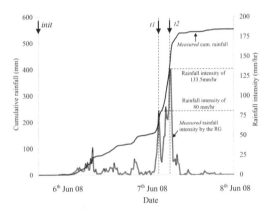

Figure 11. 3D FE mesh (a) overview, (b) section A–A, and (c) section B–B.

Figure 12. Measured rainfall intensity and cumulative rainfalls.

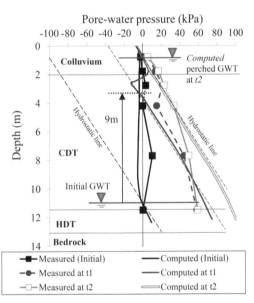

Figure 13. Comparison of measured and computed PWP profiles during the extreme rainstorm event.

the colluvium-CDT interface, while a significant rise of GWT (by 9 m) is observed. Due to the formation of the shallow perched GWT, positive PWP of up to 5 kPa was developed. This is because the permeability of the underlying CDT was one order of magnitude lower than that of colluvium at low suctions, as revealed from the *in-situ* IP tests (Fig. 8). It can be seen in Figure 13 that both measured and computed PWP in colluvium and CDT below 4 m depths distributed nearly hydrostatically. This distribution is, however, not found below the colluvium-CDT interface between 2 to 4 m depths. Given the *in-situ* geological profile (see Fig. 4(b)), the 3D analysis reveals that rainwater infiltrated near the western flank of the landslide body flowed along the shallow rock surface across the slope in the transverse direction, leading to cross-slope flow mechanism. This hence results in the significant rise of the GWT in both the measurement and simulation.

During the peak rainstorm occurred at t_2 (Fig. 13), the computed PWP increased with an increase in depth and the profile is almost hydrostatic in both the colluvium and CDT strata. This results in a thick wetting band and hence the significant building up of positive PWP, as similarly observed from the field measurement. This suggests that the slope hydrology was greatly affected by *in-situ* geological features identified at the site. In order to capture the high positive PWPs and the significant recharge of GWT in deeper depths during the rainstorm, it appears to be important,

75

for this slope, to take into account both the preferential water flow along the relict joints in the CDT stratum and the cross-slope groundwater flow along the shallow rock surface.

6.3 Dry winters

During the dry winter from October 2008 to February 2009 (see Fig. 9(a)), no rain was recorded within the study area. Suction in the top 2.5 m depths increases significantly. This is especially the case within the root zone at 0.2, 0.4 and 0.6 m depths, as recorded by the three HDSs. At deeper depths of 1.5 and 2.5 m, the measured rate of change of PWP was relatively less significant. As suctions approached 90 kPa, cavitation occurred in the three JFTs and the data measured thereafter is not shown to avoid confusion. After monitoring for two months from October to December 2008, it is evident that the measured rates of increase in suction at 0.2, 0.4 and 0.6 m depths reduced. The suctions at these depths then approached a steady-state value of 190, 160 and 165 kPa, respectively. Similar reduction of the rate of VWC change is also found at 0.5 m depth on December 2008 (Fig. 9(b)). When VWC dropped below 27%, the rate of VWC change reduces considerably. Such hydrological responses appear to be consistent with the *in-situ* SDSWRCs of the colluvium shown in Figure 7. Gradual reduction of VWC was found at both 1.5 and 2.5 m depths (Fig. 9(b)).

During the dry winter 2009/2010, substantial suction recovery was recorded in the top 0.6 m depth on November 2009, whereas suctions in deeper depths increased slightly only. When compared to the previous winter, suction induced in this winter was relatively lower. No steady-state condition is observed. This is mainly because the duration of drying period in this winter was much shorter (only one month as compared to four months in the previous winter). The rainfall events happened on November 2009 caused all the suctions disappeared. Corresponding increase in VWC at 0.5 m depth was recorded (Fig. 9(b)). The VWC responses at 1.5 and 2.5 m depths were similar to those recorded in previous winter.

6.4 Effects of vegetation on slope behaviour

The measured PWP distributions at the end of the two dry winters are depicted in Fig. 10. It should be noted that the measured data obtained from the three HDSs installed at 0.2, 0.4 and 0.6 m depths are included. The average root depth of 1.2 m for the vegetation identified at the site (refer to Fig. 4) is shown for reference. After subjecting to the four-month drying period in 2008/2009, the GWT returned to the initial elevation at 11 m. It can also be seen that the PWP induced by plant ET in colluvium was lower than the hydrostatic value. This suggests that there was net upward water flow within the colluvial stratum, predominantly due to root-water uptake. On the contrary, the distribution of PWPs in CDT almost followed the hydrostatic line, indicating that there was no net movement of water.

The observed shape of PWP profile implies that the influence depth of suction due to plant ET over four months of drying period was shallower than 2 m (i.e., about 200% of the plant root depth).

At the end of the dry winter 2009/2010, it is similar to the previous winter that suction in the top 3 m increased. However, the magnitude of suction in this winter was substantially lower. It can be further observed that except 0.2 m depth, PWPs at all depths were higher than the hydrostatic value. This suggests that downward flux prevailed even though the ground surface was subjected to ET. The observed lower suction induced in this winter is because the site was subjected to shorter duration of plant ET of only one month from October to November 2009 (see Fig. 9(a)). Another reason is that the plant ET in this winter (347 mm; see also Fig. 3) was 32% lower than that in the previous winter (507 mm).

Upon rainfalls in the two wet seasons 2008 and 2009, it is found that only marginal amount of suctions (<5 kPa) were retained within the root zone. However, when the frequent high-intensity, long-duration rainfall happened from May to July 2008, suctions at 0.5 m depth were disappeared, despite of considerable amount of plant ET (i.e., up to 5 and 8 mm/day in the wet summers 2008 and 2009, respectively). A number of case histories are reported to compare suction induced between vegetated and bare soils (Lim et al., 1996; Simon & Collison, 2002; Ng & Zhan, 2007; Kim & Lee, 2010; Leung et al., in press; Garg et al., in press). Interestingly, the hydrological effects of plant ET on induced suction identified in these previous studies are not consistent. It was sometimes reported that vegetated soil could induce higher suctions than bare soil (Lim et al., 1996; Leung et al., in press; Garg et al., in press), but in some other cases, opposite findings are observed (Simon & Collison, 2002; Kim & Lee, 2010; Ng & Zhan, 2007). The underlying reason causing this inconsistent observation is not well-understood. Unfortunately, not much information of climate and plant properties is available in these studies for further interpretation.

7 STRESS-DEFORMATION RELATIONSHIPS

7.1 Observed slope displacement profiles

Climatic effects on the horizontal displacement profiles measured by IPI$_{ce}$ and IPI$_{th}$ are compared in Figure 14. When the storm event happened in the wet summer 2008, down-slope displacement was recorded in the top 5 m of the ground at both locations. The peak movement at the ground surface at IPI$_{ce}$ was 40 mm (Fig. 14(a)). To determine the deformation profile in deeper depths, a manual inclinometer survey up to 15 m depth was conducted and two extra tilt accelerometers were installed at 7 and 10 m depths on 12th January 2009. The manual measurement shows that the slope exhibited "deep-seated" mode of movement. A remarkably large displacement of 20 mm

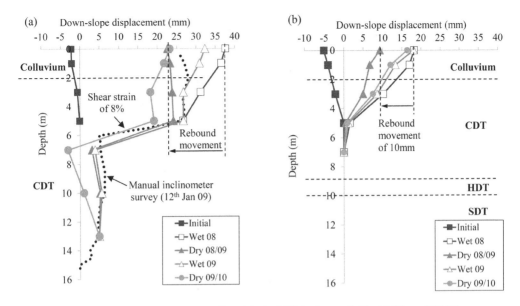

Figure 14. Measured horizontal displacement profiles of the landslide body recorded by (a) IPI$_{ce}$ and (b) IPI$_{th}$.

was recorded between 5.5 and 6 m depths, which is equivalent to an average shear strain of 8%. The relatively large shear strain indicates that the soil at these depths was likely to have been subjected to plastic deformation. The landslide body appears to undergo translational-type of down-slope movement along some rupture surfaces developed or/and re-activated at 5.5 to 6 m depth. This is consistent with the building up of positive PWP at these depths when the GWT rose significantly to an elevation close to the slope surface (see Fig. 4).

A different mode of slope deformation was recorded by IPI$_{th}$ near the thrust feature of the landslide body (Fig. 14(b)). Only the top 5 m of the slope showed substantial movement. The peak change of displacement (25 mm) was at the slope surface and the change decreased with an increase in depth. This resulted in a so-called "cantilever" mode of displacement profile. No distinctively large shear strain like the one observed at IPI$_{ce}$ was developed.

In the subsequent dry winter 2008/2009, it can be seen that at both locations, the displacements in the top 5 m depth reduced, whereas the soil below 7 m was largely stationary. The decreases in displacement mean that the soil displaced towards the up-slope direction (referred to as up-slope rebound). At IPI$_{ce}$ (Fig. 14(a)), the peak up-slope movement was 15 mm at the slope surface. This means that about 40% of the down-slope displacement recorded during the previous wet summer was rebounded. In contrast, the large plastic displacement at 5.5–6 m induced by the storm was irrecoverable. It can be identified that at both locations, the reduction of displacement was smaller in deeper depths. This is likely because the soil at deeper depths, which was subjected to higher overburden pressure, has smaller shrinkability, λ_s (see the test data shown in Fig. 5).

Down-slope displacement is again observed during the wet summer 2009 at both locations. However, a "cantilever" mode, instead of "deep-seated" mode, was recorded by IPI$_{ce}$ (Fig. 14(a)). The slope did not appear to exhibit translational movement, as observed in the previous summer. In the summer 2009, the peak displacement was less significant and it was up to 10 mm only. This is mainly because the measured changes of PWP and GWT in this summer were comparatively smaller (see Figs 4 and 9(a)).

During the dry winter 2009/2010, up-slope rebound was similarly observed at both locations (Figs 14 (a) and (b)). By comparing the initial slope displacement profile and that recorded at the end of the monitoring period, a net plastic down-slope displacement was identified. This means that the slope displacements induced in the two wet summers could not be recovered fully by the up-slope rebounds resulted in the two subsequent dry winters. This is a phenomenon known as slope ratcheting. This kind of slope behaviour was similarly found in a clay slope and a decomposed soil slope studied by Ng et al. (2003) and Cheuk et al. (2009), respectively. Through centrifuge model tests of saturated clay slopes, Take & Bolton (2011) argued that seasonal slope ratchetting was attributed to repeated mobilisation of dilatancy and softening of soil in successive wet seasons. The continuous mobilisation and accumulation of plastic shear strain of soil would eventually lead to progressive slope failure.

7.2 Effects of PWP on slope movements

In order to further understand the slope displacement responses in relation to PWP, Figure 15 correlates PWP at 0.5, 1.5, and 2.5 m depths with displacements recorded by IPI$_{ce}$ at 0, 1, and 3 m depths, respectively, during the rainstorm event from 5th to 9th June 2008.

The slope started to displace towards the down-slope direction when some critical values of positive PWP of 2, 11, and 15 kPa were recorded at 0.5, 1.5, and 2.5 m depths, respectively. When the heavy storm happened between 7th and 8th June, the slope displaced further, but interestingly, the positive PWPs recorded at all three depths remained unchanged during this event. This correlation is not expected because results from some triaxial tests (e.g. Meilani et al., 2005) conducted under constant deviatoric stress, decreasing suction path (commonly referred to as field stress path upon rainfall infiltration) showed that plastic soil deformation happened simultaneously with PWP changes (and hence effective stress changes). The field observation seems to suggest that the increase in the positive PWP in shallow depths (top 3 m) could not be the major reason leading to the large observed down-slope displacements. Instead, the slope displacement may be attributed to the translational movement of the top 5 m of the sliding mass along some rupture surfaces (refer to Fig. 14(a)). Therefore, when focusing only the responses in the top 3 m of the slope, the peak positive PWPs recorded at 0.5–2.5 m depths did not correspond to the slope displacements in Figure 15.

Correlations between the measured PWP and slope displacement during the dry winter 2008/2009 are depicted in Figure 16. As suction was recovered and increased from 10 to 90 kPa, displacement at all three

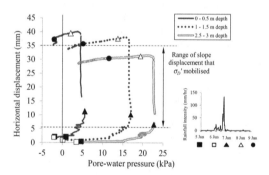

Figure 15. Relationships between PWP and horizontal displacement during the storm event.

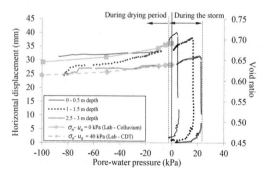

Figure 16. Relationships between PWP and slope displacement during the dry winter 2008/2009.

depths decreased (i.e., upslope rebound) at a decreasing rate. Compared to the laboratory measurements of the shrinkage of colluvium and CDT, it is interesting to observe that the void ratio of the specimen loaded at net stresses of 0 and 40 kPa reduced at very similar rates as the field measurements made at 0.5 and 2.5 m depths, respectively. It should be noted that the test data shown in this figure are the same set of those presented in Figure 5, but expressed on a linear scale for suction on the x-axis. The close responses between the field and laboratory observations suggest that the suction-induced soil plastic deformation (i.e., shrinkage) was most likely the reason causing the up-slope rebound during winters. It is worth-noting that the up-slope rebound (~10 mm in five months) occurred at a much slower rate than the down-slope displacement during the storm (>25 mm in one day; Fig. 14(a)). Such large contrast in the rate of slope movement between wet and dry seasons is the primary reason causing the net plastic down-slope movement (i.e., slope ratchetting) after two years of monitoring.

7.3 Climatic effects on stress mobilisation

7.3.1 Wet summers

Figure 17 compares the climatic effects on the responses of total horizontal stress in down-slope direction (σ_D) and total horizontal stress in down-slope direction (σ_C) at 2 m depth. It should be noted that the measurements made by both EPCs represent the changes of total horizontal stress with reference to the initial zero stress after their installation. The positive PWP recorded by the JFT installed at 2.5 m depth and rainfall intensity are also shown for references. During the two wet summers 2008 and 2009, both σ_D and σ_C increased with increase in rainfall intensity generally. It can be seen that the σ_C has a fairly good agreement with the positive PWP. This suggests that the measured changes in σ_C were likely attributed to the changes in positive PWP. The σ_D exhibited similar trends to the σ_C but the magnitude of σ_D was always higher by 7–17 kPa, depending on the rainfall intensity. Since the two EPCs were installed next to each other at the same depth of 2 m, they likely recorded comparable positive PWP. Therefore, in addition to the building up of positive PWP, substantial mobilisation of soil stress in the down-slope direction also happened during rainfalls.

In order to interpret the stress mobilisation, the modified Bishop's effective stress defined by Khalili & Khabbaz (1998) is adopted as follows:

$$\sigma' = (\sigma - u_a) + \chi(u_a - u_w)$$

$$\text{where } \chi = \begin{cases} 1, & \text{when } S = 100\% \\ \left(\dfrac{AEV}{u_a - u_w}\right)^{0.55}, & \text{when } S < 100\% \end{cases} \quad (6)$$

where σ is total stress; u_a is pore-air pressure; $(\sigma - u_a)$ is net stress; u_w is PWP; $(u_a - u_w)$ is suction; χ is Bishop's parameter; S is degree of saturation; and AEV is taken to be 2 kPa based on the field and laboratory

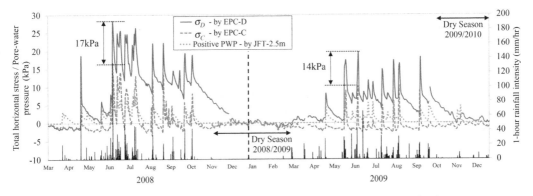

Figure 17. Measured total horizontal stresses at 2 m depth in the downslope and the cross-slope directions.

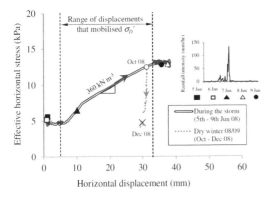

Figure 18. Observed stress mobilisation during the storm event from 5th–9th June 2008 and during the dry winter 2008/2009.

measurements shown in Figure 7. The power index of 0.55 was determined by calibrating the equation against shear strength properties of a broad range of soil types (Khalili & Khabbaz, 1998).

Figure 18 shows the process of stress mobilisation at 2 m depth as slope displaced during the storm event from 5th to 9th June 2008. The down-slope displacement expressed in x-axis is taken to be the average of the measurements made at 1 and 3 m depths. Since positive PWP and saturated VWC (i.e., $S = 100\%$) was recorded at 2.5 m depth during the storm event (see Figs 9(a) and (b)), effective horizontal stress in the down-slope direction, σ'_D, may be determined by the difference between the measured σ_D and positive PWP (i.e., setting χ in Equation (6) to be 1). It can be seen in Figure 18 that when the relatively small rainfall event happened on 6th June, σ'_D was not mobilised. This is mainly because the increase in σ_D was almost equal to the increase in positive PWP (see Fig. 17).

As the slope displaced further from 5 to 35 mm during the storm, a 350% increase in σ'_D (from 4 to 14 kPa) was recorded. By the Rankine theory and using the effective strength parameters (c' of 7.4 kPa and σ' of 33°; Table 1), it is estimated that the peak σ'_D mobilised almost 40% of the effective passive stress (38 kPa) of CDT. The mobilisation of σ'_D due to the horizontal slope displacement might be analogous to a horizontal

subgrade reaction problem. By determining the gradient of the linear portion of the σ'_D curve, it may be deduced that the coefficient of horizontal subgrade reaction, σ_h, of the silty clay is 360 kN/m³. This value is close to the lower range of σ_h (350–700 kN/m³) of clayey soil (Tomlinson & Woodward, 2007). Interestingly, the range of displacement (5–35 mm) that mobilised σ'_D corresponds to the constant responses of positive PWP in Figure 15. This relationship suggests that during the translational sliding of the top 5 m of the slope (Fig. 14(a)), deformation might have happened within the sliding soil mass. This thus led to the observed mobilisation of σ'_D at 2 m depth.

When the storm ceased on 8th June, σ'_D remained almost unchanged at the peak value of 14 kPa, as the landslide body rebounded slightly (<5 mm) towards the up-slope direction (Fig. 18). This is because the decrease in σ_D during this period was almost the same as that of positive PWP (Fig. 17).

7.3.2 Dry winters

As soil suction increased during the two dry winters 2008/2009 and 2009/2010, the σ_D decreased by up to 17 and 9 kPa, respectively, while the σ_C remained constant of 0 kPa in both dry seasons (see Fig. 17). Consistent to the suction responses measured by the HDSs and JFTs (refer to Fig. 9(a)), both the σ_D and σ_C approached a steady-state value after the prolonged drying period after December 2008. Constant negative total stresses (i.e., tensile stresses) of up to 2.5 kPa were recorded by both the EPCs. This kind of responses was similarly observed in other in-situ total stress measurements from EPCs installed by the same methods (Brackley & Sanders, 1992; Ng et al., 2003). The observed tensile stress was possibly because of shrinkage of backfilled grout upon drying.

Since suction increased and VWC dropped below the saturated value (i.e., $S < 100\%$) during the two winters (see Figs 9(a) and (b)), σ'_D in these cases is dependent upon the magnitude of suction and can be evaluated by Equation 6. As the slope rebounded towards the up-slope direction from 34 to 29 mm (see Fig. 18), σ'_D reduced from the peak value of 14 kPa to the initial value before the storm event on June 2008 happened. This suggests that the σ'_D mobilised by the plastic down-slope movement during the storm was relieved as up-slope rebound occurred in the subsequent dry winter. Moreover, it can be observed that the response of σ'_D during the dry season was stiffer (i.e., greater change in stress for the same given change in displacement) than that found during the previous wet season. This may be because the stiffness of a soil that follows an unloading path (during the dry season in this case) within a yield surface is stiffer than that following a loading path (during the wet season) at the yield surface (Ng and Xu, 2012). This indicates that the mobilisation of σ'_D during the wet season was likely a plastic process that caused the irrecoverable displacement after a cycle of wet/dry season.

8   SUMMARY AND CONCLUSIONS

This keynote paper summarises key findings of a comprehensive field monitoring work conducted in a volcanic slope situated in Hong Kong. Climatic effects on the hydrological responses (PWP, water content and GWT) as well as the mechanical behaviour (i.e., slope deformation and the associated stress mobilisation) were investigated. Based on the investigation, some conclusions may be drawn as follows:

Hydraulic properties of decomposed volcanic soil

The *in-situ* SDSWRC measurements showed that colluvium has an air-entry value (AEV) of 2 kPa. Noticeable hysteresis loop is observed. The loop size is found to be smaller in deeper depths, where soil is subjected to a higher confining pressure and has a smaller degree of weathering. On the contrary, CDT did not show hysteretic behaviour as there were negligible changes in VWC within the suction range investigated. The SDSWRCs of both colluvium and CDT obtained from the first wetting-drying cycle were comparable to those from the second cycle. This suggests that the SDSWRCs measured in the field are likely referred to as scanning curves.

Within the suctions range between 0 and 5 kPa, the field measurements show that the colluvium was more permeable than CDT. As suction increased from 1 to 4 kPa, the *in-situ* permeability of the colluvium reduces from 1×10^{-4} to 2×10^{-7} m/s. On the contrary, no significant change in the *in-situ* permeability is observed in CDT, varying within one order of magnitude between 3×10^{-6} and 4×10^{-7} m/s.

Climatic effects on slope behaviour

During rainfalls in the wet summer in 2008, suctions in the top 2.5 m of the slope were often destroyed rapidly and positive PWPs between 5 and 20 kPa were recorded. When an extreme storm event with a peak intensity of 133.5 mm/hr (equivalent to return period of 245 years) happened, transient perched GWT is identified at the colluvium-CDT interface (at 3–4 m depth) because of the large contrast of the soil permeability.

The main groundwater table (GWT), which was initially at 11 m depth, is found to rise by 6 m near the crown and the toe of the landslide body, whereas much substantial rise of up to 10 m is observed at the central portion. By comparing the results of field monitoring and a 3D anisotropic seepage analyses, the observed local recharge of the GWT may be attributed to preferential water flow along relict joints existed in the CDT stratum, as well as 3D cross-slope groundwater flow along the shallow rock surface in the transverse direction of the slope. The significant rise of the main GWT led to a nearly hydrostatic distribution of positive PWP and an increase in VWC to a saturated value of 36%.

As a result of the significant building up of positive PWP, the slope exhibited a "deep-seated" mode of displacement. A remarkably large and irrecoverable displacement of 20 mm was recorded within a stratum between 5.5 and 6 m depths (i.e., equivalent to an average plastic shear strain of 8%). It is evident that the top 5 m of the slope exhibited translational down-slope movement, whereas the slope at depths below 7 m remained largely stationary. During the plastic down-slope movement, the Bishop's effective horizontal stress was found to be mobilised by 350% (i.e., from 4 to 14 kPa) and reached a peak value equivalent to 40% of an effective passive stress. This indicates that during the translational sliding of the landslide mass, the sliding mass exhibited substantial plastic deformation that resulted in the significant stress mobilisation.

During the subsequent dry winter 2008/2009, substantial decreases of PWP and VWC were recorded within the root zone of vegetation (i.e., average depth of 1.2 m). The measured suction is found to approach a steady-state value of about 200 kPa. The depth of influence of suction is identified to be shallower than 2 m. During the 4-month drying period, substantial up-slope rebound is observed in the top 5 m of the ground, following a "cantilever" mode of deformation. However, such rebound recovered only 25% (i.e., 10 mm) of the plastic down-slope displacement resulting from the extreme storm happened in the previous wet summer. It is revealed that for a given increase in suction, the increase in up-slope displacement (4.3%–6.3%) found in the field was close to the plastic reduction of void ratio (4.1%–6.1%) of volcanic soil tested at similar overburden stress levels in the laboratory. This evidently suggests that the up-slope rebound was attributed to suction-induced plastic soil deformation (i.e., known as soil shrinkage). Due to the plastic up-slope rebound, all of effective horizontal stress built up during the previous storm event was recovered.

By comparing the initial slope displacement profile and that recorded after two cycles of wet-dry seasons between 2008 and 2010, a net plastic down-slope displacement was identified. Only 40% of the down-slope displacement resulted in wet seasons was recovered by the up-slope rebounds in dry seasons. Down-slope ratcheting was hence resulted.

ACKNOWLEDGEMENTS

The Geotechnical Engineering Office (GEO), Civil Engineering and Development Department (CEDD), the Government of the HKSAR, is acknowledged for funding the field monitoring work presented in this paper. Thanks are also given to the Head of GEO and the Director of CEDD, who permitted the use of the base photograph in Figure 1.

The research grant HKUST6/CRF/12R provided by the Research Grants Council of the Government of the HKSAR and research grant (2012CB719805) from the National Basic Research Program (973 Program) provided by the Ministry of Science and Technology of the People's Republic of China are also acknowledged for funding research assistants and providing resources for laboratory testing.

The second author would like to acknowledge the research grant provided by the EU FP7 Marie Curie Career Integration Grant under the project "BioEPIC slope" and the travel fund supported by the Northern Research Partnership (NRP).

REFERENCES

AECOM. 2012. Detailed study of the 7 June 2008 landslides on the hillside above the North Lantau Highway and Cheung Tung Road, North Lantau. GEO Report No. 272, Geotechnical Engineering Office, Civil Engineering Department, The Government of the Hong Kong Special Administrative Region

Allen, R.K., Pereira, L.S., Raes, D., & Smith, M. 1998. Crop evapotranspitation: Guidelines for computing crop water requirements. Food and Agricultural Organisation's Irrigation and Drainage Paper, No. 56.

Alonso, E.E., Gens, A. & Josa, A. 1990. A constitutive model for partially saturated soils. Géotechnique 40(3): 405-430.

Angeli, M.G., Gasparetto, P., Menotti, R.M., Pasuto, A., & Silvano, S. 1996. A visco-plastic model for slope analysis applied to a mudslide in Cortina d'Ampezzo, Italy. Quarterly Journal of Engineering Geology and Hydrogeology 29(3): 233–240.

Brackley, I.J.A & Sanders, P.J. 1992 In-situ measurement of total natural horizontal stresses in an expansive clay. Géotechnique 42(2): 443–451.

Brand, E.W. 1984. Relationship between rainfall and landslides in Hong Kong. Proc. of the 4th International Symposium of Landslides, Toronto, 1, 377–384.

Cheuk, J., Ng, A., Endicott, J. & Ho, K. 2009. Progressive slope movement due to seasonal wetting and drying. Invited paper. Proc. Seminar on "The State-of-the-art Technology and Experience on Geotechnical Engineering in Malaysia and Hong Kong". The HKIE Geotechnical Division, 25 February 2009, Hong Kong, p. 105-114.

Corlett, R.T., Xing, F.W., Ng, S.C., Chau, L.K.C., & Wong, L.M.Y. 2000. Hong Kong vascular plants: Distribution and status. Memoirs of the Hong Kong Natural History Society 23: 1-157.

Corominas, J., Moya, J., Ledesma, A., Lloret, A., & Gili, J.A. 2005. Prediction of ground displacements and velocities from groundwater level changes at the Vallcebre landslide (Eastern Pyrenees, Spain). Landslides 2(2): 83–96.

Fredlund, D.G., & Wong, D.K.H. 1989. Calibration of thermal conductivity sensors for measuring soil suction. Geotechnical Testing Journal, ASTM 12(3): 188–194.

Garg, A., Leung, A.K., Ng, C.W.W. Comparisons of suction induced by evapotranspiration and transpiration of S. heptaphylla. Canadian Geotechnical Journal In press.

Geotechnical Control Office (GCO). 1982. Mid-level studies: Report on geology, hydrology and soil properties. Geotechnical Control Office, Hong Kong

Geotechnical Control Office (GCO). 1988. GEOGUIDE 3 – Guide to rock and soil descriptions. Geotechnical Control Office, Hong Kong

Geotechnical Engineering Office (GEO). 2007. Final Geological Assessment Report for Tung Chung Foothills Study area. Landslide Assessment and Monitoring Work at Four Selected Sites – Feasibility Study. Prepared by S. W. Millis, Ove Arup and Partners Hong Kong Ltd.

IPCC. 2007. Summary for Policymakers. In: Climate Change 2007: The Physical Science Basis. Contribution of Working Group I to the Fourth Assessment Report of the Intergovernmental Panel on Climate Change. Cambridge University Press, Cambridge, United Kingdom and New York, NY, USA.

Khalili, N. & Khabbaz, M.H. 1998. A unique relationship for χ for the determination of the shear strength of unsaturated soils. Géotechnique 48(2):1–7.

Kim, Y.K. & Lee, S.R. 2010. Field infiltration characteristics of natural rainfall in compacted roadside slopes. Journal of Geotechnical and Geoenvironmental Engineernig, ASCE 136(1): 248-252.

Leroueil, S. 2001. Natural slopes and cuts: movement and failure mechanisms. Géotechnique 51(3): 197–243.

Li, A.G., Yue, Z.Q., Tham, L.G., Lee, C.F., & Law, K.T. 2005. Field-monitored variations of soil moisture and matric suction in a saprolite slope. Canadian Geotechnical Journal 42(1): 13–26.

Lim, T.T., Rahardjo, H., Chang, M.F., & Fredlund, D.G. 1996. Effect of rainfall on matric suction in a residual soil slope. Canadian Geotechnical Journal 33(2): 618-628.

Leung, A.K., & Ng, C.W.W. 2010. Back-analysis of infiltration characteristic of a saprolitic hillslope by considering permeability heterogeneity. Proc. 5th Int. Conf. on Unsaturated Soils, Sep. Barcelona, Spain, 1255–1260.

Leung, A.K. & Ng, C.W.W. 2013a. Seasonal movement and groundwater flow mechanism in an unsaturated saprolitic hillslope. Landslides 10(4): 455–467.

Leung, A.K. & Ng, C.W.W. 2013b. Analysis of groundwater flow and plant evapotranspiration in a vegetated soil slope. Canadian Geotechnical Journal 50(12): 1204–1218.

Leung, A.K. & Ng, C.W.W. Field investigation of deformation mechanisms and stress mobilisation in a soil slope. Landslides. In press.

Leung, A.K., Garg, A, Ng, C.W.W. Effects of plant roots on soil-water retention and induced suction in vegetated soil. Engineering Geology. In press.

Leung, A.K., Sun, H.W., Millis, S.W., Pappin, J.W., Ng, C.W.W. & Wong, H.N. 2011. Field monitoring of an unsaturated saprolitic hillslope. Canadian Geotechnical Journal 48(3): 339–353.

Meilani, I., Rahardjo, H. & Leong, E.C. 2005. Pore-water pressure and water volume change of an unsaturated soil under infiltration conditions. Canadian Geotechnical Journal 42(6): 1509-1531.

Ng, C.W.W. & Leung, A.K. 2012. Measurements of drying and wetting permeability functions using a new stress-controllable soil column. Journal of Geotechnical and Geoenvironmental Engineering, ASCE 138(1): 58–65.

Ng, C.W.W. & Menzies, B. 2007. Advanced Unsaturated Soil Mechanics and Engineering. Taylor & Francis, London and NY. ISBN: 978-0415-43679-3 (Hardcopy). 687p.

Ng, C.W.W. & Pang, Y.W. 2000 Experimental investigations of the soil-water characteristics of a volcanic soil. *Canadian Geotechnical Journal* **37**(6): 1252–1264.

Ng, C.W.W. & Xu, J. 2012. Effects of current suction ratio and recent suction history on small strain behaviour of an unsaturated soil. *Canadian Geotechnical Journal* **49**(2):226-243.

Ng, C.W.W., & Zhan, L.T. 2007. Comparative study of rainfall infiltration into a bare and a grassed unsaturated expansive soil slope. *Soils and Foundations* **47**(2): 207–217.

Ng, C.W.W., Lai, C..H. & Chiu, A.C.F. 2012. A modified triaxial apparatus for measuring the stress-path dependent water retention curve. *Geotechnical Testing Journal, ASTM* **35**(3): 490–495.

Ng, C.W.W., Zhan, L.T., Bao, C.G., Fredlund, D.G., Gong, B.W. 2003 Performance of an Unsaturated Expansive Soil Slope Subjected to Artificial Rainfall Infiltration. *Géotechnique* **53**(2): 143-157

Ng, C.W.W., Wong, H.N., Tse Y.M., Pappin, J.W., Sun. H.W., Millis, S.W. and Leung, A.K. 2011. Field study of stress-dependent soil-water characteristic curves and hydraulic conductivity in a saprolitic slope. *Géotechnique* **61**(6): 511–521.

Potts, D.M., Dounias, G.T. & Vaughan, P.R. 1990. Finite element analysis of progressive failure of Carsington embankment. *Géotechnique* **40**(1): 79–101.

Potts, D.M., Kovacevic, N. & Vaughan, P.R. 1997. Delayed collapse of cut slopes in stiff clay. *Géotechnique* **47**(5): 953–982.

Rahardjo, H., Lee, T.T., Leong E.C., & Rezaur, R.B. 2005. Response of residual soil slope to rainfall. *Canadian Geotechnical Journal* **42**(2): 340–351.

Sassa, K. & Canuti, P. 2008. *Landslides – Disaster risk reduction*, Springer, New York.

Simon, A. & Collison, A. 2002. Quantifying the mechanical and hydrologic effects of riparian vegetation on strembank stability. *Earth Surface Processes and Landforms* **27**(5): 527–546.

Skempton, A.W. 1970. First-time slides in over-consolidated clays. *Géotechnique* **20**(3): 320–324.

Smethrust, J.A., Clarke, D., & Powrie, W. 2012. Factors controlling the seasonal variation in soil water content and pore water pressures within a lightly vegetated clay slopes. *Géotechnique* **62**(5): 429–446.

Survey and Mapping Office (SMO). 1995. *Explanatory Notes On Geodetic Datums in Hong Kong.* Survey and Mapping Office (SMO), Lands Department, the Government of the HKSAR, HK.

Taha, H., Akbari, H., & Rosenfeld, A. 1988. Residential Cooling Loads and the Urban Heat Island: the Effects of Albedo. *Building and Environment* **23**(4): 271–283.

Take, W.A., & Bolton, M.D. 2011. Seasonal ratcheting and softening in clay slopes, leading to first-time failure. *Géotechnique* **61**(9): 757–769.

Tomlinson, M. & Woodward, J. 1994. *Pile design and construction practice.* 5th Ed., CRC Press, Taylor and Francis Group, Florida

Tommasi, P., Pellegrini, P., Boldini, D., & Ribacchi, R. 2006. Influence of rainfall regime on hydraulic conditions and movement rates in the overconsolidated clayey slope of the Orvieto hill (central Italy). *Canadian Geotechnical Journal* **43**(1): 70–86.

Tsaparas, I., Rahardjo, H., Toll, D. G. & Leong, E.C. 2003. Infiltration characteristics of two instrumented residual soil slopes. *Canadian Geotechnical Journal* **40**(5): 1012–1032.

van Genuchten, M.Th. 1980. A closed-form equation for predicting the hydraulic conductivity of unsaturated soils. *Soil Science Society of American Journal* **44**(5):892–898.

Watson, K.K. 1966. An instantaneous profile method for determining the hydraulic conductivity of unsaturated porous materials. *Water Resources Research* **2**(4): 709–715.

Wheeler, S. & Sharma, R.S. & Buisson, M.S.R. 2003. Coupling of hydraulic hysteresis and stress-strain behaviour in unsaturated soils. *Géotechnique* **53**(1): 41–54.

Zhou, C. & Ng, C.W.W. 2014. A new and simple stress-dependent water retention model for unsaturated soil. *Computers and Geotechnics* **62**: 216–222.

Volcanic Rocks and Soils – Rotonda et al. (eds)
© 2016 Taylor & Francis Group, London, ISBN 978-1-138-02886-9

Unstable geotechnical problems of columnar jointed rock mass and volcanic tuff soil induced by underground excavation: A case study in the Baihetan hydropower station, China

Q. Jiang, X.T. Feng, S.F. Pei & X.Q. Duan
State Key Laboratory of Geomechanics and Geotechnical Engineering, Institute of Rock and Soil Mechanics, Chinese Academy of Sciences, Wuhan, China

Q.X. Fan & Y.L. Fan
China Three Gorges Project Corporation, Beijing, China

ABSTRACT: In the Baihetan hydropower station of China, periodic volcanic eruptions in history have resulted in current geological coexistence of columnar jointed rock mass with high mechanical strength and volcanic tuff soil with weak mechanical property in the same stratum. The unstable geotechnical problems of columnar jointed rock mass and volcanic tuff soil, such as unloading loose, crack and collapse of columnar jointed basalt, as well as plastic deformation and collapse of volcanic tuff soil induced by underground excavation, have challenged the design and construction of large underground engineering structures. Detailed field investigation was conducted to expose their geological characteristics and corresponding mechanical properties of columnar jointed basalt. Field exploration and experiments were also implemented to estimate the properties of tuff soil with 'hard-soft-fracture' structure, which consists of basalt, tuff soil and cinerite. Our practice indicated that the accurate geological exploration of the columnar jointed rock mass and tuff soil is the key to minimize their adverse effects on the underground engineering. Moreover, a careful construction design that considers blasting control, small opening bench and timely support, is necessary.

1 INTRODUCTION

During eruptions of active volcano in history, hot lava curdled and volcanic ash was deposited layer by layer. In these igneous rock strata, the lava could cool down and form a columnar jointed rock mass under an ideal environmental temperature because of the shrinkage stresses acting on the surface of the lava (Budkewitsch et al., 1994; Lore et al., 2000; Schmincke, 2004; Goehring and Morris, 2008; Kereszturi et al., 2014). The digenesis of the volcanic ash also formed the cinerite with weak mechanical strength, and the following historical tectonic activities of lithosphere induced the shearing behaviour between the layers of magmatic rock and cinerite. The shear-damaged cinerite developed into current volcanic tuff soil because of a long period of alteration and weathering (Langmann et al., 2012; Yamashita et al., 2005; Neri and Macedonio, 1996; Durant et al., 2012; Renzulli et al., 1995). As a result, a special geological stratum is formed, which is characterized by the coexistence of columnar jointed rock mass with high mechanical strength of blocks and volcanic tuff soil with low mechanical strength.

In civil and geotechnical engineering, modern human activities, such as constructing highways, hydropower stations, and open underground repositories for nuclear waste, often have to challenge the stability issue of igneous rock with coexisting columnar jointed rock and tuff soil. In recent years, several large geotechnical hydropower and highway projects have exposed their unfavourable properties in engineering geology (Vallejo et al., 2008; Xu et al., 2010; Rotonda et al., 2010; Tommasi et al., 2015). Notably, the columnar jointed basalt (CJB) and volcanic shearing soil in the Chinese Baihetan hydropower station has exhibited serious stability issues (Shi et al., 2008, Yan et al., 2011). These coexisting stability problems induced by underground excavation often involve unloading loose, crack and collapse of columnar jointed rock mass, as well as shearing deformation and collapse of volcanic tuff soil.

This paper presents the unstable geotechnical problems of CJB and volcanic tuff soil induced by underground excavation. The case investigation in the Chinese Baihetan hydropower station not only exposes the structural, mechanical characteristics and unloading behaviour of CJB, but also reveals the geological structure and mechanical properties of volcanic tuff soil.

2 BACKGROUND

The Baihetan hydraulic station, which is the second largest hydroelectric project after the Three Gorges

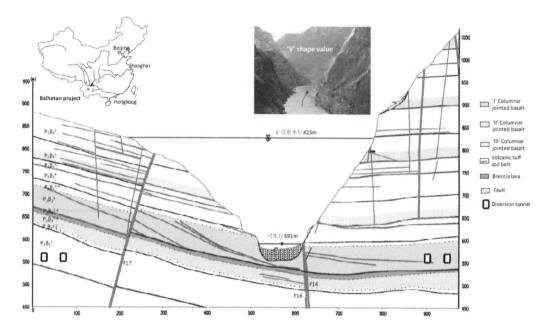

Figure 1.   Site of Baihetan hydraulic station and exposed columnar jointed basalt and volcanic tuff soil at arch dam and diversion tunnels.

station, is located at the boundary between Sichuan and Yunnan provinces. As the key node of national west–east electricity transmission project, its electric energy production is planned to be 64.095 billion kW every year on average. The main components of this hydraulic engineering are the concrete double-curvature arch dam, diversion tunnels, underground caverns, main transformer chamber, and tailrace tunnel. Among these components, the arch dam is 289 m in height. Eight electric generators, with 1000 MW capability, are fixed inside each left bank and right bank. The size of the underground powerhouse is the largest in the world, with dimensions of 438 in length, 88.7 in height and 34.0 in width. However, the geological condition of the site is imperfect because of the existence of CJB and volcanic tuff soil in magmatic rock stratum with decantation dipping. These adverse geological bodies exposed at the dam and the underground tunnels had resulted in serious geotechnical problems.

3   INVESTIGATION OF COLUMNAR JOINTED BASALT

3.1   *Field characteristics of CJB*

Field investigations in the Baihetan diversion tunnels showed that the exposed geological structure of CJB was dominated by prismatic rock blocks formed by assembly and intersection of joint sets, which compose the overall rock mass structure (Fig. 2a). Notably, the transverse sections of a typical columnar rock block were not always hexagonal (Fig. 2b). Statistical analysis of approximately 200 block sections performed in the field showed that the quadrangle, pentagonal

and hexagonal polygon frequencies in the columnar block cross-section were 32.1%, 46.7% and 17.6%, respectively, and that the combined frequency of the triangular and heptagonal polygons was less than or equal to 4% of the total shapes (Fig. 2c). Statistical analyses of field data also indicated that the average polygon edge length was approximately 0.152 m, and more than 80% of the edge lengths were in the range of 0.12–0.24 m (Fig. 2d).

Further field study on the joint patterns revealed the following three types of joint sets that form the CJB in the diversion tunnels.

(1) Columnar joints (I): This joint set is composed of rough surfaces that form the columnar rock block boundary. 'I' joints always intersect with each other to form columnar blocks, and their length is on a metre scale.

(2) Vertical internal joints (II): This joint set is found inside each columnar rock block. The plane 'II' joints are nearly parallel to the columnar block axes. 'II' joints are typically decimetres long and characterized by polished and wavy surfaces.

(3) Horizontal internal joints (III): This joint set is also found inside each columnar rock block. The plane of 'III' joints is nearly normal to the columnar block axes. 'III' joints are characterized by a smooth surface with approximately 0.01–0.03 m spacing and 0.05–0.1 m persistence.

Type 'II' and 'III' joints typically appear only inside columnar jointed rock blocks in an initially closed and tight configuration before the tunnel opening and onset of loosening. However, they can be observed in the field because of the delayed splaying after unloading excavation.

84

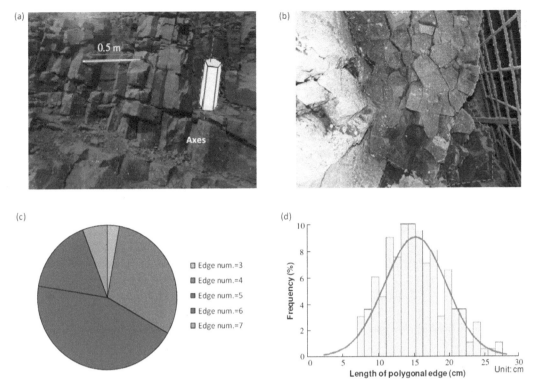

Figure 2. Structural characteristics of columnar jointed basalt (CJB) (a. Exposed CJB in diversion tunnel; b. Transverse section of CJB; c. Edge statistic; d. Length statistic of edges).

'I'
- Joint type: horizontally internal joint + uprightly internal joint + columnar joint
- Scale magnitude: meter

'II'
- Joint type: horizontally internal joint + uprightly internal joint
- Scale magnitude: decimeter

'III'
- Joint type: horizontally internal joint
- Scale magnitude: centimeter

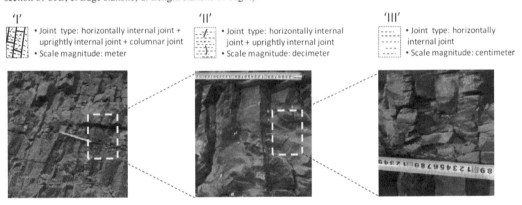

Figure 3. Multi-size joint network of columnar jointed basalt.

The investigation indicated that the CJB is composed of ordered prismatic blocks on a macro scale, but upright and sub-horizontal internal joints comprise the CJB structure at a micro scale. The different joint types are assumed to represent various *in situ* failure modes that are illustrated through differently sized scales in the rock mass. The columnar (I) joints, which are on a metre scale, control the macro deformation behavior and dictate the shape and size of the blocks that collapse during excavation. The upright internal (II) joints, which are on a decimetre scale, control the loosening and block collapse following unloading deformation triggered by excavation. The horizontal internal (III) joints, which are on a centimetre scale, control the days-delayed rock-block cracking or deformation if support is not installed in time (Fig. 3).

3.2 *Anisotropic deformability*

The field and laboratory investigation results presented earlier indicate that the CJB is inherently anisotropic, which should manifest in their anisotropic mechanical behaviour. Using *P*-wave velocity tests in the field is advantageous to check this property, because these experiments are not only

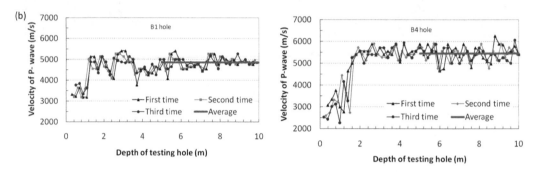

(a)

Travel direction of wave		Travel direction paralleling to the axis of prismatic block			Travel direction normal to the axis of prismatic block		
		V_{B1}	V_{B2}	V_{B3-B4}	V_{B3}	V_{B4}	V_{B1-B2}
Velocity (m/s)	First time	4844	4859	4842	5539	5414	5429
	Second time	4871	4790	4814	5514	5398	5379
	Third time	4851	4803	4802	5522	5520	5417
Average P-velocity (m/s)		4830			5459		

Figure 4. Anisotropic characteristics of P-wave velocity (a. Statistic result of P-wave velocity; b. Curve of P-wave velocity along the borehole).

non-destructive methods that do not require sample extraction, but they can also represent the field stress and original moisture content state of the rock mass tested. Therefore, through *in situ* P-wave velocity tests, the CJB elastic deformability in various directions can be determined in the field under the existing environmental conditions.

To obtain a reliable wave velocity using *in situ* ultrasonic velocity measurements in boreholes, the exploratory boreholes should be oriented parallel or orthogonal to the CJB transverse isotropic plane. First, the orientation of prismatic block axes was measured to determine the optimum orientation of the drill holes. The exploratory boreholes B1 and B2 were drilled in the tunnel floor with axes parallel to the columnar basalt axis. Both boreholes were 10 m in length and had a distance of 1 m from each other. The exploratory boreholes B3 and B4 were drilled in the sidewall; both were 10 m long with 1 m spacing between them. Ultrasonic tests were performed after drilling the exploratory boreholes (from B1 to B4). The first single borehole P-wave velocity tests were performed in boreholes B1, B2, B3 and B4. P-wave velocity tests were then performed across two boreholes for the borehole couples B1–B2 and B3–B4. As a result, the P-wave velocity along the transverse isotropic plane direction for the CJB can be obtained using data from V_{B3}, V_{B4} and V_{B1-B2}, and the P-wave velocity along the direction normal to the transverse isotropic plane can be derived using data from V_{B1}, V_{B2} and V_{B3-B4}.

The aforementioned testing methodology, which was performed at position K0 + 550 m in the No. 2 diversion tunnel, provided six sets of P-wave velocity profiles for the different CJB directions. All P-wave

velocity profiles exhibited an initial low-velocity segment that was obtained in close proximity to the excavation face. These low-velocity segments represented the excavation-induced damaged zone and disturbed zone with stress release. Thus, these segments near the free face were excluded from further analyses of the CJB intrinsic mechanical anisotropy.

The average velocities indicate that the average P-wave velocity of an intact CJB in the direction parallel to the columnar block axes is approximately 4830 m/s but approximately 5459 m/s normal to the block axis. When the general relationship between the dynamic modulus with the P-wave velocity is considered, the anisotropic coefficient (δ) obtained for the dynamic elastic modulus between the directions parallel and normal to the CJB axis is 0.78 (Eq. 1).

$$\delta_d = \frac{E_{d,\parallel}}{E_{d,\perp}} = \left(\frac{V_{p,\parallel}}{V_{p,\perp}}\right)^2 \tag{1}$$

is the anisotropic coefficient of CJB's dynamic-elastic modulus; $E_{d,\parallel}$ and $V_{p,\parallel}$ are the dynamic Young's modulus and P-wave velocity parallel to the columnar basaltic block axis, respectively; $E_{d,\perp}$ and $V_{p,\perp}$ are the dynamic Young's modulus and P-wave velocity normal to the columnar basaltic block axis, respectively.

3.3 *Anisotropic strength of basaltic block*

The anisotropic strength of CJB was determined using point load tests performed in the field and uniaxial compression tests performed at the laboratory.

Approximately 100 basaltic block specimens were selected from the No. 4 diversion tunnel for the point load tests. Based on the assumption that the irregular basalt blocks were anisotropic in strength, loading was

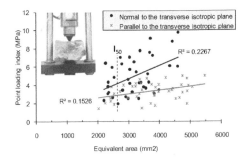

Figure 5. Experimental result of point load test in different directions.

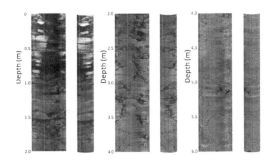

Figure 6. Typical panoramic picture and three-dimensional virtual borehole core images captured by digital borehole camera.

applied parallel and normal to the transverse isotropic plane of the CJB. The tests were performed in accordance with the suggested methods of IRSM (1985) and ASTM (2002) standards. The basaltic block point load values obtained for the normal and parallel loadings to the CJB transverse isotropic plane were 39 and 44, respectively (Fig. 5). This experimental data regression analysis indicated that the size-corrected point load index (I_{50}^1) on the anisotropic plane was approximately 3.94 MPa, and the size-corrected point load index (I_{50}^2) on the transverse isotropic plane was approximately 2.95 MPa. The anisotropic signature is evident. In general, the anisotropic coefficient for compressive strength (δ_s) is quantitatively estimated at approximately 0.75 using Eq. 2 (ASTM, 2002).

$$\delta_s = I_{50}^2 / I_{50}^1 \qquad (2)$$

3.4 Unloading crack of columnar jointed basalt

The unloading opening would inevitably induce the excavation damage zone. Measuring the depth and distribution of the fractured zone of the surrounding CJB is highly important to understand the damage degree and supporting design. In this part, the digital borehole camera was used to observe the unloading cracks inside the CJB by pushing the optical probe into the borehole.

The panoramic picture and three-dimensional virtual borehole core images captured by the digital borehole camera indicated that heavy cracks can be observed in the 0–2 m range, and several cracks can also be observed in the 2–4 m range along the axes of hole from sidewall to inner rock of the tunnel. The cracks inside the CJB are considered as unloading fractures because these cracks cannot be found in the deep part of the borehole. Moreover, most cracking planes are parallel to the free surface of the tunnel. This phenomenon indicated that the unloading direction of the surrounding columnar jointed rock mass was pointed to the free surface after excavation.

3.5 Unloading failure behaviour of columnar jointed basalt

Investigation of *in situ* unstable behaviour of CJB during tunnel excavation indicated that its unloading

failure pattern belonged to the 'structure-stress' induced failure mode (Hoek et al., 1995, Martin et al., 1999). These specific failure patterns included the following types.

(1) Slide because of separation of columnar joints (Fig. 7a). The stress unloading between columnar joints led to the separation of columnar joints, and thus the fall of prismatic blocks appeared from the sidewall.

(2) Disintegration of columnar jointed block (Fig. 7b). Disturbance of opening blasting and stress unloading induced the loosening of vertical internal joints, and thus the decomposition of columnar blocks happened.

(3) Break of columnar block (Fig. 7c). Distribution of opening blast and stress unloading also led to the stretching and loosening of horizontal internal joints. Thus, a resultant failure pattern was observed in which the long columnar block was divided into several short parts.

(4) Dilatancy of joint shear (Fig. 7d). Given the redistribution of rock stress after excavation, the existing stress concentration around the surface of the tunnel often led to the shearing behaviour of joints inside the CJB. This shearing effect produced the volumetric increase of the surrounding rock. As a result, the volumetric expansion of sheared CJB can lead to the cracking of sprayed concrete.

4 INVESTIGATION OF VOLCANIC TUFF SOIL

4.1 Occurrence characteristic

In general, the current volcanic tuff soil is the historical production of altered cinerite that was generated from the erupted volcanic ash and had undergone strong shearing tectonism and argillisation. The volcanic tuff soil in the Baihetan project exhibited several characteristics.

(1) Spatial extension in large scale. This weak volcanic tuff belt, which is parallel to the stratum attitude, can stretch to several kilometres or several tens of kilometres (Fig. 8). Thus, several

Figure 7. *In situ* failure patterns of columnar jointed basalt (a. Slide; b. Disintegration; c. Break; d. Dilatancy).

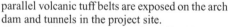

Figure 8. Surface belt of volcanic tuff soil on slope of Baihetan bank.

Figure 9. *In situ* ternary structure of volcanic tuff soil.

parallel volcanic tuff belts are exposed on the arch dam and tunnels in the project site.

(2) Composited rock structure. The tuff soil belt is a ternary structure, i.e., hard basalt, softening soil and fractured cinerite (Fig. 9). The basalt is approximately several tens of metres thick and the cinerite is approximately several decimetres or one metre thick, but the tuff soil belt is approximately several centimetres thick. This composited soil-rock structure has significant weak shearing strength along the extending plane of the tuff soil belt.

(3) Multi-shearing history of tuff soil. The microscopic structure and arrangement orientation of forming ingredients indicated that stratification

occurred, which is induced by the reverse slip of hanging side and frictional boulder clay (Fig. 10). Thus, the tuff soil belt has a tendentious shearing behaviour during unloading excavation in the tunnel because of its sliding behaviour.

4.2 Unloading failure behaviour of tuff soil

Given its weak strength and complicated structure, the tuff soil exposed several kinds of failure behaviour during unloading excavation of the tunnel.

(1) Plastic spill of softened tuff soil (Fig. 11a). The tuff soil belt with small dipping angle would produce plastic flow because of the unloading effect of tangential stress on the sidewall. In addition, the concentrated normal stress on the plane soil belt intensified the crashing effect and ledto its large deformation.

Figure 10. Microscopic structure and arranging orientation of volcanic tuff.

(2) Structure-stress collapse (Fig. 11b). Serious stress aggregation and deformation discontinuity can occur at the special 'hard-soft-fracture' ternary structure. Thus, the fracture cinerite and basalt would collapse because of the excessive concentrated stress.

(3) Crack of shotcrete (Fig. 11c). The large plastic deformation would induce local volumetric expansion, and therefore the separation between rock mass and supported shotcrete. As a result, the sprayed concrete on the tuff soil belt would crack and fall.

4.3   Mechanical experiments for tuff soil

Some reasonable experiments for tuff soil are essential to estimate its strength because its stability prediction is very important. In this part, several laboratory and *in situ* tests have been conducted.

(1)   Ring-shear test for the tuff soil.

The ring-shearing experiment was conducted using Japanese DPR-1 equipment. The specimen has a hollow annular shape with the dimensions of 75 mm external diameter, 50 mm inner diameter and 20 mm height. The shearing speed for the tuff soil was set to 0.02 mm/min (Fig. 12). The shearing stress–shearing displacement curves of tuff soil under different normal stresses can be obtained (Fig. 13). If the strength of softened tuff soil is supposed to obey the Mohr–Coulomb strength criterion, then the strength parameters of tuff soil in the laboratory can be calculated according to the peak strength of gained experimental curves. The detailed experimental regression indicated that the internal friction angle of the specimen was approximately 26 and its cohesive strength was approximately 15–48 kPa under the condition of natural moisture content.

(2)   Triaxial compression test for tuff soil.

To understand the loading mechanical response and strength parameters, a series of triaxial compression tests for the tuff soil were conducted in different conditions of moisture content. The testing specimens were obtained by cutting the *in situ* tuff soil into standard sample with dimensions of 5 cm diameter and 10 cm height. The experimental results indicated that both the internal frictional angle and cohesive strength of the tuff soil decreased with increasing moisture content (Fig. 14). The test also showed that the cohesive strength and frictional angle of the tuff soil were approximately 300–400 kPa and 5°–12°, respectively, under the condition of natural moisture content.

(3)   Shearing test for complex rock-soil body.

The complex rock-soil specimens, which were obtained by drilling on the field rock, were used in the shear test on the rock mechanical system (RMT150C). During the shearing experiments, several loading steps of normal stress were applied (Fig. 15). These shearing tests for complicated rock-soil specimens indicated that the cohesive strength and frictional angle of the complicated rock-soil body were approximately 30 kPa and 23°–30°, respectively.

(4)   Triaxial compression for complex rock-soil body.

A triaxial compression test for complex rock-soil specimens were also conducted to determine its compressively mechanical behaviour. The specimens with 10 cm diameter and 20 cm height was composed of hard basalt, soft soil and hard basalt along the axial way. The tuff soil inside the complex specimen had solidified firstly before the compressive experiments. The experimental results indicated that the integral cohesive strength and frictional angle of this complicated specimen were approximately 840 kPa and 11.2°, respectively. Moreover, the typical failure of this kind of specimens was compressive crush of the tuff soil, and its critical plastic strain was approximately 1% in general (Fig. 16).

(5)   *In situ* experiment for tuff soil belt.

In situ shearing experiments were also conducted based on the Chinese national test standard. The technical steps in field shearing experiments for the tuff soil belt included cutting a cubic basalt specimen with the bottom of tuff soil, concert plastic for the specimens and installation of normal and shearing jacks for shearing test (Fig. 17). The experiments showed that the cohesive strength and frictional angle of this complicated specimen were approximately 90 kPa and 20.3°, respectively, under different normal stresses (i.e., 0.6, 0.9, 1.2 and 1.5 MPa).

4.4   Comparative analysis of shearing strength

These five kinds of experiments, which showed different values of strength parameters, provided us with abundant data to estimate the shearing strength of tuff soil under different conditions. A comparative discussion of the tuff soil included the following points:

(1)   The cohesive strength of tuff soil gained from the compressive experiments was larger than that

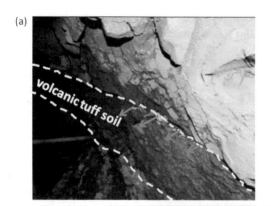

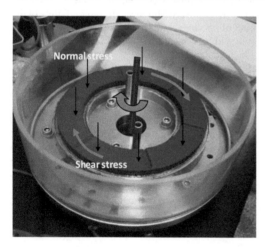

Figure 11. Exposed failure patterns of volcanic tuff soil in tunnel.

Figure 13. Typical shearing displacement–shearing stress curves of tuff soil based on ring-shear test in laboratory.

Figure 12. Tuff soil specimen for ring-shear test.

gained from the shearing tests, but the internal frictional angle was the inverse (Eq. 3). This difference may have originated from the different failure mechanics between the shearing and compressive tests.

$$\begin{cases} C_{compressive} > C_{shearing} \\ \phi_{compressive} < \phi_{shearing} \end{cases} \qquad (3)$$

(2) The cohesive strength gained from the complex rock-soil was larger than that gained from the soft soil specimens under the same normal stress condition, but the internal friction angle was close to each other under different experimental cases (Eq. 4). The essential reason was that the rough contacting face increased its occlusive force between the hard basalt and soft tuff soil. This local occlusive force between the contacting face was exhibited as macro cohesive strength during the experiments.

$$\begin{cases} C_{complex} > C_{soil} \\ \phi_{complex} \approx \phi_{soil} \end{cases} \qquad (4)$$

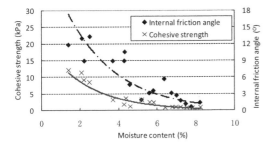

Figure 14. Relationship between strength parameters and moisture content of tuff soil under triaxial compression test.

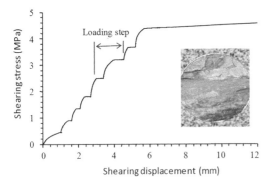

Figure 15. Typical shear displacement–shear stress curve of complicated rock-soil specimen.

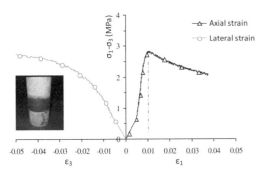

Figure 16. Typical stress–strain curves of complex rock-soil specimens under triaxial compression test.

(3) The cohesive strength gained from low moisture specimens was larger than that gained from high moisture specimens, but the internal friction angle was also close to each other under the same kind of experimental method (Eq. 5).

$$
\begin{cases}
C_{\text{high moisture}} < C_{\text{low moisture}} \\
\phi_{\text{high moisture}} \approx \phi_{\text{low moisture}}
\end{cases}
\tag{5}
$$

5 DISCUSSION AND CONCLUSIONS

In general, the periodic volcanic eruption in history would result in a special geological coexistence of columnar jointed rock mass with high mechanical

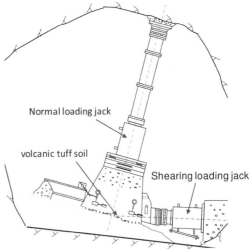

Figure 17. Sketch map of *in situ* shearing experiment for tuff soil belt.

strength and volcanic tuff soil with weak mechanical property in the same stratum. The unstable geotechnical problems of columnar jointed rock mass and volcanic tuff soil, such as unloading loose, crack and collapse of columnar jointed rock mass, as well as shearing deformation and collapse of volcanic tuff soil, which was induced by underground excavation, often challenge the design and construction of large underground engineering structures.

The field investigation indicated that the shape of most transverse sections of the Baihetan CJB were not hexagonal but rather quadrangle or pentagonal. The columnar joint and other internal joint nets played a crucial role in determining the mechanical properties of the Baihetan columnar jointed rock mass and led to the multi-failure patterns during unloading excavation.

The tuff soil in the Baihetan project was characterised by large spatial extension, ternary rock structure and multi-shearing history. This special 'hard-soft-fracture' structure with basalt, tuff soil and cinerite induced the special plastic spill of softened tuff soil and collapse of rock mass.

Our practices indicated that exploring the geological distribution for the columnar jointed rock mass and tuff soil was the key to prevent its adverse effect on the underground excavation. In addition, a careful construction design that considers blasting control, small opening bench and timely support, is necessary.

ACKNOWLEDGEMENT

The authors gratefully acknowledge the financial support from National Natural Science Foundation of China (Grant No. 41172284 and No. 51379202). In particular, authors also wish to thank Prof. X.D. Zhu, Prof. A.C. Shi and Prof. X.B. Wan for their kindly help in *in-situ* investigation.

REFERENCES

ASTM, 2002. Standard Test Method for Determination of the Point Load Strength Index of Rock. Current edition approved Nov. Annual Book of ASTM Standards, West Conshohocken, United States, Vol. 04.08, D5731–02.

Budkewitsch, P. & Robin, P.Y. 1994. Modelling the evolution of columnar joints. J. Volcanol. Geotherm. Res. 59: 219–239.

Durant, A.J., Villarosa, G., Rose, W., et al. 2012. Long-range volcanic ash transport and fallout during the 2008 eruption of Chaitén volcano, Chile. Physics and Chemistry of the Earth 45–46: 50–64.

Goehring L. & Morris, S.W. 2008. Scaling of columnar joints in basalt. J. Geophys. Res., 113 (B10): 1–18.

Hoek, E., Kaiser, P.K. & Bawden, W.F. 1995. Support of underground excavations in hard rock. A.A. Balkema, Rotterdam, p. 215.

ISRM, 1985. InternationalSociety for Rock Mechanics Commission on Testing Methods: Suggested Methods for Determining Point Load Strength. Int. J. Rock. Mech. Min. Sci. Geomech. Abstr., 22(2): 51–60.

Kereszt/uri, G., Németh, K., Cronin, S.J., et al. 2014. Influences on the variability of eruption sequences and style transitions in the Auckland Volcanic Field, New Zealand. Journal of Volcanology and Geothermal Research 286: 101–115.

Langmann, B., Folch, A. & Matthias M.V. 2012. Volcanic ash over Europe during the eruption of Eyjafjallajökull on Iceland, AprileMay 2010. Atmospheric Environment 48: 1–8.

Lore, J., Gao, H. & Aydin, A. 2000. Viscoelastic thermal stress in cooling basalt flows. J. Geophys. Res. 105: 23695–23709.

Martin, C.D., Kaiser, P.K. & Mccreath D.R. 1999. Hoek–Brown parameters for predicting the depth of brittle failure around tunnels. Can. Geotech. J., 36: 136–151.

Neri, A. & Macedonio, G. 1996. Physical Modeling of Collapsing Volcanic Columns and Pyroclastic Flows. Monitoring and Mitigation of Volcano Hazards, pp. 389–427.

Renzulli, A., Nappi, G., Falsaperl, S., et al. 1995. Annual report of the word volcanic eruptions in 1992. Bulletin of Volcanic Eruptions No. 32 for 1992.

Rotonda, T., Tommasi, P. & Boldini, D. 2010. Geomechanical Characterization of the Volcaniclastic Material Involved in the 2002 Landslides at Stromboli. Journal of Geotechnical and Geoenvironmental Engineering, 136(2): 389–401.

Schmincke, H. 2004. Volcanic Edifices and Volcanic Deposits. Volcanism, Springer-Verlag Berlin Heidelberg, pp. 127–154.

Shi, A.C., Tang, M.F. & Zhou, Q. 2008. Research of deformation characteristics of columnar jointed basalt at Baihetan hydropower station on Jinsha river. Chinese Journal of Rock Mechanics and Engineering, 27(10): 2079–2086. (Abstract in English).

Tommasi, P., Verrucci L. & Rotonda, T. 2015. Mechanical properties of a weak pyroclastic rock and their relationship with microstructure. Can. Geotech. J., 52(2): 211–223.

Vallejo, L., Hijazo, T. & Ferrer, M. 2008. Engineering geological properties of the volcanic rocks and soils of the canary Islands. eprints. ucm.es/.../2008_Articulo_publicado _SOILS&ROCKS_01.pdf.

Xu, W.Y., Deng, W.T., Ning Y., et al. 2010. 3D anisotropic numerical analysis of rock mass with columnar joints for dam foundation. Rock and Soil Mechanics, 31: 949–955. (Abstract in English).

Yamashita, H., Saito, T. & Takayama, K. 2005. Numerical investigations of volcanic eruption and prodution of hazard maps. Shock Waves, ISBN 978-3-540-22497-6. Springer-Verlag Berlin Heidelberg, p. 1013.

Yan, D.X., Xu, W.Y., Zheng, W.T., et al. 2011. Mechanical characteristics of columnar jointed rock at dam base of Baihetan hydropower station. J. Cent. South Univ., 18: 2157–2162.

Session 1:   Structural features of volcanic materials

Volcanic Rocks and Soils – Rotonda et al. (eds)
© 2016 Taylor & Francis Group, London, ISBN 978-1-138-02886-9

A micro- and macro-scale investigation of the geotechnical properties of a pyroclastic flow deposit of the Colli Albani

M. Cecconi
Department of Engineering, University of Perugia, Italy

M. Scarapazzi
Geoplanning Servizi per il Territorio S.r.l., Rome, Italy

G.M.B. Viggiani
Department of Civil Engineering and Computer Science Engineering, University of Rome Tor Vergata, Italy

ABSTRACT: The paper presents the results of a recent investigation of the geotechnical properties of a pyroclastic flow deposit of the Colli Albani volcanic complex, locally known as *Pozzolanelle*. The investigation focused on the micro-structural features of the material, as observed in thin sections and SEM micrographs, on its physical and mechanical properties at the scale of the laboratory sample, and on its macro-structural features at the scale of the *in situ* deposit.

This paper examines a pyroclastic flow deposit belonging to the IV cycle of the Tuscolano-Artemisia eruptive phase of the Colli Albani volcanic complex (Rome, Italy), locally known as *Pozzolanelle*. The work presents the results of a recent investigation of the geotechnical properties of this deposit at different scales, focusing on: *i*) the main micro-structural features of the material, as detected from the analysis of a relatively large number of thin sections, *ii*) the mechanical behaviour observed in triaxial tests carried out in a range of mean effective stress from 50 to 600 kPa, and *iii*) the structural features of the deposit in situ.

Figure 1(a) shows the typical succession of pyroclastic flow deposits of the Colli Albani volcanic complex as exposed on the sub-vertical cut in a quarry at Fioranello, South East of the city of Roma; the deposit of *Pozzolanelle* is the upper unit, delimited at the bottom by the dotted white line. According to Giordano et al. (2006), the *Pozzolanelle* are at the lower limit of the Tuscolano-Artemisio lithosome (De Rita et al. 2000).

The analysis of thin sections and SEM micrographs permitted to identify the main minerals and a number of features of the grains and of the pores, such as size, shape, and orientation; it also provided evidence that inter-granular pores are partially filled by altered material and secondary minerals, indicating that bonding is partly diagenetic.

Figure 2 shows a typical thin section of undisturbed *Pozzolanelle*, at two values of the magnification factor. The material consists of a compact matrix containing darker clasts, scoria with various degrees of vesiculation and frequent crystals, recognizable from their

Table 1. Main physical properties of *Pozzolanelle*.

w/c (%)	G_s (−)	γ (kN/m^3)	γ_d (kN/m^3)	n (−)	S_r (%)
16.3 ± 1.9	2.74	11.92 ± 0.06	9.93 ± 0.03	0.64 ± 0.01	25 ± 5.2

straight edges. The main minerals are leucite, biotite and clinopyroxene. Cross- and star-wise microlites of leucite are visible in Fig. 2(a) and (b), at higher magnification factor. Pores appear as white areas with no defined edges. Inter-granular pore features, such as size, shape, and orientation are very variable. SEM micro-graphs on dry specimens of Pozzolanelle confirmed the observation that intergranular pores are partially filled by altered material and secondary minerals, indicating that bonding may be at least in part diagenetic, i.e., due to lithification by formation of hydrated aluminosilicates (zeolites).

Table 1 reports the average values of the initial physical properties of the material as determined at the scale of the laboratory sample.

Conventional drained triaxial tests were carried out on undisturbed samples at increasing confining pressures between 50 and 600 kPa. As typical for geotechnical materials, the mechanical behaviour of intact Pozzolanelle gradually changes from brittle and dilatant to ductile and contractant with increasing confining pressure (Fig. 3). The peak deviator stress from triaxial tests are slightly above the strength envelope previously obtained from direct shear tests. The stress–dilatancy relationships observed during triaxial

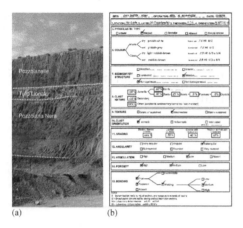

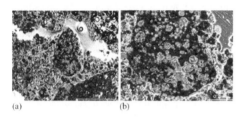

Figure 1. (a) Sub-vertical cut in a quarry at Fioranello showing the sequence of pyroclastic flow deposits from the Colli Albani; (b) compiled data sheet for the deposit of *Pozzolanelle* (Rome).

Figure 2. Thin section of *Pozzolanelle* at MF of: a) 2×; b) 10×.

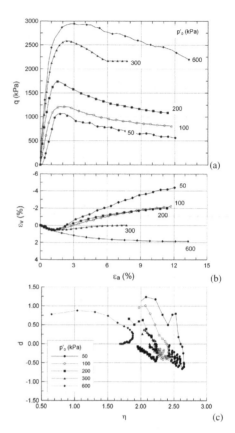

Figure 3. Results of drained TX compression tests on intact samples of Pozzolanelle: (a) deviator stress, q, vs. axial strain, ε_a; (b) volume strain, ε_v, vs. axial strain, ε_a; (c) dilatancy, $d(=-\delta\varepsilon_v^p/\delta\varepsilon_a^p)$ vs. stress ratio, $\eta(=q/p')$.

compression (Fig. 3c) indicate that, at high confining stress, a peak of stress ratio occurs in a contractant regime, while, at low confining stress, the peak of stress ratio always precedes the point in the test where dilatancy is minimum. The condition of zero dilatancy, which in classical critical state models defines the friction of the material, is attained under different values of stress ratio. This behaviour was explained by Cecconi et al. (2002) as due to progressive grain crushing and de-structuring.

The geological complexity of pyroclastic rocks and soils makes it difficult to classify these deposits using conventional systems. This motivated the Authors to develop an original descriptive model for volcanic soils (Cecconi et al., 2010), based on an operationally tested data sheet, that can be compiled in situ for each eruptive unit observed on outcropping formations, or exposed in artificial cuts, or on cores obtained by sampling as well as on retrieved samples in the laboratory. Classification (Fig. 1b) indicates that Pozzolanelle is a welded pyroclastic rock with a massive and quite homogeneous structure at the scale of the deposit, presenting limited local variations in grading. Its texture, describing the relation between the clasts and the fine matrix, is intermediate, with few contacts between randomly oriented clasts. Clast shape is sub angular, vesiculation low to absent. Porosity is medium to high.

True bonding is medium and may be related to the formation of secondary minerals.

REFERENCES

Cecconi M., De Simone A., Tamagnini C., Viggiani G.M. (2002). A constitutive model for granular materials with grain crushing and its application to a pyroclastic soil. Int. J. Numer. Anal. Meth. Geomech. 26: 1531–1560.

Cecconi, M., Scarapazzi, M. and Viggiani, G. MB. (2010). "On the geology and the geotechnical properties of pyroclastic flow deposits of the Colli Albani,". Bull. Eng. Geol. Environ., 69, 185–206.

De Rita, D., Fabbri, M., Giordano, G., Rodani, S., (2000). Proposta di organizzazione della stratigrafia delle aree vulcaniche secondo i principi delle unità stratigrafiche a limiti inconformi e sua informatizzazione, Bollettino Della Società Geologica Italiana, 119, 749–760.

Fisher, R.V. (1966). Rocks composed of volcanic fragments. Ea. Sci. Rev.

Giordano, G., De Benedetti, A.A., Diana, A., Diano, G., Gaudioso, F., Marasco, F., Miceli, M., Mollo, S., Cas, R.A.F., Funiciello, R. (2006). The Colli Albani mafic caldera (Roma, Italy): stratigraphy, structure and petrology. Journal of Volcanology and Geothermal Research, Elsevier, 155, 49–80.

Volcanic Rocks and Soils – Rotonda et al. (eds)
© *2016 Taylor & Francis Group, London, ISBN 978-1-138-02886-9*

New geotechnical classification proposed for low density pyroclastic rocks

M. Conde
Geological Engineering, CEDEX, Madrid, Spain

A. Serrano
Technical University of Madrid (UPM), Spain

Á. Perucho
Civil Engineering, Geotechnical Laboratory, CEDEX, Madrid, Spain

ABSTRACT: A new qualitative classification for low density pyroclasts is proposed, based on the particles size, their magmatic composition, their welding or lithification degree and the presence or absence of a matrix filling the macropores.

1   INTRODUCTION

An extensive study has been done in the last years with the main objective of advancing in the understanding of the stress-strain behavior of low density pyroclasts, determining the most influential factors in it, such as its structure or alteration (CEDEX 2013, Conde 2013,). Particular attention has been paid to examine aspects considered paramount such as: macroporosity (size and morphology of the pores andtheir relative size in relation to the particles), alteration of the particles and the partial or complete filling of the macropores in altered materials, with deposits of crystals or particles due to the circulation of fluids.

This paper presents a proposal for a new geotechnical classification for low density pyroclastic rocks, based on the lithotypes defined in a previous classification from Hernández-Gutiérrez & Rodríguez-Losada (Gobierno de Canarias 2011).

2   STUDIED MATERIALS

A great number of specimens from the Canary Islands were studied and tested at laboratory. Samples were from lapilli, scoria, basaltic and sialic ashes and pumice.

A study of the macroporositywas performed through a hand microscope (magnification 10x-150x). The macroporosity of each sample was defined according to the classification previously defined by the authors (CEDEX 2007; Santana et al. 2008). The samples tested in this work include all the defined types of macroporosity:

- Reticular porosity was observed in non-altered pumice samples classified in this study as PM-W.
- Mixed porosity was observed in non-altered lapilli samples classified as LP-W;
- Vacuolar porosity was found in scoria classified as SC-W;

- Matrix porosity was found in non-altered basaltic and sialicashes (BA-L and SA-L) and altered samples of types LP-L-M, SA-L-MP and PM-L-M. It is remarkable that although neither non-altered basaltic ashes (BA-L) nor non-altered sialic ashes (SA-L) show a real matrix their porosity may be classified as so, due to the small size of their particles.

3   CONCLUSIONS

The new qualitative classification for low density pyroclasts is proposed as defined in Table 1.

This new classification defines a lithotype for each low density pyroclast, taking into account the size of the particles, the magmatic composition of the particles, their degree of welding or lithification and the presence or not of a fine grain matrix surrounding the particles and filling the pores. It is considered to be useful from the civil engineering point of view.

REFERENCES

CEDEX. Caracterización geotécnica de los piroclastos canarios débilmente cementados. *Final Report* (April 2007).
CEDEX. Estudio del comportamiento geomecánico de los piroclastos canarios de baja densidad para su aplicación en obras de carreteras. *Final Report* (April 2013).
Conde M, Caracterización geotécnica de materiales volcánicos de baja densidad, *PhD Thesis*, Universidad Complutense de Madrid, 2013.
Gobierno de Canarias. *Guía para la planificación y realización de estudios geotécnicos para la edificación en la comunidad autónoma de Canarias.* GETCAN 011 (2011).
Santana, M., de Santiago, C., Perucho, A. and Serrano, A. Relación entre características químico-mineralógicas y propiedades geotécnicas de piroclastos canarios. *VII CongresoGeológico de España. Geo-Temas 10*, 2008.

Table 1. Geotechnical classification proposed for low density pyroclastic rocks.

	Description	Particles size (mm)	Welding or lithification	Matrix	Lithotype	Photo (Scale: circle diameter=7±0.5 cm)	Porosity type
BASALTIC	Basaltic ashes (Ash tuff) (BA)	<2mm	Lithified	NO	BA-L		Matrix
			Slightly lithified		BA-SL		
			Loose		BA-Lo		
	Basaltic ashes with particles of different nature (BA)	Matrix:<2mm Particles: different sizes	Lithified	YES	BA-L-MP		Matrix
	Lapilli (LP)	2-64mm	Welded	NO	LP-W		Mixed
			Slightly welded		LP-SW		
			Loose		LP-Lo		
			Lithified	YES	LP-L-M		Matrix
	Scoria (Bombs, Blocks) (Pyroclastic breccia) (SC)	>64mm	Welded	NO	SC-W		Vacuolar
			Slightly welded		SC-SW		
			Loose		SC-Lo		
SALIC	Sialic ashes (Ash tuff) (SA)	<2mm	Lithified	NO	SA-L		Matrix
			Slightly lithified		SA-SL		
			Loose		SA-Lo		
	Salic ashes with particles of different nature (SA)	Matrix:<2mm Particles: different sizes	Lithified	YES	SA-L-MP		Matrix
	Pumice (Pumice-flow deposits) (PM)	>2mm	Welded or Lithified	NO	PM-W/L		Reticular
			Slightly welded/ lithified		PM-SW/SL		
			Loose		PM-Lo		
			Lithified	YES	PM-L-M		Matrix

In gray: materials that have not been tested due to lack of samples.

Volcanic Rocks and Soils – Rotonda et al. (eds)
© *2016 Taylor & Francis Group, London, ISBN 978-1-138-02886-9*

Alteration of volcanic rocks on the geothermal fields of Kuril-Kamchatka arc

J.V. Frolova
Faculty of Geology, Lomonosov Moscow State University, Russia

ABSTRACT: The paper focuses on the alteration of volcanic rocks in near-surface zone in geothermal fields of Kuril-Kamchatka arc (Far East, Russia). The mechanism of transformation of hard volcanic rocks with brittle failure into plastic clays with ductile deformation is described in details on macro- and micro-scale. Special attention is given to the variation of the physical and mechanical properties which occur due to geothermal alteration.

1 INTRODUCTION

Geothermal energy is presently used in many countries for electricity production and heat supply. In Russia the most promising is the Kuril-Kamchatka island arc which is located in the northwestern segment of Circum Pacific Belt. The hydrothermal systems are hosted in volcanic formations of Neogene-Quaternary age. Thermal waters with different temperature and chemical composition act on the rocks changing their mineralogy, pore-space morphology and finally changing their physical and mechanical properties (Wyering et al. 2014, Frolova et al. 2014). In the near-surface horizon of geothermal fields, which can be the foundation for power plants construction, volcanic rocks transform into argillified rocks, hydrothermal clays or opalites. Hydrothermal clays are the most problematic horizon. They often form a very heterogeneous cover several meters thick which is characterized by high porosity, plasticity, hygroscopicity, and compressibility, increased weakness and occasionally swelling. Slope geological processes frequently develop in connection with hydrothermal clays.

The main target of this research is to describe the transformation of volcanic rocks into clayey soil. Special attention is given to the variation of the physical and mechanical properties which accompanies geothermal alteration. Transformation of volcanic rocks into hydrothermal clays were studied in detail in several thermal fields. In this research we consider the alteration at the Low-Koshelevsky thermal field as an example.

2 GEOLOGICAL AND GEOTHERMAL CONDITIONS

Koshelevsky hydrothermal system occupies the southernmost position on the Kamchatka Peninsula and is located on the slope of a volcano of the same name.

It is a high-temperature, steam-dominated system presently under investigation (Rychagov et al. 2012). The host rocks consist of effusive and volcaniclastic formations of Neogene-Quaternary age. Several thermal fields are known within the Koshelevsky volcano; two of those are very large i.e. – Low- and Upper-Koshevevsky – and differ in their geochemical and thermodynamic conditions. The cover of hydrothermal clays is characteristic for Low-Koshelevsky field (T = 70–100°C). Totally 24 samples of andesites were taken from natural outcrops and three trial pits and 3 samples of hydrothermal clay were taken from trial pit (depth 35–55 cm, 80–100 cm, and 120–140 cm).

3 APPLIED METHODS

In the laboratory each sample was separated into several specimens for physical and mechanical measurements. Several tests were carried out for each property, and finally the mean value was calculated for each sample. All laboratory tests were performed in accordance with the standards (ISRM 2007). Special testing was made for hydrothermal clayey soils. The following properties were determined: in-situ moisture content, hygroscopic moisture, plasticity, swelling, and shear strength parameters (direct shear test in undrained conditions).

4 HYDROTHERMAL ALTERATION AND MECHANICAL PROPERTIES

In the Low-Koshelevsly thermal field andesites are exposed in the surface. Under the action of fluids (with pH = 3.5–5.5; T up to 95–100°C), they are gradually destroyed and transformed into clays. Alteration starts in fracture networks exposed to thermal water and steam. The walls of fractures are rapidly replaced by clay minerals. Gradually the fractures propagate and expand, and new blocks of andesite are altered by argillization. In the end only cores of andesites (which are intensely fractured, ferruginizated, and argillized) remain surrounded by a clayey mass. It can be assumed that the progressive fracturing occurs under the pressure of swelling in the surrounding clay mass. Thus, the clay mass is very heterogeneous, and contains hard relicts of andesites. The clay soil is mainly smectite or mixed-layer minerals with some silica minerals and occasionally pyrite. It has pseudomorphic texture inherited from andesite. Alteration process is followed by leaching with formation of secondary pores in the rock matrix and than the initiation of microcracks. Electron microscopy has shown that dense volcanic glass is replaced by cristobalite and porous smectites (mixed-layers), firstly with scaly and then with cellular microstructure, that reflects on rock properties. Subsequently the microcrysts in the groundmass are partially dissolved.

Alteration of plagioclase phenocrysts basically begins from microcrakes and flaws. Then, the inner (central) part of the crystal, which is more anorthic or andesine composition is replaced, whereas the outer part of the crystal (more albite composition) is more stable to alteration and succumbs to alteration at later stages. Finally the crystal can be totally substituted but it maintains its initial shape.

Hydrothermal alteration effects on rocks physical and mechanical properties. By the intensity of alteration the studied andesites are subdivided on fresh or slightly altered, moderately, and intensely altered. Intensity of alteration correlates with the temperature of thermal fluids (70°C for slightly altered, 84–92°C – for moderately altered, and 95–100°C – for intensely altered). Alteration processes results in mechanically weaker rock. Development of smectite-filled microcracks decreases density of andesite from 2.6 to 2.1 Mg/m^3, and increases porosity from 2–5 to 25–30%. As a result uniaxial compressive strength decreases from 130–150 (fresh andesites) to 25–30 MPa (intensely altered andesites), velocity of the P-wave changes from 4.5–5.0 to 2.5–3.5 km/s. Dynamic elastic modulus decreases from 50–55 to 15–20 GPa and static modulus is approximately 2–2.5 times less than dynamic. The saturation of rocks with water influences in different way for fresh and altered andesites. Strength is almost the same in dry and saturated states for unaltered andesite whereas it decreases by 40% due to saturation for intensely altered andesite.

The total transformation of andesites to clays is accompanied by a strong reduction in dry density (andesites 2.5–2.6 Mg/m^3, clays 1.0–1.1 Mg/m^3), increase in porosity (from 2–8% to 61–63%) and hygroscopicity. Cohesion decreases from 15–20 MPa (andesites) to 0.02–0.05 MPa (clays) and friction angle changes from 55–57° down to 10–21°. Hydrothermal clays are characterized by high plasticity (plasticity index equal to 29–30) and slightly swelling. In-situ they belong to the stiff clays.

5 CONCLUSION

Geothermal areas are basically located within volcanic formations. At geothermal fields volcanic rocks interact with thermal water and steam and gradually transform in clayey soils, which can be the foundation for power plants construction and affect the selection of sites and design for buildings and pipelines. Alteration processes results in strong decrease in density and mechanical properties which correlate with intensity of alteration. The cover of hydrothermal clays is a very heterogeneous horizon characterized by high porosity, plasticity, hygroscopicity, increased weakness, softening and occasionally swelling in saturated state.

ACKNOWLEDGEMENTS

This research was supported by Russian Foundation for Basic Research (Grant # 13-05-00530).

REFERENCES

Frolova J.V., Ladygin V. M., Rychagov S.N., Zukhubaya D. Z. 2014. Effects of hydrothermal alterations on physical and mechanical properties of rocks in the Kuril–Kamchatka island arc. Engineering Geology 183: 80–95.

International Society on Rock Mechanics. 2007. In: Ulsay, R., Hudson, J. (eds.), *The complete ISRM suggested methods for rock characterization, testing and monitoring*: 1974–2006. ISRM

Rychagov, S.N., Sokolov, V.N., Chernov, M.S., 2012. Hydrothermal clays in geothermal fields on the South Kamchatka: The new approach and results of the studies. *J. Geochemistry* (4): 378–392.

Wyering, L.D., Villeneuve, M.C., Wallis, I.C., Siratovich, P.A., Kennedy, B.M., Gravley, D.M., Cant, J.L. 2014. Mechanical and physical properties of hydrothermally altered rocks, Taupo Volcanic Zone, New Zealand. *J. of Volcanology and Geothermal Research* 288: 76–93.

Volcanic Rocks and Soils – Rotonda et al. (eds)
© *2016 Taylor & Francis Group, London, ISBN 978-1-138-02886-9*

Hydrothermally altered rocks as a field of dangerous slope processes (the Geysers Valley, Kamchatka peninsula, Russia)

I.P. Gvozdeva & O.V. Zerkal
Lomonosov Moscow State University, Moscow, Russia

ABSTRACT: The Geysers Valley is located in the territory of Kronotskiy State Natural Biosphere Reserve and is a tourist attraction. Geologically it is a canyon river valley composed by Pleistocenic lacustrine-volcanic hydrothermally altered deposits. Presently hydrothermal activity continues actively. The study rock strata are interbeded pumice litoclastic and pelitic tuffs. The article describes composition, structure and properties of them. It is shown the studied rocks are very sensitive to humidification their strength and deformation properties decrease sharply at water saturation. Moreover they erode readily. Experiments of alternating wetting – desiccation and freezing – thawing samples showed that some tuffs are destroyed after the first test cycle. Thus weak zones can form in zones of fracturing and higher permeability that subsequently become zones of separation and sliding. The map of slope processes distribution demonstrates the variety of slope processes formed in hydrothermally altered tuffaceous sediments in the Geysers Valley.

1 INTRODUCTION

Two significant slope failures occurred in the Geysers Valley last time – the famous tourist attraction of Kamchatka peninsula (Russia). A large landslide with estimated volume of 16.3 million cubic meters formed in lower part of the valley in Jun – 2007. It changed landscape strongly, destroyed one geyser field, effected on regime of some geysers and dammed the river. The second event occurred in January – 2014. A rockfall took place in upper part of the valley. This rockfall caused a mudflow whose deposits covered some geyser spots along river channel. In both cases the displacement involved hydrothermally altered tuffs forming the slopes of the Geyser River. Also a big landslide happened in 1981 during the cyclone Elsa is described. The landscape of the valley indicates permanent slope processes.

2 GEOLOGICAL AND GEOMORPHOLOGICAL SETTING

Geysers Valley is located in the eastern part of the Uzon-Geysernaya volcano-tectonic depression formed, according radiocarbon dating, about 40 ka ago. It is a canyon with a length of 4 km directed from north-east to south-west. Vertical drop from the source to the mouth is 750 m. Depth of canyon is 300 m in the upper part and 500–550 m in the bottom part. Slopes of the valley are composed of Pleistocenic lacustrine-volcanic hydrothermally altered of pumice litoclastic and pelitic tuffs north-west dipping. Thus rocks are fallen into the river in the left side and into the slope in the right slope. Cross bedding of deposits divide the strata into several units: Geysernaya (Q_3^4grn), Pemsovaya (Q_3^4) and "Yellow tuffs" (Q_3^4js).

Geysers Valley is situated in an active tectonic zone. It is connected with crossing north-eastern, latitudinal and ring faults. Echelon fractures create zones of increased permeability to thermal water and fluids. Hydrothermal activity is high in the valley and manifested in the form of geysers, fluids and mud pots.

3 ANALYSIS AND TESTING METHODOLOGY

Sixty seven tuffs samples for the laboratory study were selected in the middle and bottom parts of the Geysers Valley in natural outcrops and landslide deposits. All samples were studied petrographically. Secondary minerals were also identified using X-ray diffraction. Microprobe analysis was conducted for a portion of the samples to study the morphology of pore space and chemical alteration that occurred during the hydrothermal process. Furthermore for each sample the following properties were measured or calculated: bulk density (ρ), grain density (ρs), open (no) and total (n) porosity, gas permeability (K_g), hygroscopic moisture (Wg), water absorption (W), velocity of ultrasonic P- and S-waves (Vp, Vs), elastic modulus (E), Poisson's ratio (ν), uniaxial compressive (σ_c) and tensile (σ_t) strength, softening coefficient (C_{soft}), angle of internal friction (ϕ), cohesion (C). All measurements were performed in accordance with testing procedures suggested by the International Society for Rock Mechanics (ISRM, 2007).

4 ROCK CHARACTERISATION

4.1 *Petrology and hydrothermal alteration*

Studied tuffs belong to hydrothermal argillites with high-silica zeolites. Secondary minerals are Mg-Na-smectites, zeolites (clinoptilolite, mordenite, heulandite) and silica minerals (tridymite and cristobalite). By the first approximation all tuffs can be subdivided into two large groups: pelitic and coarse-grained. Pelitic tuffs consist of hydrothermally altered volcanic glass, coarse-grained tuffs contain pumice and lithoclasts locating in fine-grained matrix. Smectites replace volcanic glass without changing their microstructure. Zeolites fill pores and fractures. Degree of hydrothermal alternation depends on the hypsometric position. It decreases up the slope.

4.2 *Petrophysical properties*

A large amount of smectites define physical and mechanical properties of tuffs: a low density (ρ – 1.15–1.28 g/cm^3), high porosity (n – 50–59%) and hygro-scopic moisture (Wg – 2.5% or higher). Pelitic tuffs are stronger than coarse-grained. Uniaxial compressive strength (σ_c) of pelitic tuffs is 16 MPa whilst that of coarse-grained is 7 MPa only. All tuffs lose their strength after water saturation. Softening coefficient (C_{soft}) is 0,32–0,34. Gas permeability has zero value for pelitic tuffs and can reach 16 mD for coarse-grained tuffs. Laboratory weathering test showed different behavior of thin- and coarse-grained tuffs during cyclical freezing – thawing and wet – drying. Coarse tuff withstood 25 cycles of freezing and moisture with little or no visible changes. Pelitic have started to break down after the first cycle. Cracking occurred on mud crack surface.

5 DISTRIBUTION OF SLOPE PROCESSES IN THE GEYSERS VALLEY

Traditional methods of GIS mapping were used to assess the slope processes hazard in the valley. Interpretation of remote sensing data was provided using high resolution satellite images. From the thematic processing of remote sensing data of high resolution and subsequent field verification of interpretation results. The map of slope processes distribution was prepared. All landslides can be separated in two principal groups: 1. Large landslides with complex mechanism of displacement that forming the modern relief. The volume involved in the displacement is more than a million cubic meters; 2. Different scale landslides with the volume of rocks involved in the displacement varying from a few thousand to hundred thousand cubic meters. By the mechanism of displacement these landslides subdivided into rockfalls, block landslides, and earthflow, complicating the modern relief. Five hundred thirty phenomena of slope processes have been allocated the observed area. Among them 249 landslides occurred within the volcano-sedimentary strata. Identified slope processes have different mechanism of displacement, including 74 rockfalls, 30 block landslides, 48 complex landslides, 295 earthflows, 20 from them transformed to mudflows, and 60 taluses.

6 CONCLUSIONS

Investigations of hydrothermally altered tuffs in the Geysers Valley and correlation of testing results with remote testing data have showed that ancient and modern slope processes and phenomena are widespread in this rocks. There are slope processes with different mechanism of displacement, including block landslides, complex landslides, earthflows and mudflows. Recently there have been several major events in the studied area that changed the landscape of the valley and damaged or destroyed the thermal manifestations. Geological, geomorphological and climatic factors contributed to their development, but the type of process depends mainly on the composition, structure and physical and mechanical properties of rocks. Hydrothermally altered tuffs contain a lot of clay minerals and their strength properties decrease with humidification. They destroy also with cyclic freezing. History of slope hazard in the region confirms that the most significant events occurred after snowmelt or rainfalls. Analysis of remote sensing data showed an inherited character of slope processes, in particular landslides.

REFERENCES

Leonov, V.L. at al 1991. Caldera Uzon and Geyser Valley. In S.A.. Fedotov & J.P. Masurenkov, *Active Volcanoes of Kamchatkae*: 94–141. Moscow: Nauka Publishers.

Leonov, V.L. 1982. Geology of the Shumnaya river and Uzon-Geysernaya depression, Kamchatka, *Volcanology and Seysmology, v.2, 100–103.*

Sugrobova N.G. & Sugrobov V.M. 1985. Change of regime of thermal manifestations influenced by cyclone Elsa, *Voprosy Geography Kamchatka, v/9, 88–94 (in rus.)*

Volcanic Rocks and Soils – Rotonda et al. (eds)
© *2016 Taylor & Francis Group, London, ISBN 978-1-138-02886-9*

Effects of compaction conditions on undrained strength and arrangements of soil particles for Shikotsu volcanic soil

S. Yokohama
Faculty of Engineering, Hokkaido University, Japan

ABSTRACT: In order to prevent ground disasters on volcanic soils, the mechanical properties have to be cleared in detail. In this study, the mechanical characteristics of a volcanic soil which have deposited in Japan were investigated by a series of laboratory tests. The arrangements of soil particles of the compacted specimens were also observed with a microscope. Based on the test results, the effects of compaction condition such as the difference of water content on the undrained strength was discussed. From the results of cross-sectional observation, it was found that the trends of arrangements of soil particles were influenced by the water contents at compaction work. Finally, the undisturbed specimens were sampled on site of slope failure. The undrained strength and soil structure were also discussed in this study.

1 INTRODUCTION

In the cold-snowy regions such as Hokkaido, Japan, the mechanical properties of volcanic soils deteriorated by freeze-thaw actions. Ground disasters, such as for example failure of slopes or embankments constructed by volcanic soils, have occurred due to torrential rain in spring-summer season. In order to prevent such ground disasters, the mechanical properties of the various soils have to be investigated in detail. In this study, a series of triaxial compression tests on volcanic soil were conducted. Furthermore, the arrangements of soil particles were also observed with a microscope using both of the reconstituted compacted specimens and undisturbed specimen sampled from a slope failure site. Based on the results of the observation, the effects of inherent anisotropy on the mechanical properties of the compacted volcanic soil were discussed.

2 UNDRAINED STRENGTH

Figure 1 showed the results obtained CU tests as the relationship between the deviator stress q $(\sigma_1' - \sigma_3')$ and axial strain ε_a at the effective confining pressure of 150 kPa. From this figure, it was observed that the values of deviator stress q at drier condition (D) and the optimum water content condition (OWC) were higher than that of wetter condition (W). Figure 2 showed the relationship between the deviator stress at ε_a of 5%, $q_{at\,5\%}$ and initial water content at initial. From this figure, it was found that the values of $q_{at\,5\%}$ at D- and OWC-conditions were higher than these of W-condition. Furthermore, it seemed that the effect of effective confining pressure σ_c' on $q_{at\,5\%}$ was not remarkable. Based on these results, it was implied that the undrained strength of K soil was influenced by

difference of the compaction state such as the water content.

3 ARRANGEMENTS OF SOIL PARTICLES FOR UNDISTURBED AND RECONSTITUTED SPECIMEN

Figure 3 showed the schematic figure of a failure of slope, which was constructed artificially by using K soil. The data for dry densities on the surface of failure plane were also denoted. The undisturbed specimens were sampled at the points indicated in this figure. It was seen that the range of the dry densities was from 0.782 Mg/m^3 to 1.020 Mg/m^3. Although the effects of disturbance at sampling on the values of dry density might not be ignored, it could be thought that the dry density of the slope had been reduced due to rainfall and temperature change in soil until the slope failure occurred. In order to clarify the soil fabric of undisturbed specimen, a series of cross-sectional observations were produced.

Figure 4 defined the orientation angle θ of a soil particle. θ indicated the orientation angle between the elongated axis of the soil particle and X-axis shown in Figure 4. Figure 5 (a) and (b) are the rose diagrams indicating the orientation angle θ at $S < 100\,\mu$m and $S > 100\,\mu$m for undisturbed specimen, respectively. In both figures, the distributions of θ at OWC-condition observed from the reconstituted compacted specimen are also illustrated. From Figure 5 (a), it was found that the values of θ for $S < 100\,\mu$m with 70° to 120° for the undisturbed specimen were smaller than these of the compacted specimen at OWC-condition. On the other hand, Figure 5 (b) indicated that the differences of θ distributions between undisturbed and compacted specimen were very small. From these observation

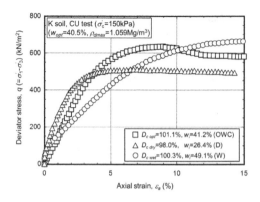

Figure 1. Deviator stress – axial strain relationships.

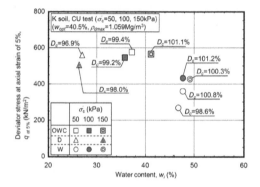

Figure 2. Deviator stress at ε_a of 5% of K soil.

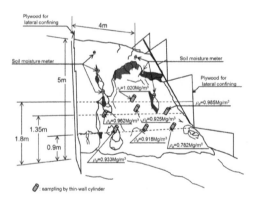

Figure 3. Slope failure and dry densities.

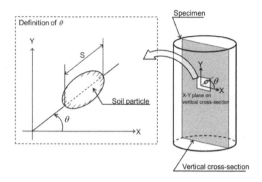

Figure 4. Definition of orientation angle of a soil particle.

(a) [1387particles]

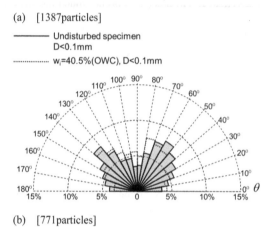

(b) [771particles]

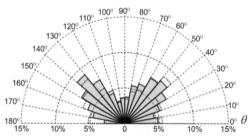

Figure 5. Rose diagrams for orientations of soil particles for undisturbed specimen, (a) S < 100 μm (b) S > 100 μm.

results, it was implied that the differences of θ distribution for S < 100 μm between undisturbed and reconstituted specimens might produce the difference of undrained mechanical properties.

4 SUMMARY

– The undrained strengths of K soil were influenced by the difference of the water content at compaction.

– The arrangements of soil particles were influenced by the water content at compaction. The trends of arrangement patterns at drier and wetter conditions were different from that of the optimum water content condition.

– The undisturbed specimens were sampled on the surface of the failure slope. The arrangement pattern of undisturbed specimen differed from the reconstituted compacted specimens.

Session 2:   Mechanical behaviour of volcanic rocks

Volcanic Rocks and Soils – Rotonda et al. (eds)
© 2016 Taylor & Francis Group, London, ISBN 978-1-138-02886-9

Microstructural features and strength properties of weak pyroclastic rocks from Central Italy

M. Cecconi
Department of Engineering, University of Perugia, Italy

T. Rotonda & L. Verrucci
Department of Structural and Geotechnical Engineering, Sapienza University, Roma, Italy

P. Tommasi
CNR – Institute for Environmental Geology and Geo-Engineering, Roma, Italy

G.M.B. Viggiani
Department of Civil Engineering, University of Roma Tor Vergata, Roma, Italy

ABSTRACT: The paper focuses on the results of an experimental study of the microstructural features and mechanical properties of three pyroclastic deposits from Central Italy: one weak pyroclastic rock from a volcanic complex nearby Rome; and the two materials forming the slab on which the historical town of Orvieto is founded. Possible relationships between microstructural features and strength are examined.

1 INTRODUCTION

Volcanic deposits cover about 8000–9000 km^2 of Central and Southern Italy. Most of them are of pyroclastic origin and exhibit a mechanical behaviour stiffer and brittle at low confining stress (rock-like) and progressively soften and loose brittleness as confining stress increases (soil-like). The rock-like behaviour is exhibited within a range of confining stress which is often controlled by microstructural features which increase the continuity of the solid skeleton (i.e. different types of intergranular bonds) and the low strength of many constituents, rather than by the sole initial void ratio and stress history, as for most part of soils. In the present study, the results of experimental investigations on three materials from different ignimbrite deposits are compared, focusing the attention on their microstructural features and the way in which these may affect their shear strength properties.

2 MICROSTRUCTURAL AND PHYSICAL FEATURES

The first two pyroclastic materials belong to a thick ignimbrite deposit from the eastern volcanic plateaux of northern Latium (Volsini complex, Fig. 1a). The ignimbrite consists of a rock-like facies (tuff, T_{Or}) and a slightly coherent facies (pozzolana, P_{Or}). They are both composed of pumices/scoriae, lithic fragments and phenocrystals immersed in an aphanitic mass. Through SEM and polarized light analyses, at

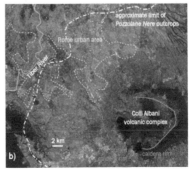

Figure 1.   a) Outcrops areas of the *Orvieto-Bagnoregio Ignimbrite*; b) approximate limits of *Pozzolana Nera* outcrops.

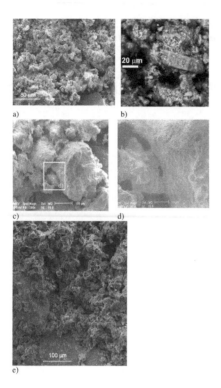

a)

b)

c)

d)

100 µm

e)

Figure 2. SEM image of P_{OR} groundmass at $300 \times$ (a) and $1000\times$ (b); P_{N_RM} material (c, d); T_{OR} (e), showing a completely zeolitized ground mass, with apparent pores, surrounding two lithic fragments.

Table 1. Mean values of physical properties.

Mat.	w/c (%)	G_s (–)	D_{50} (mm)	U (–)	γ_d (kN/m^3)	n (%)
P_{OR}	14.3	2.60	5.0	29	10.90	57
P_{N_RM}	13.0	2.69	4.0	34	14.64	45
T_{OR}	–	2.33	–	–	11.12	51

w/c: water content; G_s: specific gravity; D_{50}: mean grain size; U: uniformity coefficient; γ_d: dry unit weight; n: porosity.

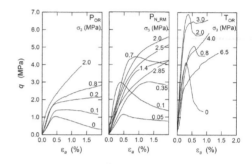

Figure 3. Deviatoric stress q vs. axial strain ε a in the TC tests for P_{OR}, T_{OR}, P_{N_RM} specimens at different confining pressure (numbers near the curves).

The main physical properties of the tested materials are shown in Table 1.

medium magnification the groundmass of the pozzolana (Fig. 2a, b) appears to be formed by glassy particles welded over very small areas and with large inter-particle pores. Observations at higher magnification indicate that many of the glass particles are aggregates of welded minuscule glassy fragments separated by extremely small voids. The groundmass of the tuff (Fig. 2e) is instead pervaded by zeolite minerals of post-depositional origin. They have partially replaced the original glass particles of the pore walls and reduced the pore volume. Zeolite crystals are stronger and stiffer than the glass particles and bonding between zeolite crystals is stronger than that between glass particles in the pozzolana.

The third tested material is the Pozzolana Nera (P_{N_RM}) unit, product of the Colli Albani volcanic complex, about 20 km southeast of Rome (Fig. 1b). The texture of the material is intermediate between granular and matrix-sustained with randomly oriented clasts. The absence of zeolites, was confirmed by X-ray powder diffraction analyses, which allowed to define the mineralogical composition. SEM analyses on P_{N_RM} (Fig. 2c, d) revealed that the microstructure of such material consists of sub-angular grains of very variable size with a rough and pitted surface. The contact between particles reveal the presence of physical bridges bonding the grains, but no significant difference in the mineralogical composition of bonds and grains was revealed by chemical micro-analyses.

3 MECHANICAL BEHAVIOUR

Strength properties of the three materials were investigated through triaxial compression and uniaxial compression (Fig. 3). P_{OR} and T_{OR} specimens were tested in dry conditions, while P_{N_RM} specimens were directly saturated in the triaxial cell, and then subjected to isotropic and triaxial compression under drained conditions.

The uniaxial compressive strength, σ_f, of the materials varies from 0.8 to 7.0 MPa. Porosity is the main physical parameter characterizing the initial state of the specimens and, although a significant void change is expected during loading, n can be used as unifying criterion for mechanical classification, especially when different lithotypes are compared. In uniaxial conditions and at very low confinements the strength drop is more pronounced for the stronger materials (T_{OR}), while the weaker ones are characterized by a gentle descent after the peak. The behaviour turns into hardening over a confinement threshold which varies significantly among the different lithotypes, ranging from 0.1 (P_{OR}) to 3.0 MPa (T_{OR}). The discrepancy is mainly due to the significantly different strengths at low confinements. P_{N_RM}, though having uniaxial strength close to that of P_{OR}, at intermediate confining pressure (0.5–2 MPa) presents stress-strain features similar to those of the stronger T_{OR}.

Volcanic Rocks and Soils – Rotonda et al. (eds)
© *2016 Taylor & Francis Group, London, ISBN 978-1-138-02886-9*

Compressibility of geothermal reservoir rocks from the Wairakei–Tauhara fields with insights gained from geotechnical laboratory testing and scanning electron microscope imaging

M.J. Pender & B.Y. Lynne
University of Auckland, Faculty of Engineering, Auckland, New Zealand

ABSTRACT: Geothermal fluid has been extracted for electricity generation from the Taupo Volcanic Zone in the central North Island of New Zealand since the mid-1950s. This has induced regional subsidence of more than 1 m; but there are a few localized bowls with much greater subsidence. A comprehensive geotechnical investigation with recovery of undisturbed samples from depths of up to 774 m was undertaken with testing of core samples from within and outside subsidence bowls to determine material properties. More than 130 K_o triaxial compression tests were done. Given the volcanic origins of most of the material, the observed wide variation in the measured constrained modulus values is not surprising. Scanning electron microscope images (SEM) reveal differences between the texture of the soft materials and the very stiff materials, these correlate with the constrained modulus values. It was SEM imaging that provided a window into the complex nature of these subsurface environments and identified processes driving the physio-chemical conditions acting and either hardening or softening rocks at specific depths. As geothermal environments are constantly changing and each environment leaves a footprint which is preserved within the rock, SEM can be used to track these changes placing them into a spatial and temporal context. While compressibility measurements identified soft zones within the formations, SEM observations documented subsurface processes responsible for weakening the various lithologic units; these changes in the subsurface conditions are ultimately responsible for the subsidence visible at the surface.

1 INTRODUCTION

Geothermal fluid has been extracted for power generation from the Wairakei field in the Central North Island of New Zealand since the mid-1950s. This extraction has reduced the fluid pressures in the geothermal reservoirs by up to 2 MPa and has induced regional subsidence of 1–2 metres. In addition, there are a number of bowl-shaped regions with greater subsidence, including one where the central subsidence has reached 15 m. More details of the Wairakei-Tauhara geothermal field and the underlying geology are given in Bromley et al. (2013) and Rosenberg et al. (2009).

Extensive investigation drilling had been undertaken previously at Wairakei, with the aim of geothermal resource proving and provided little geotechnical information. To obtain a better understanding of the subsidence mechanism a comprehensive geotechnical investigation was undertaken during 2008 and 2009 in which nine boreholes across the Wairakei–Tauhara field, both within and outside subsidence bowls, were drilled having a total length of 4391 m, and the deepest to 774 m. A total of 3928 m of core was recovered plus 269 undisturbed samples. The investigations were aimed at full core recovery with periodic "undisturbed samples" and presented significant challenges due to the depth and geothermal conditions.

The prediction of subsidence requires, along with other information, the compressibility of the layers affected by extraction of steam and hot water. This paper reviews results from 130 K_o triaxial compression tests on samples selected from the 269 samples recovered. K_o compression is a one-dimensional compression test in a special triaxial cell designed to mimic the in-situ strain conditions during the subsidence. A very wide range of compressibility values was measured, as the materials ranged in consistency from soil to very stiff rock. Scanning electron microscope (SEM) images and mineralogical testing revealed differences between the texture of the soft materials and the very stiff materials (Lynne et al. 2013).

One important outcome from the investigation core logging, laboratory testing, SEM, and mineralogical testing was that localised subsidence bowls are associated with considerable thicknesses of materials with a soil-like consistency.

2 K_o TRIAXIAL TESTING

Conventional oedometer testing was discounted as bedding errors are significant for stiff materials. The K_o cell, proposed by Davis and Poulos (1968) and Campanella and Vaid (1972), provides an alternative.

This is a modified triaxial cell with the loading ram the same diameter as the specimen, and a thick metal cell wall with the cell fluid isolated during the loading, so that no cell fluid enters or leaves the cell. It was necessary to build cells specifically for the project to accommodate the sample diameter of 62 mm, which was dictated by the coring equipment, and to sustain the cell pressures required to reconsolidate the samples to the confining pressures that exist at the large depths from which they were recovered. The K_o testing was done at ambient temperature, rather than the 200°C and greater temperatures at depth in the geothermal field.

The constrained modulus (CM) results indicate that the materials are highly variable and that the property values for the various formations do not separate into distinct ranges.

3 IMAGING METHODS

Scanning electron microscopy (SEM) images and semi-quantitative analyses determined by electron dispersive spectroscopy (EDS), combined with standard thin section petrography, provided critical information on the mineralogy of the samples, mineral compositions, and fine-scale textural relationships (e.g., dissolution textures, micro-fracturing, and mineral morphologies). Samples were examined under the SEM using a Phillips (FEI) XL30S field emission gun. Operating conditions were 5 keV accelerating voltage, a spot size of 3 μm, and a working distance of 5 mm. EDS analyses were performed to obtain semi-quantitative compositional data on the mineral phases. EDS operating conditions were 20 keV accelerating voltage, with a spot size of 5 μm, and a working distance of 5 mm. Standard petrographic analyses provided additional information on mineral distributions and microscale textures.

Fifty two samples were analyzed which show the effects of hydrothermal alteration vary significantly over relatively short distances.

SEM analysis and constrained modulus values on samples of core allowed direct comparison of results. SEM was used to examine subsurface processes affecting both the mineralogy and overall rock compressibility. Differences in rock modulus can be related to the amount of clay minerals or crystals present within individual samples, Lynne et al. (2013). An important conclusion from the SEM imaging is that the materials with low constrained modulus values have conspicuous amounts of clay minerals present, whereas the composition and fabric from the materials with large constrained modulus values reveal an abundance of hard crystalline material such as feldspar.

4 CONCLUSIONS

Selected results from more than 130 K_o and constant cell pressure drained triaxial tests are reported in this paper. A significant feature of the results obtained is the substantially lower constrained moduli of some materials beneath subsidence bowls compared with those of materials at similar depths outside the bowls. These local variations are sufficient to explain why the bowls have formed.

SEM clearly differentiates between minor and major crystal degradation, the type of degradation, and crystal – clay relationships, which can be correlated to measured CM values. Samples that revealed no clay minerals attached to crystal surfaces produced significantly higher CM values than those samples where clay minerals were attached to and altering the crystals.

Subsidence within a geothermal field can be induced through multiple processes. The combination of techniques used in this study proved useful tool for unraveling complex geothermal processes altering the subsurface rock stiffness. By examining hydrothermal alteration processes and coupling this with CM values, insights were obtained into the stiffness of cored rocks and fluid-rock interactions within the Wairakei-Tauhara geothermal system.

REFERENCES

Bromley, C. J., Brockbank, K., Glynn-Morris, T., Rosenberg, M., Pender, M. J., O'Sullivan, M. J., Currie, S. 2013. Geothermal subsidence study at Wairakei-Tauhara. Geotechnical Engineering 166 (GE2), 211–223.

Campanella, R.G. & Vaid, Y.P. 1972. A Simple K_o Cell. Can. Geotech. Jour. Vol. 9, pp. 249–260.

Davis, E.H. & Poulos, H.G. 1963. Triaxial Testing and Three-dimensional Settlement Analysis. Proceedings Fourth Australian-New Zealand Conference on Geomechanics, pp. 233–243.

Davis, E.H. & Poulos, H.G. 1968. The use of elastic theory for settlement prediction under three-dimensional conditions. Geotechnique, Vol 18, No. 1, pp. 67–91.

Lynne, B.,Y., Pender M. J., Glynn-Morris T. and Sepulveda, F. 2013. Combining scanning electron microscopy and compressibility measurement to understand subsurface processes leading to subsidence at Tauhara Geothermal Field, New Zealand. Engineering Geology. 166 (2013), 26–38.

Rosenberg, M.D., Bignall, G., Rae, A.J., 2009. The geological framework of the Wairakei–Tauhara geothermal system, New Zealand. Geothermics 38, 72–84.

Volcanic Rocks and Soils – Rotonda et al. (eds)
© *2016 Taylor & Francis Group, London, ISBN 978-1-138-02886-9*

Geomechanical characterization of different lithofacies of the Cuitzeo ignimbrites

A. Pola & J.L. Macías
Geophysical Institute, Universidad Nacional Autónoma de México, Mexico

G.B. Crosta & N. Fusi
Dipartimento di Scienze Geologiche e Geotecnologie, Università di Milano–Bicocca, Milano, Italy

J. Martínez-Martínez
Laboratorio de Petrología Aplicada (Unidad CSIC-UA), Departamento de Ciencias de la Tierra y del Medio Ambiente, Universidad de Alicante, Alicante, España

ABSTRACT: Volcanic rocks, especially ignimbrites exhibit complex mechanical behaviors due to their large variation in welding, porosity, textural and granulometrical characteristics. It is well known that the mechanical properties of rocks (mainly the uniaxial strength) are greatly controlled by variables such as grain and pore size and their shapes. In order to understand such variations the authors combined field and laboratory study of ignimbrites applying different analyses. The authors studied a well-exposed ignimbrite succession located at the southern edge of the Cuitzeo lake in Michoacán, México. The distribution of each ignimbrite lithofacies was spatially and temporally assigned, while only the most representatives ones were physical-mechanically characterized. Strength, deformation, as well as the failure behavior of selected lithofacies were directly related to micro-textural and granulometrical characteristics by a series of uniaxial and pre- and post-failure non-destructive analyses, including ultrasonic pulse velocity and X-ray image tomographies. Results were compared and evaluated allowing to explore the variation in the mechanical properties, directly related to the characteristics derived from their diagenesis. The results emphasized the importance of the physical-mechanical characterization of each ignimbrite lithofacies to construct conceptual and numerical models of different slope instabilities occurred in the zone, particularly along the Guadalajara-Mexico City highway.

1 INTRODUCTION

The characterization of the physical and mechanical properties of the rocks is the basis to construct conceptual models useful in numerical modelling of any natural phenomena (e.g. pyroclastic density currents; landslides, volcanic instabilities, flow of water and heat transport in rock medium). Particularly, the characterization of volcanic rocks has been always problematic, basically because they have their own history of emplacement, evolution and interaction with the environment and host rock. In addition, there is a huge gap in the knowledge of how external factors affect or interact with the intrinsic properties of rocks. For example, in a volcanic environment, especially in hydrothermal systems, the alteration could drastically change the geometry and morphology of the grain framework and pore network of rocks; in turn these changes control the mechanical and hydrologic behavior of the resulting altered-material.

The main objective of the present research is to investigate the physical and mechanical (e.g. strength and deformational) behavior of several Miocene ignimbrite lithofacies, characterized by different textural and granulometrical characteristics. Another motivation is to investigate the interaction of the physical-mechanical properties in these rocks, for example, how the grain size affects the uniaxial strength. The construction of a general lithostratigraphic section and mapping of the entire zone was necessary to assign the temporal and spatial distribution of each lithofacies.

The study zone is located in the northern part of Michoacán state, along the limit of Guanajuato state, in west-central Mexico. Particularly, the geomorphological characteristics of the zone correspond to deposits emplaced by explosive eruptions. In fact, the mappable units in the area consist of pyroclastic surge and fall deposits, ignimbrites, volcaniclastic deposits and lava flows.

Several samples of different Ignimbrite lithofacies were taken providing an excellent setting to study the evolution of the physical-mechanical properties of the porous and weak materials. The results obtained in this study together with the general stratigraphy of the southern shore of the Cuitzeo Lake represent the first complete study in the area that can help to construct conceptual and numerical models.

2 METHODS AND THEORETICAL BACKGROUND

Several thematic maps, were constructed by processing hypsometric images constructed by DEM data with 20 m of resolution, scale (1:50,000), and image constructed by the fusion of two SPOT images, in order to construct preliminaries maps of the zone.

During fieldwork different stratigraphic sections were described, from which rocks from three different lithofacies (spl, mpl, and mpa) were collected for different types of laboratory analyses.

Mineralogical and petrographical changes were examined by means of optical microscopy and X-ray powder diffraction (XRD). Physical properties including unit weight (γ), effective porosity (η_e), and ultrasonic pulse velocities (V_p, Vs and αs) have been determined.

Effective porosity (η_e) was initially obtained following the procedure recommended by ISRM (2007) by means of comparing dry unit weigh (γ_{dry}) and under void saturated unit weigh (γ_w) of cylindrical core samples.

Rock structure and rock fracturing of each lithofacie was also studied by 3D image visualization, generated by a series of 2D images, obtained from X-Ray Computed Tomography (BIR Actis 130/150 Micro CT/DR system; 40–60 µm of resolution).

V_p and Vs were measured by using a precise ultrasonic device, consisting of signal emitting-receiving equipment (NDT 5058PR) and an oscilloscope (TDS 3012B), which acquired and digitalized the waveforms. Ultrasonic waveform was registered by direct transmission of a 1 MHz signal in cylindrical samples, using Panametrics transducers (model 5660B, gain 40/60 dB, bandwidth 0.02–2 MHz). Velocities (V_p and V_s), spatial (α_s) and temporal attenuation (α_t) were calculated according to the equation described in Martínez-Martínez et al. (2011).

Mechanical properties (e.g. tensile and uniaxial strengths) were performed following standard procedures (ASTM D2938, 1995) on a 250 kN GDS VIS servo-controlled hydraulic testing frame: uniaxial compressive strength (UCS) testing on cylindrical samples (30 mm in diameter) was performed at a constant displacement rate (4 mm/h). Splitting tensile tests were carried out at a constant displacement rate (6 mm/h) on circular disk samples (30 mm in diameter).

Dynamic Young's Modulus (ED) and Poisson's ratio (υ_D) were determined using both V_p and V_s from the equations included in standards (ASTM D3148D, 2002), while static Young's modulus was obtained from the average modulus of linear portion of axial stress-axial strain curve, following those described in ASTM D3148D (2002).

3 CONCLUSIONS

Values of all properties, as well as fracture mode have shown that there is a large dependence on the textural characteristics of each sample, particularly on pumice content. The average values of γ vary from 13.78 (spl-a) to 15.74 kN/m^3 (mpl). In particular, γ values, seem to be influenced by the proportion of matrix content not only in each lithofacies, but also in each specimen.

X-ray images reconstruction shows that principal differences in average values of physical parameters are directly associated to the geometry and morphology of the grains and pore network, which in turn strongly affect the mode of failure of each specimen. The geometry and arrangement of clasts in each specimen largely contribute in determining the total porosity value. In particular, the abundance of mm-to-cm size pumice clasts largely influences the values of porosity in the mpa lithofacies.

The high degree of anisotropy, particularly observed in mpa lithofacies, suggests large variations in mechanical properties: in general the UCS for all lithofacies tends to increase linearly with physical properties (e.g. Vp, bulk density, porosity), but shows larger dispersion in mpa samples than in the other samples. In addition, the decrease in the mechanical properties (e.g. UCS, TS) indicates that the strength of each lithofacies is largely influenced by the intrinsic characteristics of each specimen, inherited by its own history of emplacement mechanism.

Finally, the results presented in this work provided the physical and mechanical properties of four different Ignimbrite lithofacies of the Cuitzeo zone. It should be remembered that the entire sequence is composed by at list sixteen lithofacies. However, it is highlighted that further studies related to geomechanical characterization and detailed stratigraphy needs to be performed. Additionally, all of the existing values need to be evaluated to explore the variation in rock behavior. Finally, this evaluation will emphasize the importance of the physic-mechanical characterization of the zone to construct conceptual and numerical models of different phenomena as slope instabilities occurred in the zone, particularly along the Guadalajara-Mexico City highway.

REFERENCES

ASTM. Standard D 2938, 1995. Standard test method for unconfined compressive strength of intact rock core specimens.

ASTM. Standard D 3148, 2002. Standard test method for elastic moduli of intact rock core specimens in uniaxial compression.

ISRM, 2007. The Complete ISRM Suggested Methods for Rock Characterization, Testing and Monitoring: 1974–2006. Suggested Methods Prepared by the Commission on Testing Methods, International Society for Rock Mechanics, R. Ulusay and J.A. Hudson (eds.), Compilation Arranged by the ISRM Turkish National Group, Ankara, Turkey, Kozan Ofset, 628 p.

Martínez-Martínez, J., Benavente, D., Garcia del Cura, M.A., 2011. Spatial attenuation: the most sensitive ultrasonic parameter for detecting petrographic features and decay processes in carbonate rocks. *Engineering Geology* 119 (3–4): 84–95.

Volcanic Rocks and Soils – Rotonda et al. (eds)
© 2016 Taylor & Francis Group, London, ISBN 978-1-138-02886-9

Triaxial and shear box tests on a pyroclastic soft rock

A. Scotto di Santolo
Università Telematica Pegaso, Napoli, Italy

F. Ciardulli & F. Silvestri
Università di Napoli Federico II, Napoli, Italy

EXTENDED ABSTRACT

The paper presents some experimental results on the mechanical behavior of a pyroclastic rock (*Ignimbrite Campana*) characterizing the subsoil of *Sant'Agata de' Goti*, a typical pre-medieval village in Southern Italy lying on a steep cliff. Therein, the risk of instability induced by the singular morphology is enhanced by the presence of a network of underground anthropogenic cavities, sometimes facing out the cliff (Scotto di Santolo et al. 2015). This study was therefore focused on the experimental characterization of the shear strength of the intact rock and of joints, by means of a laboratory investigation addressed to define the constitutive parameters necessary to analyze with rational methods the stability conditions of the jointed rock mass along the slopes and around the cavities.

The laboratory investigation was carried out on specimens (prismatic and cylindrical) collected from a block of LYT fallen in one cavity. The testing program is reported in Table 1.

Table 2 reports the average physical properties of the samples. The compression wave velocity (V_P) was measured before all the tests through ultrasonic pulse probes, in order to indirectly evaluate the homogeneity of the samples. The mean measured value of V_P is significantly higher than that measured in the field (672–1060 m/s), since this latter is affected by the presence of discontinuities.

The unconfined compression strength, *UCS*, is equal to 2.64 MPa, i.e. about 65% of the mean value measured on dry specimens. The reduction of strength with increasing saturation is in agreement with that observed by means of unconfined compression tests on a fine yellow tuff (Aversa & Evangelista, 1998).

All the intact specimens show a brittle behavior, with the peak shear stress attained at relatively low values of shear displacement or axial strain, and growing with the normal stress or confinement. The lower the normal stress, the more pronounced the dilative behavior observed. Note that all samples reached about the same ultimate stress, corresponding to the critical state condition only for the highest normal stress (300 kPa). For instance, Figure 1 shows the summary of the direct shear tests carried out on intact samples at normal stress of 70, 150 and 300 kPa.

After the full-scale 20 mm shear displacement was reached, the box was opened and some pictures were taken, so that the slope of the failure surface and the thickness of the shear band were measured. The observed fracture patterns changed as the normal load increased; moreover, the failure surfaces followed different patterns inside the samples, according to the heterogeneities present in the natural rock.

The saw-tooth jointed samples were thereafter reloaded with the same normal stresses (tests B in Table 1) and then sheared again to failure. The results (Figure 2) show that the shear behavior of the natural joint is dependent on the normal stress. At lower normal stress, the joint slides along the inclined surface with little damage of the asperities and shows relatively high values of dilation and friction. At high normal stress, the constraint of normal displacement leads to large asperity failure, resulting in low friction. The dilatancy peak shows up at higher sliding for lower normal stress, and only the test at 300 kPa shows a stationary response, as for the intact rock. As a result, the

Table 1. Testing program.

Test type	σ'_c or σ'_v	Specimen condition
UNIAX	0	Intact rock
TX-CID	150	Intact rock
TX- CID	300	Intact rock
DS	150	A2*: Intact rock – No gap
DS	70	A1: Intact rock
DS	150	A2: Intact rock
DS	300	A3: Intact rock
DS	70	B1: Natural joint
DS	150	B2: Natural joint
DS	300	B3: Natural joint
DS	150	Cd: Planar smooth – Dry joint
DS	150	Cw: Planar smooth – Wet joint

Table 2. Mean physical properties of the samples.

Gs	γ	w	γ_d	n	S_r	V_P
(-)	kN/m³	%	kN/m³	%	%	m/s
2.46	12.5	15	11.6	54	31	1670

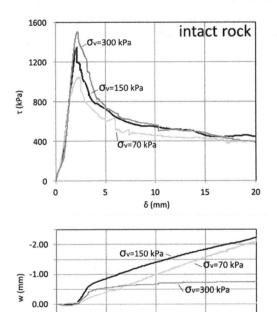

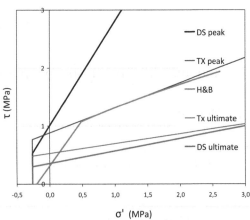

Figure 3. Comparison between peak and ultimate strength envelops measured in DS and TX tests on intact rock.

Table 3. Peak strength parameters from TX and DS tests.

	φ' (°)	c' (MPa)	R^2	m_b	σ'_c (MPa)	R^2
TX	23	0.9	0.969	2.9	2.68	0.913
DS	61	1.0	0.780			

Figure 1. Direct shear test results on intact rock.

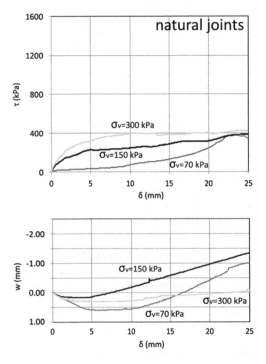

Figure 2. Direct shear test results on natural joints.

ultimate strength of the joints is practically equal to that of the intact samples.

Figure 3 compares the Mohr-Coulomb peak and ultimate failure envelopes measured on intact rock by DS and TX tests. As expected, the peak strength evaluated by direct shear is higher than that measured in triaxial tests, due to the failure surface imposed by the first apparatus. For the ultimate condition, the differences are quite lower due to the localization of the shear band. The peak strength parameters of the intact rock according to the Mohr-Coulomb and Hoek & Brown criteria are reported in Table 3. Although inferred from few data, they represent a first important step for the geotechnical characterization of the rock mass.

REFERENCES

Aversa, S. & Evangelista, A. 1998. Mechanical behaviour of a volcanic tuff: yield, strength and "destructuration" effects. *J. Rock Mechanics and Rock Engineering,* 31(1), 25–42.
Scotto di Santolo, A., de Silva, F., Calcaterra, D., Silvestri, F. 2015. Preservation of cultural heritage of Sant'Agata de' Goti (Italy) from natural hazards, *Engineering Geology for Society and Territory, vol. 8,* Springer International Publishing: Switzerland.

Volcanic Rocks and Soils – Rotonda et al. (eds)
© 2016 Taylor & Francis Group, London, ISBN 978-1-138-02886-9

Failure criterion for low density pyroclasts

A. Serrano
Technical University of Madrid (UPM), Spain

Á. Perucho
Civil Engineering, Laboratorio de Geotecnia, CEDEX, Madrid, Spain

M. Conde
Geological Engineering, CEDEX, Madrid, Spain

1 INTRODUCTION

Studies have been carried out since 1970 in different countries on a variety of volcanic rocks: e.g. in Spain (e.g. Uriel & Serrano 1973, Serrano et al. 2002, González de Vallejo et al. 2006, Serrano et al. 2007); in Italy (e.g. Aversa & Evangelista 1998, Tommasi & Ribacchi 1998, Cecconi & Viggiani 1998, Rotonda et al. 2002, Cecconi et al. 2010); in Japan (Adachi et al. 1981); or in New Zealand (Moon 1993). In those studies it was observed that some of the volcanic materials, mainly the low density ones, have a very similar and peculiar stress-strain behavior. At low pressures they behave like a rock, with high modulus of elasticity and low deformations, while at high pressures their structure is broken and their deformability greatly increased behaving then more like soils.

The stress-strain behavior of low density pyroclasts has been studied for several years by the authors looking for the most influential factors on their mechanical behavior (CEDEX 2013, Conde 2013) and trying to find a failure criterion for these materials, based on a few parameters. The study has been mainly based on a large amount of strength tests performed at Laboratorio de Geotecnia (CEDEX) on samples from the Canary Islands.

This paper presents the failure criterion proposed for these materials.

2 PERFORMED TESTS

Around 250 specimens from this and previous studies were tested to study the strength and deformability of these materials.

Identification and strength tests (uniaxial, triaxial and isotropic compression) were performed on these samples. A study of the macroporosity of all the samples through a hand microscope was also done.

3 SUMMARY AND CONCLUSIONS

A general failure criterion for low density pyroclasts based on a few parameters has been proposed. This

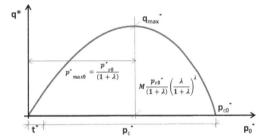

Figure 1. Strength criterion in Cambridge variables.

model has been adjusted to empirical results from around 250 specimens and it is proposed as the failure criterion for these materials as defined in equation 1,

$$q = Mp_0(1 - p_0)^\lambda \qquad (1)$$

This failure criterion (see Figure 1) depends on four parameters: two explicit ones, λ and M, and two hidden ones, ζ and p_{c0}^*. M is a friction parameter while λ is a shape parameter, $p_{c0}^* = p_c^* + t^*$, p_c^* the isotropic compressive strength, t^* the isotropic tensile strength and $\zeta = t^*/p_{c0}^*$ a tensile strength coefficient of the rock.

Variables with physical dimensions are represented with an asterisk, dimensionless variables without asterisk (these are obtained dividing by p_{c0}^*) and subindex 0 refers to translation of origin of normal pressures so that $p_0^* = p^* + t^*$).

A very good fit is obtained to most test data from this study and from published data and values of the model's parameters have been achieved from the adjustments.

As parameter λ presents a value very close to 1 in most samples it was decided to adjust also a simplified model with $\lambda = 1$ to the test results, as defined in equation 5. A very good fit is also obtained for all samples tested in this study as well as for the data taken from literature, indicating that this simplified criterion is a good option for these materials.

A summary of the obtained parameter values is shown in Table 1 for different lithothypes.

Table 1.   Summary of values obtained from the study.

MATERIAL	Sample	γ_d (kN/m^3)	λ	M	p_{c0}^* (MPa)	$\zeta = t^*/p_{c0}^*$ (%)
Welded lapilli (LP-W)	7074	9.28	1.00	2.97	1.12	1.8
	7078	10.85	0.98	2.53	1.81	17.1
Slightly welded lapilli (LP-SW)	7068	9.68	0.83	2.93	1.60	0.3
	7072	10.86	0.90	2.81	1.88	0.0
	7075	8.85	0.71	1.66	0.48	16.7
Weathered lapilli (LP-L-M)[+]	7066	11.70	0.99	1.71	7.81[+]	0.5
	7076	12.07	1.00	1.64	12.22[+]	1.9
	7077	12.78	1.00	1.16	11.86[+]	6.5
	7080	10.50	1.00	1.23	12.52[+]	4.2
Welded pumice (PM-W)	6638	5.25	0.86	2.15	0.28	25.0
Weathered pumice (PM-L-M)	4386	10.13	0.99	2.75	3.58[+]	0.3
	6633	9.24	0.98	2.02	7.17	5.9
Basaltic welded ashes (BA-L)	5141	11.53	1.00	2.62	5.26[+]	2.3
Salic slightly welded ashes (SA-SL)	3855	9.22	1.00	1.83	4.31[+]	4.9
Red Tuff		12.10	1.00	2.99	9.27	6.2
Yelow Tuff		11.00	0.89	1.77	8.00	22.8
Pozzolana Nera		14.62	0.78	2.17	7.24	5.1
Fine grain Tuff		12.91	0.82	1.94	25.27	11.8
Fine grain (Tr) Tuff		12.73	0.78	1.81	20.95	14.2

+ Collapse pressure deduced from adjustments, not obtained from tests;
++ Data from literature Data from literature as indicated in Table 2 (Aversa et al. 1993, Evangelista et al. 1998, Aversa & Evangelista 1998, Tommasi & Ribacchi 1998; Cecconi & Viggiani 1998 and 2001).

They could be used in first stages of projects in volcanic materials when there may not be test results but the material is identified and classified. However, the obtained ranges are quite wide, reflecting the fact that there are other factors influencing the strength of these materials but their lithotype or structure, as their density or their particles strength.

More research is needed in the tensile and low confining pressures areas of the failure criterion.

REFERENCES

Adachi, T., Ogawa, T. and Hayashi, M. Mechanical properties of soft rock and rock mass. *Proc. 10 TH ICSMFE 1*, 527–530 (1981).

Aversa, S. and Evangelista, A. The mechanical behaviour of a pyroclastic rock: failure strength and "destruction" effects. *Rock mechanics and Rock engineering*, 31, 25–42 (1998).

Cecconi, M. Scarapazzi, M. and Viggiani, G. On the geology and the geotechnical properties of pyroclastic flow deposits of the colli albani. *Bulletin of Engineering Geology and the Environment, 69*, 185–206 (2010).

Conde M. Caracterización geotécnica de materiales volcánicos de baja densidad. *PhD Thesis.* Univ. Polit., Madrid (2013).

Conde, M., Serrano, A. & Perucho, A.. New geotechnical classification proposed for low density pyroclastic rocks. *Workshop on volcanic rocks and soils, Ischia* (2015).

González de Vallejo, L.I., Hijazo, T., Ferrer, M., and Seisdedos, J. (2006), Caracterización Geomecánica De Los Materiales Volcánicos De Tenerife., *ed. M. A. R. G. N° 8*, Madrid: Instituto geológico y minero de España.

Moon, V. G. Geotechnical characteristics of ignimbrite: a soft pyroclastic rock type. *Eng. Geology 35*(1–2): 33-48 (1993).

Rotonda, T., Tommasi, P. and Ribacchi, R. Physical and mechanical characterization of the soft pyroclastic rocks forming the orvieto cliff. *Workshop on Volcanic Rocks, Eurock 2002, Funchal, Madeira* (2002).

Serrano, A, Olalla, C, and Perucho, Á. Evaluation of non-linear strength laws for volcanic agglomerates. *Workshop on Volcanic Rocks, Funchal, Madeira.* Eurock 2002.

Serrano, A., Olalla, C., Perucho, A. and Hernández, L. Strength and deformability of low density pyroclasts. *ISRM International Workshop on Volcanic Rocks. Ponta Delgada, Azores.* (2007).

Serrano, A., Perucho, A. & Conde, M. Empirical correlation between the isotropic collapse pressure and identification parameters for low density pyroclasts. *Workshop on volcanic rocks and soils, Ischia* (2015).

Tommasi, P. and Ribacchi, R. Mechanical behaviour of the orvieto tuff. The Geotechnics of Hard Soils-Soft Rocks. 2do International *Symposium on Hard Soils-Soft Rocks/Naples/Italy*, 2: 901–909 (1998).

Uriel, S. and Serrano, A. Geotechnical properties of two collapsible volcanic soils of low bulk density at the site of two dams in canary island (Spain). *8th Congress I.S.S.M.F.E.*, Vol. I: 257–264. Moscú (1973).

Volcanic Rocks and Soils – Rotonda et al. (eds)
© *2016 Taylor & Francis Group, London, ISBN 978-1-138-02886-9*

Relationship between the isotropic collapse pressure and the uniaxial compressive strength, and depth of collapse, both derived from a new failure criterion for low density pyroclasts

A. Serrano
Technical University of Madrid (UPM), Spain

Á. Perucho
Civil Engineering, Laboratorio de Geotecnia, CEDEX, Madrid, Spain

M. Conde
Geological Engineering, CEDEX, Madrid, Spain

ABSTRACT: A theoretical relationship between the isotropic collapse pressure and the uniaxial compressive strength is obtained for low density pyroclasts, based on a failure criterion proposed by the authors, and it is compared to empirical results. Also based on the proposed failure criterion, a theoretical collapsed depth of ground has been deduced, due to the internal stresses overcoming the failure criterion of the material.

1 INTRODUCTION

A comprehensive study for the geotechnical characterization of low density volcanic pyroclasts coming from different areas of the Canary Islands has been performed at CEDEX's Laboratorio de Geotecnia. As a result, a failure criterion has been proposed for these materials (Serrano et al 2015) as well as a new classification (Conde et al. 2015). From the proposed failure criterion, a theoretical relationship between the isotropic collapse pressure and the uniaxial compression strength is deduced for these materials and it is compared to empirical results. Furthermore, a theoretical collapsed depth of the ground is deduced, due to the internal stresses overcoming the failure criterion of the material.

2 STRENGTH CRITERION

The strength criterion obtained by the authors for low density pyroclasts is proposed (Serrano et al. 2015):

$$q^* = M p_0^* \left(1 - \frac{p_0^*}{p_{co}^*}\right)^\lambda \tag{1}$$

This failure criterion depends on four parameters: two explicit ones, λ and M, and two hidden ones, ζ and p_{co}^*. M is a friction parameter that modifies the initial slope of the criterion and the maximum value of the strength; λ is a shape parameter that changes the position of the maximum strength point; p_{co}^* and ζ are strength parameters. More details are given in Serrano et al. 2015.

3 RELATIONSHIP BETWEEN THE ISOTROPIC COLLAPSE PRESSURE AND THE UNIAXIAL COMPRESSIVE STRENGTH

A theoretical relationship between the isotropic collapse pressure and the uniaxial compressive strength is obtained for low density pyroclasts, based on the failure criterion proposed by the authors. In Figure 1 the ratio between the isotropic collapse pressure, p_c^*, and the uniaxial compressive strength, σ_c^*, is shown depending on the other parameters of the strength criterion: ζ, λ and M. This chart is useful to obtain the isotropic collapse pressure for low density pyroclasts once determined the uniaxial compressive strength if the other three parameters are known. The influence of parameter λ is very low as shown in this figure.

A main consequence of this study is that it is not possible to obtain a constant ratio between the isotropic collapse pressure and the uniaxial compressive strength for different low density pyroclasts as it depends on the values of other parameters.

4 DEPTH OF COLLAPSE IN LOW DENSITY PYROCLASTS

Also based on the proposed failure criterion, a theoretical collapsed depth of ground has been deduced, due to the internal stresses overcoming the failure criterion of the material.

The following data are considered:

– Primary data: ground specific weight, γ^*, isotropic collapse pressure, p_c^*, and isotropic tensile strength, t^*.

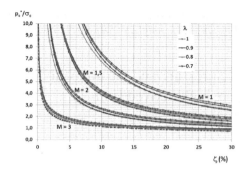

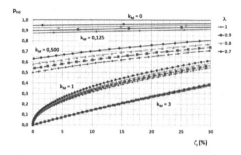

Figure 1. Ratio between the isotropic collapse pressure (p_c^*) and the uniaxial compressive strength (σ_c^*).

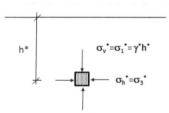

Figure 2. Scheme of principal stresses at depth h^*.

– Secondary data: absolute isotropic collapse pressure, p_{c0}^*, and a tenacity coefficient, ζ, so that:

$$p_{c0}^* = t^* + p_c^* \; ; \; \zeta = \frac{t^*}{p_{c0}^*}$$

It is verified: $p_c^* = p_{c0}^*(1 - \zeta)$.
An intrinsic length, L^*, is defined as

$$L^* = \frac{p_c^*}{\gamma^*} \; ; \; \text{and} \; L_0^* = \frac{p_{c0}^*}{\gamma^*} \qquad (2)$$

It is verified: $L^* = L_0^*(1 - \zeta)$.
And a dimensionless height may be defined as

$$h = \frac{h^*}{L_0^*} \qquad (3)$$

At a depth h^* the following expressions are verified (Figure 2):

$$\sigma_v^* = \sigma_1^* = \gamma^* h^* \; ; \; \sigma_h^* = \sigma_2^* = \sigma_3^* = k_0 \sigma_1^* = k_0 \gamma^* h^*$$

It is supposed that geostatic conditions are verified so that $\sigma_1^* \geq \sigma_2^* = \sigma_3^*$.

4.1 Cambridge dimensionless stresses

The stresses can be expressed in Cambridge variables as:

$$q_h^* = \sigma_1^* - \sigma_3^* = \sigma_v^* - \sigma_h^* = (1 - k_0)\sigma_v^* = d\gamma^* h^*$$

$$p_h^* = \frac{\sigma_1^* + 2\sigma_3^*}{3} = \frac{\sigma_v^* + 2\sigma_h^*}{3} = s\gamma^* h^*$$

where: $d = 1 - k_0$ and $s = \frac{1 + 2k_0}{3}$
And they can be transformed in dimensionless stresses by dividing them by the strength p_{c0}^* and

Figure 3. Canonic collapse pressure, p_{h0}, as a function of ζ for different values of k_M and for different values of λ.

simplified by taking into account the intrinsic length defined in (2):

$$q_h = d\frac{\gamma^* h^*}{p_{c0}^*} = \frac{dh^*}{L_0^*} = dh$$

$$p_h = s\frac{\gamma^* h^*}{p_{c0}^*} = s\frac{h^*}{L_0^*} = sh$$

And a canonic collapse pressure, p_{h0}, can be defined as:

$$p_{h0} = sh + \zeta \qquad (4)$$

A global earth pressure coefficient, k_M, is considered and defined as:

$$k_M = \frac{3(1 - k_0)}{M(1 + 2k_0)} = \frac{d}{Ms}$$

In Figure 3 the canonic collapse pressure, p_{h0}, is obtained for any values of other parameters λ, ζ and k_M, This chart can be used to obtain the canonic collapse pressure and then, through equations (4) and (3), the depth of collapse can be deduced for certain values of the parameters of the ground. As it can be observed, the influence of the parameter λ is higher for intermediate values of k_M, and almost null for $k_M = 3$.
This topic is considered to be of relevance for tunnels and other underground constructions on these types of materials.
The theoretical depths of collapse have been deduced for different types of low density pyroclasts from the Canary Islands.

REFERENCES

CEDEX. Caracterización geotécnica de los piroclastos canarios débilmente cementados. *Final Report* (April 2007).
CEDEX. Estudio del comportamiento geomecánico de los piroclastos canarios de baja densidad para su aplicación a obras de carreteras. *Final Report* (April 2013).
Conde, M., Serrano, A. & Perucho, A. New geotechnical classification proposed for low density pyroclastic rocks. *Workshop on volcanic rocks and soils, Ischia* (2015).
Serrano, A., Perucho, A. & Conde, M. Failure criterion for low density pyroclasts. *Workshop on volcanic rocks and soils, Ischia* (2015).

Volcanic Rocks and Soils – Rotonda et al. (eds)
© *2016 Taylor & Francis Group, London, ISBN 978-1-138-02886-9*

Correlation between the isotropic collapse pressure and the unit weight for low density pyroclasts

A. Serrano
Technical University of Madrid (UPM), Spain

Á. Perucho
Civil Engineering, Laboratorio de Geotecnia, CEDEX, Madrid, Spain

M. Conde
Geological Engineering. CEDEX, Madrid, Spain

ABSTRACT: A theoretical relationship between the isotropic collapse pressure and the dry density for these materials was previously obtained by the authors. Some formal changes are performed to that relationship and it is adjusted to empirical results of low density volcanic pyroclasts.

1 INTRODUCTION

A theoretical relationship between the isotropic collapse pressure and the dry density for low density pyroclasts was previously obtained by the authors (Serrano et al. 2010).

In this paper some formal changes are performed to that relationship and it is adjusted to empirical results of low density volcanic pyroclasts from the Canary Islands and to some published data of volcanic tuffs to estimate the values of a new parameter defined as structural parameter L.

2 RELATIONSHIP BETWEEN THE ISOTROPIC COLLAPSE PRESSURE AND THE SPECIFIC WEIGHT

2.1 Previous studies

In Figure 1 a relationship between the isotropic collapse pressure and the dry specific weight proposed by Serrano et al. (2010) is shown. An expression theoretically deduced for both reticular and vacuolar types of porosity was proposed in the following form:

$$p_c = L\gamma \left(\frac{\gamma}{\gamma_s - \gamma}\right)^\alpha \qquad (1)$$

where p_c is the isotropic collapse pressure, γ is the specific weight, γ_s is he specific gravity, α is a coefficient that was assumed as $\alpha = 1$ from the previous experimental adjusting, and L is a parameter with length dimension that, according to the theoretical assumptions made, mainly depends on the shapes and sizes distributions of pores and particles of the material and on the ratio between the particles and contacts strength and the specific gravity, γ_s. From fitting the proposed expression to some test results from low density pyroclasts a mean value of $L = 433$ m was obtained.

2.2 New proposal

Terms from equation (1) can be rearranged so that the following expression is obtained, assuming $\alpha = 1$:

$$p_c = L\gamma_s \frac{(1-n)^2}{n} \qquad (2)$$

where n is the porosity of the material.

The following three main factors affecting the isotropic strength of these materials can be seen:

– The structure of the material, represented by parameter L (Serrano et al. 2010).
– The type of material, represented by parameter γ_s, higher for basic materials and lower for acid ones.
– The porosity, represented by a function of the total porosity: $(1 - n)^2/n$.

2.3 Adjustment to experimental data

Equation (2) can be adjusted to experimental results of low density pyroclasts where the isotropic collapse pressure, the specific gravity of solids and the porosity of the samples have been determined, so that structural parameter L is empirically obtained. Figure 2 shows the isotropic collapse pressure versus the dry specific weight, similar to the one in Figure 1 but with more new data incorporated on it.

2.4 Adjusted values of parameter L

Mean values and ranges for the structural parameter L are shown in in Table 1 for each type of porosity and in Table 2 for the different groups of pyroclasts considered.

From the analysis of tables 1 and 2 three important conclusions spring out:

1. There is not an only value for L that could be used for all the pyroclasts.

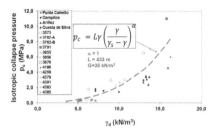

Figure 1. Relationship between the isotropic collapse pressure (p_c) and the dry specific weight (γ_d) using tests results from Canary Islands pyroclasts (Serrano et al. 2010).

Table 1. Mean values, standard deviation and ranges of variation of structural parameter L for the different porosities.

Porosity	Range of L (m)	Mean (m)	Stand. dev. (m)
Matrix	425.3–1776.3	1012.0	453.2
Mixed	108.6–343.4	252.2	72.9
Reticular	96.4–283.4	172.4	68.8
Vacuolar		388.6	

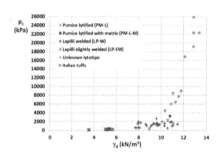

Figure 2. Isotropic collapse pressure (p_c) versus dry specific weight (γ_d). Tests results from pyroclasts from Canary Islands and from Italian tuffs (Aversa et al. 1993, Evangelista et al. 1998, Aversa & Evangelista 1998, Tommasi & Ribacchi 1998).

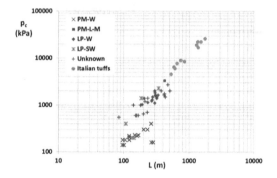

Figure 3. Structural parameter L versus the isotropic collapse pressure, p_c, for the studied materials. Logarithmic scale.

Table 2. Mean values, standard deviation and ranges of variation of structural parameter L for the different materials studied.

Lithotype	Range of L (m)	Mean (m)	Standard dev. (m)	Type of porosity
Fine grain tuffs (assumed SA-L-MP)	1316.4–1776.3	1456.3	173.5	Matrix
Red and yellow tuffs (assumed SA-L-MP)	533.7–847.4	665.5	114.2	Matrix
Lithified pumice and with matrix (PM-L-M)		425.3		Matrix
Welded lapilli (LP-W)	190.6–336.4	278.5	44.2	Mixed
Slightly welded lapilli (LP-SW)	108.6–343.4	184.0	93.3	Mixed
Welded or lithified pumice (PM-W/L)	96.4–283.4	172.4	68.8	Reticular
Slightly welded scoria (SC-SW)		388.6		Vacuolar
Campitos pyroclasts	85.3–513.1	277.3	135.0	Unknown

2. The value of L is clearly dependent of the type of porosity of the material.

3. The higher values of L are attained for pyroclasts with matrix-type porosity while the lower ones are obtained for reticular-type porosity materials. Intermediate values are obtained for mixed-type porosity materials.

In Figure 3 the values of the parameter L versus the isotropic collapse pressure, p_c, are plotted in logarithmic scale. A correlation is observed for values of parameter L higher than around 300 meters or for values of isotropic collapse pressure, p_c, higher than around 1500 kPa. Figure 3 shows that there is no correlation for lower values of L and p_c. This correlation seems to verify only for pyroclasts with matrix-type porosity.

REFERENCES

Aversa, S., Evangelista, A., Leroueil, S. & Picarelli, L. Some aspects of the mechanical behaviour of "structured" soils and soft rocks. *International symp. on Geotechnical engineering of hard soils and rocks,* Athens, 1, 359–366 (1993).

Aversa, S. and Evangelista, A. The mechanical behavior of a pyroclastic rock: failure strength and "destruction" effects. *Rock mechanics and Rock engineering,* 31, 25–42 (1998).

Evangelista, A., Aversa, S., Pescatore, T.S. & Pinto, F. Soft rocks in southern Italy and the role of volcanic tuffs in the urbanization of Naples. The Geotechnics of Hard Soils-Soft Rocks. *2do International Symposium on Hard Soils-Soft Rocks/Naples/Italy,* 3, 1243–1267 (1998).

Serrano, A., Perucho, Á. and Conde, M. Isotropic collapse load as a function of the macroporosity of volcanic pyroclasts. *3rd International Workshop on Rock Mechanics and Geo-Engineering in Volcanic Environments* (2010).

Tommasi, P. and Ribacchi, R. Mechanical behaviour of the orvieto tuff. The Geotechnics of Hard Soils-Soft Rocks. *2do International Symposium on Hard Soils-Soft Rocks/Naples/Italy,* 2: 901–909 (1998).

Volcanic Rocks and Soils – Rotonda et al. (eds)
© 2016 Taylor & Francis Group, London, ISBN 978-1-138-02886-9

Underground caverns in volcanic rocks: Geological aspects and associated geotechnical behaviour of pyroclastic rocks

P. Vaskou & N. Gatelier
Geostock, Rueil-Malmaison, France

ABSTRACT: Hydrocarbon storage caverns are unlined openings preferentially excavated in hard rock for stability and hydraulic containment purpose. Volcanic rocks are good candidates for the development of medium to large caverns but they are often excluded during site selection for hydrogeological reasons related to the rock structure. Based on laboratory tests and in situ measurements, the geomechanical behaviour pyroclastic rocksare presented and compared to granite in which many caverns have already been excavated worldwide.

1 INTRODUCTION ON VOLCANIC TUFF

Volcanic tuff represents a series of rock from volcanic origin, generally having a petrographic composition of andesite or dacite. The high viscosity to the magma induces blast events, with ashes and other volcanic debris blown apart by explosing gases and followed by the settlement of ashes in layers.

Tuffs are formed by sedimentation and possibly welding of hot ashes (pyroclastic flow). It is commonly read that welded tuffs have a very high compressive strength and the rock mass is made of thick and very slightly fractured layers. It is also often considered that mechanical and rock mass properties are induced by the formation process which gives homogeneous and thick layers. These properties are analysed and compared to granite properties.

2 EXAMPLE OF JURASSIC EAST CHINA VOLCANIC TUFF

2.1 Geological context

A very large zone of tuff is present at the East of China, from Ningbo to Hong Kong. This zone is dated from Yanshan orogenesis (late Jurassic). A project of storage cavern, launched at the end of the 90's close to the city of Ningbo allowed investigating the volcanic tuff with surface geological observations, core drillings, borehole televiewer, sampling and laboratory testing, as well as observations and measurements during the construction of tunnels and two caverns. Recently, another project in the same area gave the opportunity to complete the knowledge of East China volcaniclastics through new investigation boreholes, laboratory tests on core samples, televiewer and in situ stress measurements.

With these two projects, a comprehensive analysis of the geomechanical behaviour from various types of tuff, from cinerite to tuff and agglomerate was made possible, based on a significant number of field observations and laboratory tests. The quantity of data and results allow comparing the geomechanical properties of tuff with the ones from a very classical rock type in which many caverns have been excavated in the world, the granite.

2.2 Matrix properties

Uniaxial compression tests and compression tests under confining pressure in the range 2–8 MPa are implemented to derive basic elastic and failure properties of the rock matrix. The fine/medium grained volcanic rocks havehigh strength values (average UCS a bit less than 200 MPa for tuff) which seem to be higher values as compared to typical granite.

So, we compare more in detail strength and other geomechanical parameters of pyroclastic rocks and granitoid selected for its homogeneity, both on geological basis and mechanical behaviour. At the matrix scale, the comparison of the mechanical parameters of volcanic tuff (cinerite and tuff) and granitoid show higher values for the volcanic rocks. Some differences in the failure mode have been also noticed and seem to be linked with the brittleness index, the welded tuff having a more brittle behaviour.

Comparison has also been carried out with welded tuff studied by others, namely welded tuff from the Newberry volcano (Oregon, USA) and the welded Topopah Spring tuff (Nevada, USA). Interestingly, the older Chinese tuff is stiffer and more competent than the welded tuff from Newberry volcano but comparable to the least porous Topopah Spring welded tuff. The lowest range of porosity for the Topopah tuff (10–13%) is still significantly higher than the porosity of the Chinese welded tuff (typically <1%). The role played by the diagenesis, with a decrease of the porosity with time and a subsequent increase of strength, is certainly a key factor.

Based on the results obtained and comparison with others geological environment, the strength properties of the rock matrix for the studied Chinese volcanic rocks are therefore considered as fully able to host a stable opening at depth with a large cross-section.

2.3 *Rock mass properties*

The rockmass mechanical properties at the scale of the site were estimated based on empirical approach whereas hydraulic conductivities were calculated through Lugeon test profiles in the same boreholes.

The cinerite is very homogeneous with rock classification rated as Good or better whereas the larger range of rock quality occurred for the tuff with rock classification ranging from Very Poor to Very Good. For an average granitoid, 50% of the rockmass is rated as Poor whereas 50% is rated as Good.

The estimated Q-index distribution of cinerite/tuff and agglomerate is comparable to an average fresh granitoid, rock type in which a large volume of rock has already been excavated for underground storage caverns, and proven to be stable. It therefore supports the feasibility to excavate large underground excavations in the Chinese tuff.

A preliminary estimate of the basic mechanical rock mass parameters is presented following Barton empirical approach. The predicted rockmass strength for the volcaniclastics ranges from 45 to 55 MPa about twice the rockmass strength of the granitoid. Based on the hydrofracturing tests, stress-induced stability problems in the rockmass are not the expected failure mechanism and unstable structural wedges are the most probable events.

3 COMPARISON WITH RECENT ANALOGUES

3.1 *Objective and selection of analogues*

We have tried to compare recent analogues from recent volcanic environment from Japan, Chile and French Carrabean. For each of them, we have selected pyroclastic rock types with andesite or dacite composition. All of them are (geologically speaking) extremely recent, say less than or about a century.

3.2 *Comparison between recent and ancient tuff*

The most spectacular difference between ancient and recent tuff is a huge difference of porosity as well as strength. Recent rocks have a relatively high porosity compared with the porosity of ancient tuff: a 5% porosity is measured at the Montagne Pelée 1902 tuff whereas ancient Jurassic pyroclastic rocksfrom East China exhibit a very low porosity ranging from 0.2 to 0.7%. Strength is also very different with very high values measured in ancient tuff and this difference is most likely due to or related to the porosity. The observation of thin sections evidences that secondary crystals took place inside the primary porosity and due to the high silica content, minerals forming the cement providing a strong framework to the tuff matrix. The process is rather similar with larger grain size rocks such as agglomerates.

Both matrix and rock mass were analysed and compared, for volcanic tuff and granitoid. It appears that granitoid, considered as a good rock for underground excavations and in particular large caverns, when fresh, has a quality which can easily be exceeded by ancient pyroclastic material.More in detail, matrix of fine grained to medium grained tuff (cinerite to tuff) exhibits higher strength values as compared to granitoid (granite, granodiorite and quartz diorite) and regarding rock mass scale, tuff is at least in the range of granite whereas agglomerate is largely higher.

4 CONCLUSIONS

This study evidences the very high geomechanical characteristics of volcanic tuffs, and in particular the medium- to fine-grained material, say tuff and cinerite. Standard and welded tuffs exhibits similar strength, higher than granitoid, whereas agglomerate shows lower strength values, rather similar to granitoid. Regarding rock mass, ancient pyroclastic rocks exhibit a high to very high quality, at least equal to but often higher than granitoid rock masses.

Another conclusion of the comparison with recent analogues is that the common acceptation that welded tuffs have very high strength due to welding is infirmed, since standard and welded tuffs show rather similar mechanical strength values. It seems that the age of the material could have a strong influence on the strength, most likely due to initial high porosity, later filled with siliceous material after water percolation through the rock. The results of the tests found in the literature carried out on Newberry volcano samples and Topopah Spring tuff (dated Miocene) seem to corroborate this hypothesis, with both UCS and ages in between recent (Quaternary) tuffs from active volcanoes (Martinique, Japan) and ancient (Jurassic) Chinese tuffs.

The investigation campaigns carried out both in the late 90's and more recently have shown that the pyroclastic rocks (cinerite/welded tuff and agglomerates) are fully adapted to host a large storage cavern for storing LPG or crude oil. The mechanical strength of the intact pyroclastic rock samples is even higher than fresh igneous rock (granite).

Volcanic tuff (tuff, welded tuff, cinerite and agglomerate as well) when ancient enough, represents an adequate rock type and rock mass for unlined cavern construction, even very large caverns, at least as homogeneous and as strong as granite which is considered as a standard and classical material for this purpose. However, tuff is an exception inside the volcanic rocks, in which the excavation of large unlined underground openings is difficult due to heterogeneities, average low strength and relatively low quality rock masses.

Session 3:   Mechanical behaviour of volcanic soils

Volcanic Rocks and Soils – Rotonda et al. (eds)
© *2016 Taylor & Francis Group, London, ISBN 978-1-138-02886-9*

One-dimensional compression of volcanic ash of Mount Etna

V. Bandini, G. Biondi, E. Cascone & G. Di Filippo
University of Messina, Italy

EXTENDED ABSTRACT

The activity of Mount Etna has always been characterized by spectacular lava fountaining events, ranging
from strombolian to plinian and causing copious ash
fallout varying from tens to hundreds of kilometers
away from the vent. In August 2011 a large amount of
volcanic ash, hereafter referred to as Etna Sand, was
erupted and fell down, blown by the wind, on the Ionian
coast North of the city of Catania. Due to its highly
crushable and compressible nature, the Etna Sand may
result to be problematic from an engineering and construction viewpoint and mechanical characterization,
as well as the constitutive modeling, involves practical
and theoretical difficulties.

Accordingly, a physical and mechanical characterization of Etna Sand was performed. The activities
concerned with X-ray diffractometry, electron scanning miscroscopy (*SEM*), particle size and shape
dynamic image analysis and, finally, one-dimensional
compression tests carried out on both dry and wet
specimens prepared at different initial void ratios.

In the paper, some of the experimental results are
presented and discussed focusing on the effects of
particle breakage and on the non linearity of the $e-\log$
σ'_v relationship.

Etna Sand particles are glass shards, which might
be classified as angular particles, and consist of many
gas bubbles or only a portion of a single gas bubble.
Cristobalite and anaorthite are the main constituents
of the material.

For irregular particles, as those of the tested material, the measured size depends on both the measuring
technique and the definition of the particle size. Thus,
particle size analyses were carried out by using a particle analyser (*QicPic*) which overcomes the limitations
of the traditional low tech approaches by means of a
dynamic image analysis technique based on the use of
a dispersing unit equipped with a laser and a camera
detector. Figure 1 shows the mean particle size distributions obtained through the *QicPic*, in terms of the
diameter of a circle of Equal Projection area (d_{EQPC})
and in terms of minimum (F_{min}) and maximum (F_{max})
Feret diameters. Etna Sand particles are not spherical and are characterised by an irregular perimeter
with the presence of re-entrant sections. These features
imply the d_{EQPC} distribution to be very close to the
F_{min} distribution. In Figure 1 the particle size distribution obtained by sieving is also plotted for comparison.

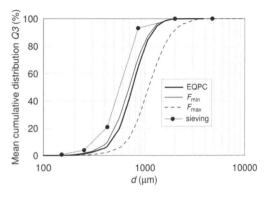

Figure 1. Mean cumulative particle size distributions $Q3$
for *EQPC*, F_{min} and F_{max} diameters.

Consistently with other results available in the literature, the particle size distribution related to F_{min} is the
closest to that obtained by means of the hand sieving.

The one-dimensional compression tests were carried out in a 38 mm diameter oedometer cell. Specimens were reconstituted at different initial densities,
by either pluviation or tamping, and were compressed
under wet and dry conditions. Four tests (ES_OED1-4)
were carried out on saturated specimens created with
the initial grading of Etna Sand and compressed up
to 16 MPa. A further one-dimensional compression
test (ES_CRS1) was carried out on a dry specimen,
characterised by the same initial grading, using a constant strain rate cell which allowed applying on a
38 mm diameter specimen a maximum vertical stress
of 45 MPa. Finally, two more tests were carried out on
dry specimens prepared both as loose (DES_OED5)
and as dense (DES_OED6) as possible, by using the
damaged material tested up to 9240 kPa.

To evaluate the effect of one-dimensional compression on particles crushing of Etna Sand, the particle
size and the shape parameters distributions at the
beginning and at the end of each tests were compared.
Figure 2 shows the distribution of three shape parameters: the sphericity (S), calculated as the ratio of the
equivalent circle perimeter over the real perimeter of
the two-dimensional image of the particles, the convexity (C), describing the compactness of a particle as
calculated from the ratio of the projected particle area
over the gross area including any re-entrant section,

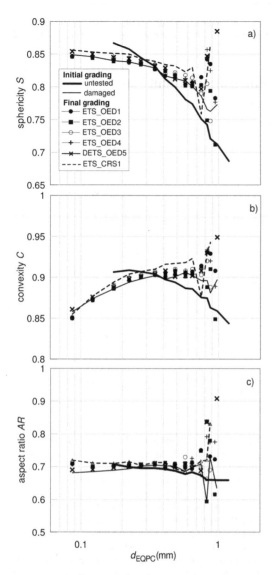

Figure 2. Shape parameters at the beginning and at the end of the one-dimensional compression tests: a) sphericity, b) convexity and c) aspect ratio.

and the aspect ratio $AR = F_{min}/F_{max}$. The initial grading is referred to both the original material (specimens ES_OED1-4) and to the damaged material adopted to prepare the specimens DES_OED5 and DES_OED6.

The values of both S and C of the original material vary in the 0.7–0.85 (Fig. 2a) and 0.85–0.92 (Fig. 2b) ranges, respectively, moving from the coarser particles towards the smaller ones. The initial aspect ratios were of about 0.65–0.70 (Fig. 2c) for the entire range of particle diameters.

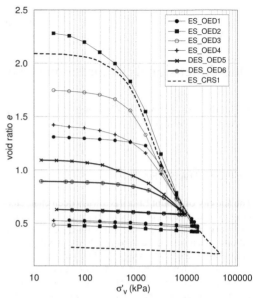

Figure 3. One-dimensional compression test results on Etna Sand: e–log σ'_v plot.

At the end of the oedometer tests the coarser particles ($d > 0.3$–0.4 mm) tend to become more spherical and more convex, especially for $d = 0.8$–1.0 mm. The smaller particles, including the new particles created by the grain crushing, are characterized by lower sphericity and convexity than the original ones (Figs. 2a,b).

The effect of particle breakage is also evident in Figure 2c, where a major change in the aspect ratio again for $d = 0.8$–1 mm is shown.

The results of the one-dimensional compression tests in terms of void ratio e against the vertical effective stress σ'_v are plotted in Figure 3. It can be observed that:

– the e–logσ'_v relationship for Etna Sand is non-linear over the entire range of applied stresses, since the compression index C_c increases as specimens are loaded in the elevated stress level range (1–10 MPa) and decreases in the high stress level range ($\sigma'_v > 10$ MPa);
– at higher stresses ($\sigma'_v > 10$ MPa), all the e–log σ'_v curves obtained in the tests are approximately coincident, denoting a unique response of the material, and allowing to infer that the limiting compression curve (LCC) for the Etna Sand, dependent only on soil mineralogy, is reached.

Volcanic Rocks and Soils – Rotonda et al. (eds)
© *2016 Taylor & Francis Group, London, ISBN 978-1-138-02886-9*

Geotechnical characterization of Mount Etna ash for its reuse preserving human health

G. Banna
Unit of Medical Oncology, Cannizzaro Hospital, Catania, Italy

P. Capilleri, M.R. Massimino & E. Motta
Department of Civil Engineering and Architecture, University of Catania, Italy

EXTENDED ABSTRACT

In the last decades Mount Etna (Sicily, Italy) has partially modified its activity from effusive to eruptive. More precisely, since 1971 significant eruptive activity has been observed. This was due to the development of new craters. The erupted volcanic material is characterized by pyroclastic elements of different sizes, water vapor and carbon and/or sulfur dioxide. The pyroclastic material of Mt. Etna is characterised by: blocks (>64 mm), lapillus (2–64 mm), coarse ash (62 micron – 2 mm), fine ash (<62 micron). From the eruption of 1991–1993 to that of 2001 more than 150 paroxysmal activities occurred, involving above all the great Southeast crater (Figure 1). New important activities occurred from January 2011 to the end of 2013. The paroxysmal activities from 2011 to 2013 were due to a new crater in the Southeast area. In the 2014 about 30 events of moderate intensity occurred. Overall, an exponential increasing of the paroxysmal activity has been observed in the XXI century. Large amounts of ash have repeatedly covered the city of Catania (Figure 2). This has caused problems on the road and air traffic, problems on the agricultural sector and human health damage. As regards the human health damage, a great attention should be devoted to cancer, due to the significant toxicity of this material.

The storage of Mt. Etna ash has been a great problem for this area. Due to the great quantity of erupted volcanic ash, the recent trend is not considering the volcanic ash as a waste, but as a resource in different fields of Civil Engineering. Volcanic ash could be particularly used in Geotechnical Engineering, avoiding environmental and human health damage. Since many

years volcanic material has been used as construction materials, for example in roadway embankments and river dikes. Nevertheless, the rational use of this material as construction materials was not investigated taking into account human health damage. Moreover, most studies have focused on old volcanic deposits; while the geotechnical properties of fresh volcanic products, such as Mt. Etna ash, have not been carefully investigated (Orense et al., 2006).

The present paper deals with human health damage and geotechnical characterization of Mt. Etna ash. In particular, the effects of inhalation and/or ingestion of volcanic elements on acute or chronic disease in humans were highlighted (Table 1), despite the actual Italian regulation (D.L. 152/2006), which allows us to use products of volcano activities for engineering

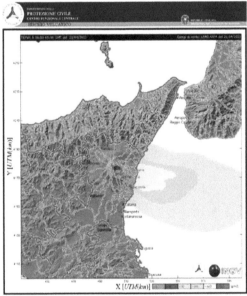

Figure 2. Areas interested by ash emission (g/m^2), 21 April 2013, h 21.00–24.00 GMT. Magenta: 1–10 g/m^2; Blue: 11–50 g/m^2; light blue; 51–100 g/m^2; green: 101–1000 g/m^2; yellow: 1001–9000 g/m^2; red: 9001–10000 g/m^2.

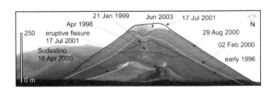

Figure 1. The new craters since 1996 to 2001 (INGV-CT).

Table 1. Possible acute and late effects of volcanic elements on human health.

Possible effects (Agent)	Inhalation (gas, silica, cristobalite, fluoro-edenite)	Ingestion (water/soil)
Acute	Acute respiratory disease (asthma and bronchitis) Exacerbations of pre-existing lung and heart disease	–
Chronic	Lip, oral cavity and pharynx cancers Malignant mesothelioma Lung cancer Chronic obstructive pulmonary disease (COPD) Silicosis	Thyroid cancer Kaposi sarcoma

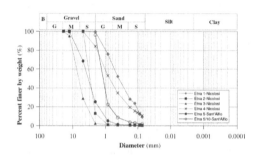

Figure 3. Grain size distribution curves of the tested materials.

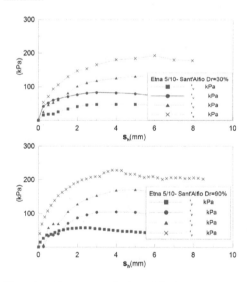

Figure 4a. Shear stress vs horizontal displacement curves at different relative densities and constant vertical stresses for "*Etna 5/10 – Sant'Alfio*" samples.

purposes classifying this material as not dangerous according to codes CER 200303; CER 170504.

Grading and index properties (Figure 3) and strength behaviour (Figure 4) of some samples of ash due to Mt. Etna eruption were also detected.

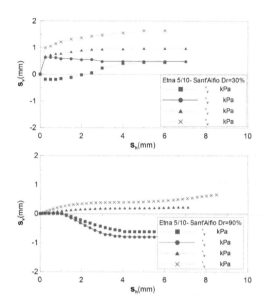

Figure 4b. Vertical displacement vs horizontal displacement curves at different relative densities and constant vertical stresses for "*Etna 5/10 – Sant'Alfio*" samples.

The results indicate that the tested materials are of gravelly-sandy nature. The peak friction angle range is 31°–42°; the friction angle at the critical state is in the range 29°–36°. Medical studies on the effects of inhalation of volcanic elements (Rapisarda et al., 2003) suggest us to avoid their reuse with a direct contact with air. Thus, reuse in embankments, earth retaining walls or other geotechnical structures are indicated. However, medical studies on the effects of their ingestion (Malandrino et al., 2013) leads to a reuse of these materials, including engineering remedial for protection of aquifers.

REFERENCES

Malandrino P, Scollo C, Marturano I, Russo M, Tavarelli M, Attard M, Richiusa P, Violi MA, Dardanoni G, Vigneri R, Pellegriti G. (2013). Descriptive epidemiology of human thyroid cancer: experience from a regional registry and the "volcanic factor". *Front Endocrinol (Lausanne)*. Jun 4;4:65.

Orense R.P., Zapanta A., Hata A. & Towhata I. (2006). Geotechnical characteristics of volcanic soils taken from recent eruptions. *Geotechnical and Geological Engineering*, 24: 129–161, Springer, DOI 10.1007/s10706-004-2499-y.

Rapisarda V, Amati M, Coloccini S, Bolognini L, Gobbi L, Duscio D. (2003). The in vitro release of hydroxyl radicals from dust containing fluoro-edenite fibers identified in the volcanic rocks of Biancavilla (eastern Sicily). *Med Lav*. Mar–Apr; 94(2):200-6.

Volcanic Rocks and Soils – Rotonda et al. (eds)
© *2016 Taylor & Francis Group, London, ISBN 978-1-138-02886-9*

Geotechnical issues concerning the material removal and reuse of pyroclastic soils

G. Caprioni, F. Garbin, M. Scarapazzi & F. Tropeano
Geoplanning S.r.l., Rome, Italy

G. Bufacchi & M. Fabbri
Consultants, Italy

Q. Napoleoni
Department of Civil Engineering Building and Environment (DICEA), Sapienza University of Rome, Rome, Italy

A. Rignanese
Parco Industriale della Sabina S.p.A., Italy

ABSTRACT: This paper presents the results of a work developed on pyroclastic fall deposits from Sabatino volcano complex, typical of the subsoil of the area placed in the North of Rome (Italy). The particular physical and mechanical properties of the materials are related to the geological origin of the deposits, to their formation environment and mechanisms. The geotechnical description follow the scheme recently proposed on pyroclastic flow deposits. The case study concerns the material removal and reuse of pyroclastic soils for road embankments; especially the compaction issues are treated because, sometimes, these soils are not included in the specifications standards despite that overall they have good mechanical characteristics. The paper presents the results of an experimental investigation by means of laboratory and on site tests and suggest a way of managing these problems.

Volcanic Rocks and Soils – Rotonda et al. (eds)
© 2016 Taylor & Francis Group, London, ISBN 978-1-138-02886-9

V_S measurements in volcanic urban areas from ambient noise Rayleigh waves

M.R. Costanzo, R. Mandara & R. Strollo
Department of Earth, Environmental and Resources Science, University of Napoli Federico II, Italy

C. Nunziata
Department of Earth, Environmental and Resources Science, University of Napoli Federico II, Italy
International Seismic Safety Organization (ISSO)

F. Vaccari
Department of Mathematics and Geosciences, University of Trieste, Trieste, Italy
The Abdus Salam International Centre for Theoretical Physics, SAND Group, Trieste, Italy

G.F. Panza
Department of Mathematics and Geosciences, University of Trieste, Trieste, Italy
The Abdus Salam International Centre for Theoretical Physics, SAND Group, Trieste, Italy
International Seismic Safety Organization (ISSO)
Institute of Geophysics, China Earthquake Administration, Beijing, People's Republic of China

ABSTRACT: An original approach is proposed for V_S measurements in highly urbanized area, like Napoli and Ischia island, that is based on the non-linear inversion of group velocities of fundamental-mode Rayleigh waves extracted with the frequency-time analysis from the cross-correlation of synchronous ambient noise recordings at two sites. The obtained V_S models reach depths (up to 1.5 km) that are prohibitive for controlled source experiments and allow the definition of the seismic basement for reliable microzoning studies.

1 INTRODUCTION

The increasing demand of realistic modeling of hazard scenarios has accelerated the research for new seismic methods for rapid, accurate and low cost shear wave velocity (V_S) measurements, mostly in urban areas.

We present some examples of V_S measurements extending to depths of hundreds of meters in the highly urbanized areas of Napoli and Ischia island, affected both by seismic and volcanic hazard. They are obtained by using the FTAN and Hedgehog methods applied to noise cross-correlation functions (NCF) computed between two receivers, which is a reliable estimate of the Green's function (surface wave packet) (e.g. Bensen et al. 2007). Frequency-time analysis (FTAN method), based on group velocity dispersion, is an implemented multifilter technique able to separate the different modes that compose a signal, in particular the fundamental mode of surface waves (Levshin et al. 1992). Average group velocity dispersion curve, computed for the path travelled by the seismic wave, is inverted with Hedgehog method, that is an optimized Monte Carlo non-linear search of velocity-depth distributions, which allows to define V_S models with a resolution that depends on the measurement accuracy (Panza, 1981).

2 EXPERIMENTS IN URBAN AREAS

We show results of cross-correlation measurements performed in the eastern part of the urban area of Napoli (Ponticelli, Poggioreale and Barra quarters), including the Volla-Sebeto depression, along alignments of ~500–1,100 meters (Fig. 1). The subsoil mainly consists of fluvio-lacustrine deposits and volcanic soils and rocks.

Noise recordings (2–4 hours) were processed to get vertical and radial components of the NCFs which were later analyzed with FTAN method in order to extract the group velocity dispersion curve of the fundamental-mode Rayleigh surface waves. Then average dispersion curves, with error bars, were inverted with Hedgehog method to get V_S models with depth. The possible presence of aquifers was investigated by several tests on the most suitable V_P/V_S ratio, by varying it between 1.8 and 3.0, with 0.05 step. As a general result we can report a high V_P/V_S (2.0–2.9). Once established the preferred V_p/V_s ratio, for each path, the value was kept fixed during the inversion procedure. We fixed V_P to 1.5 km/s, in the water table.

Significant lateral variations can be observed if we map the horizons with $600 < V_S < 800$ m/s (related to rocks from fractured to compact) and $V_S > 800$ m/s

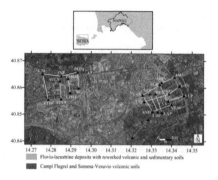

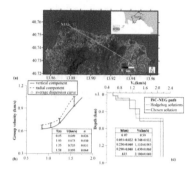

Figure 1. Location on geological simplified map (modified after Corniello et al. 2008) of the NCF paths at Napoli, crossing the quarters of Barra (orange line), Ponticelli (green lines) and Poggioreale (cyan lines).

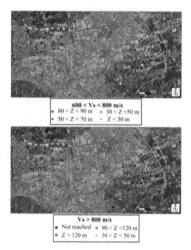

Figure 2. Depth of the seismic horizons with $600 < V_S < 800$ m/s and $V_S > 800$ m/s. The border of the Sebeto-Volla Plain (azure area) is shown together with a proposal of enlargement of it, as inferred from our experiments (dashed cyan line).

(seismic bedrock in Italian building code) together with the borders of the Volla-Sebeto depression (e.g. Corniello et al. 2008) (Fig. 2).

The correlation of our results with geological data is impressive, so that we can suggest an enlargement of the depression in the northern part of the investigated paths. The absence of the seismic horizon with $V_S > 800$ m/s represents another important result of our study that has to be taken into account for seismic response evaluation.

As regards the NCF experiments at Ischia (Strollo et al. 2015 and references therein), which is mainly characterized by tuffs, lavas and landslide deposits, we present the results along a path (ISC-NEG) (interstation distance of 5.8 km) crossing the town of Casamicciola, where seismic activity concentrates (Fig. 3). The obtained V_S model (up to 1.5 km of depth) is characterized by velocities increasing from ~0.6 to 2.2 km/s at ~0.6 km of depth b.s.l. Taking into account

Figure 3. Noise cross-correlation measurements at Ischia: (a) location of the ISC-NEG path; (b) average dispersion curve with error bar; (c) V_S models (referred to the average altitude of the measurement sites) obtained from the non-linear inversion of the average dispersion curve. The chosen solution (red bold line) is also reported with the uncertainties of the inverted parameters in the adjacent table.

the stratigraphies of some deep wells in the southwestern part of the island, the V_S increment to 2.2 km/s can be attributed to the presence of trachytic lavas, representing the top of the geological basement of Ischia.

3 CONCLUSIONS

Our experiments in the complex geological setting of volcanic, highly urbanized areas like Napoli and Ischia island evidence the power of the approach based on the rigorous FTAN and Hedgehog methods applied to ambient noise cross-correlation. Results of similar high quality cannot be obtained with standard professional procedures. The obtained V_S models reach depths that are prohibitive for controlled source experiments and allow the reliable definition of the seismic basement, a key information for trustworthy evaluation of the site seismic response.

REFERENCES

Bensen, G.D., Ritzwoller, M.H., Barmin, M.P., Levshin, A.L., Lin, F., Moschetti, M.P., Shapiro, N.M., Yang, Y., 2007. Processing seismic ambient noise data to obtain reliable broad-band surface wave dispersion measurements. *Geophys. J. Int.* 169: 1239–1260.

Corniello, A., de Riso R., Ducci, D., 2008. Carta idrogeologica della provincia di Napoli (1/250.000). 3° Convegno Nazionale della Protezione e Gestione delle Acque Sotterranee per il III Millennio.

Levshin, A.L., Ratnikova, L.I., Berger, J., 1992. Peculiarities of surface wave propagation across Central Eurasia. *Bull. Seism. Soc. Am.* 82: 2464–2493.

Panza, G.F., 1981. The resolving power of seismic surface waves with respect to crust and upper mantle structural models. In Cassinis, R. (ed), *The solution of the inverse problem in geophysical interpretation. Plenum Publishing Corporation*, pp. 39–77.

Strollo, R., Nunziata, C., Iannotta, A., Iannotta D., 2015. The uppermost crust structure of Ischia (southern Italy) from ambient noise Rayleigh waves. *J. Volcanol. Geotherm. Res.* 297: 39–51.

Volcanic Rocks and Soils – Rotonda et al. (eds)
© *2016 Taylor & Francis Group, London, ISBN 978-1-138-02886-9*

Collapse-upon-wetting behaviour of a volcanic soil

E. Crisci
Dipartimento di Ingegneria Civile, Edile e Ambientale (DICEA), Università degli studi di Napoli Federico II, Italy

A. Ferrari
Laboratory for Soil Mechanics (LMS), Ecole Polytechnique Fédérale de Lausanne (EPFL), Switzerland

G. Urciuoli
Dipartimento di Ingegneria Civile, Edile e Ambientale (DICEA), Università degli studi di Napoli Federico II, Italy

ABSTRACT: Experimental results of volumetric response of a volcanic soil were collected in standard oedo-metric tests, with the aim to evaluate the volumetric response of this soil when subjected to loading and wetting paths. Microstructural features of reconstituted samples used are investigated. Test results are then interpreted within the framework of a constitutive model using effective stress.

1 INTRODUCTION

Volcanic soils often behave as collapsible soils when wetted, because of their high-porosity open-fabric which results from their deposition. These soils cover 0.84% of the world's land surface and are often related to engineering issues e.g. building foundations problems and rainfall-induced landslides (Eichenberger et al., 2013).

A particularly interesting phenomenon happened in 2001 in Via Settembrini, Naples (Southern Italy), when, after a heavy rainfall, some buildings suffered important settlements (until 27 cm). The pyroclastic soil, though having a good resistance to the vertical stress of the buildings, was weakened by the wetting (Gens, 2010), with disastrous consequences.

Experimental results of the volumetric response of a volcanic soil were collected in oedometric tests, with the aim to evaluate the volumetric response of this soil when subjected to loading and wetting paths.

2 TESTED MATERIAL AND EXPERIMENTAL PROGRAMME

The soil used in the experimentation is a Vesuvio pyro-clastic soil, collected in Monteforte Irpino (southern Italy), constituted of ashes and pumices of Ottaviano's eruption. It is a cohesion-less high-porosity open-fabric ($e = 2.57$) soil, *in situ* in unsaturated condition (saturation around 50%). The grain-size distribution presented a gravel-size fraction of 9.5%, a sand-size fraction of 44.9% and a silt size fraction of 43.4%. The soil was classified as a slightly clayey, gravelly sand and silt.

In order to reproduce the average in situ characteristics of the soil, specimens were prepared by an *ad hoc* compaction technique at a water content of 0.40 and reaching an average initial void ratio of 2.57.

Microstructural features of reconstituted samples were investigated by the means of Scanning Electron Microscope (SEM) images and a Mercury Intrusion Porosimetry (MIP) test. The specimen showed PSD with dominant pore radii in the range 10–20 μm, which corresponds sufficiently well to the inter-grain pores observed in SEM images.

Several oedometric tests on the prepared specimens were carried out. In each test the oedometer cell was initially covered with a plastic cover, to avoid water loss and to keep the initial water content constant. Then specimens were loaded with conventional loads; the partially saturated samples were inundated at a given stress (Tab. 1) in order to measure strains due to wet-ting. Afterwards, the loading carried on normally. The collapse measured in the wetting paths increases for low stress value, and then decreases for high stress, in agreement with literature evidences (e.g. Alonso & Gens 1987; Ferrari et al., 2013; Munoz-Castelblanco, 2011). After the saturation, the soil behaves as a saturated soil.

Table 1. Characteristics of reconstituted specimen.

n°	e_o	Wet at σ_v:
	–	kPa
1	2.590	1.1
2	2.597	12
3	2.590	20
4	2.560	40
5	2.593	80
6	2.616	500

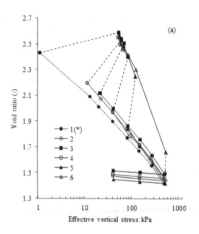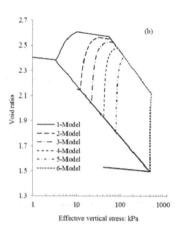

Figure 1. (a) Odometric tests on prepared specimens; (b) AMEG-s model application compared to experimental results; (*) Numbers referrer to experimental test specimens presented in table 1.

3 EXPERIMENTAL RESULTS INTERPRETATION

Soil behaviour was interpreted by the means of the advanced constitutive model for environmental geomechanics, ACMEG-s (Nuth & Laloui, 2007; Nuth & Laloui, 2008). ACMEG-s is a Cam-Clay-type elasto-plastic model using Bishop generalized effective stress (Bishop, 1961) σ'_{ij}:

$$\sigma'_{ij} = (\sigma_{ij} - u_a \delta_{ij}) + S_r(u_a - u_w)\delta_{ij} \qquad (1)$$

where σ_{ij} is the total stress, u_a is the air pressure (or the general gas that could be present in the soil), u_w is the water pressure, S_r is the degree of saturation, and δ_{ij} is the Kronecker's delta. The difference $(u_a - u_w)$ represents the matric suction (s). The product between the degree of saturation and the matric suction is computed by means of the retention curve, interpreted as Equation 2 (Van Genuchten, 1981):

$$S_r = S_{r,res} + \left(\frac{1 - S_{r,res}}{1 + (\alpha \cdot s)^n}\right)^m \qquad (2)$$

where n, m and α are material parameters and $S_{r,res}$ is the residual degree of saturation.

By the use of the parameters calibrated on the experimental results, an application of ACMEG-s in oedometric condition with constant water content is produced in order to test its capacity to correctly predict volumetric deformations induced by loading – wetting paths. Few wetting paths are simulated at different value of total vertical stress, kept constant during suction reduction (Fig. 1b).

4 CONCLUDING REMARKS

The experimental test results allowed the calibration of an advanced constitutive model based on the effective stress concept extended to partially saturated conditions (ACMEG-s). The modeling of the experimental

results highlights the advantage of using the effective stress concept when modeling the collapse-upon-wetting behavior. From one side the water retention behavior of the volcanic soil is intrinsically linked to the mechanical behavior through the product between the matric suction and the degree of saturation; on the other side, the mechanical behavior in fully saturated conditions is recovered after the wetting is completed.

REFERENCES

Alonso, E. E., Gens, A. & Hight, D. W. 1987. Special problem soils. General report. In: *Proceedings of the 9th European conference on soil mechanics and foundation engineering, Dublin:* 1087–1146.

Bishop, A. W. & Blight, G. E. 1963. Some aspects of effective stress in saturated and partly saturated soils. *Géotechnique*, 13 No. 3, 177–197.

Eichenberger, J., Ferrari, A. & Laloui, L. 2013. Early warning thresholds for partially saturated slopes in volcanic ashes. *Computers and Geotechnics*, 49, 79–89.

Ferrari, A., Eichenberger, J. & Laloui, L. 2013. Hydromechanical behaviour of a volcanic ash. *Géotechnique*, 63, No. 16, 1433–1446.

Gens, A. 2010 Soil–environment interactions in geotechnical engineering. *Géotechnique*, 60.1: 3–74.

Munoz-Castelblanco, J., Delage, P., Pereira, J.M. & Cui Y.J. 2011. Some aspects of the compression and collapse behaviour of an unsaturated natural loess. *Geotechnique Letters*, 1–6.

Nuth, M. & Laloui, L. 2007. New insight into the unified hydromechanical constitutive modelling of unsaturated soils. In: *Proceedings of the 3rd Asian conference on unsaturated soils* (UNSAT-Asia), (eds Z. Z. Yin, Y. P. Yuan and A. C. F. Chiu). Nanjing, China: Science Press, 109–25.

Nuth, M. & Laloui, L. 2008. Effective stress concept in unsaturated soils: clarification and validation of a unified framework. *Int. J. Numer. Anal. Methods Geomech.* 32, No. 7.

van Genuchten, M. T. 1980. A closed-form equation for predicting the hydraulic conductivity of unsaturated soil. *Soil Sci. Soc. Am. J.* 44, No. 5, 892–898.

Volcanic Rocks and Soils – Rotonda et al. (eds)
© *2016 Taylor & Francis Group, London, ISBN 978-1-138-02886-9*

Experimental investigation and constitutive modelling for an unsaturated pyroclastic soil

Sabatino Cuomo, Mariagiovanna Moscariello & Vito Foresta
Laboratory of Geotechnics, Department of Civil Engineering, University of Salerno, Italy

Diego Manzanal & Manuel Pastor
Department of Applied Mathematics and Computer Science, Universidad Politecnica de Madrid, Spain

EXTENDED ABSTRACT

Unsaturated granular soils involved in flow-like land-slides (Cascini et al., 2010, 2013) require specific experimental tests to assess the role of several distinct hydromechanical coupled processes. Among those, one can mention the pore water pressure build-up upon shearing in contractive soils, the so-called volumetric collapse upon wetting and the increase of the shear strain rate once any soil mechanical instability behaviour incepts.

The Soil Water Characteristic Curve (SWCC) regulates the soil water content and soil hydraulic conductivity, and thus the water exchange, for any RVE (Representative Volume Element) in a boundary value problem. On the other hand, the stiffness and volumetric soil behaviour (contractive or dilative) affect the pore water pressure. Finally, the soil shear strength whether high enough inhibits the accumulation of plastic strains, localised or diffuse, thus preventing the build-up of pore water pressure.

It is worth noting that in unsaturated conditions, all the above mentioned soil mechanical properties depends on the matric suction, which is the difference of the air pressure (usually assumed equal to zero for practical applications) to the negative pore water pressure.

In the paper an unsaturated volcanic (air-fall) pyroclastic soil was dealt with (Bilotta et al., 2005; Sorbino and Nicotera, 2013). Oedometric tests with different initial void ratios were considered (Bilotta et al., 2006). As well, compression triaxial tests were carried out in drained (suction controlled) conditions. For all the tests, undisturbed specimens were utilized and the suction spanned within the range from 0 to 60 kPa. The experimental results were firstly discussed, with special emphasis on the observable relationships between the pore water pressure and the soil (volumetric and shear) strains.

The experimental tests were also described, with special reference to the Water Retention Curve (WRC) and the Critical State Line (CSL), used for the calibration of the constitutive model.

Table 1. Oedometeric tests performed on soil specimens (Sarno site) at initial suction ($u_a - u_w$) of 50 kPa, and at a fixed value of net vertical stress ($\sigma_v - u_a$) the suction was decreased until the null value.

Test	v	$(\sigma_v - u_a)_{coll}$ (kPa)*	ε_v^{coll}
ESA12_03	3.698	41	0.044
ESA13_03	3.537	95	0.068
ESA11_04	3.194	120	0.036
ESA11_03	3.194	162	0.056
ESA9_03	2.993	295	0.044
ESA6_04	3.212	400	0.019
ESA10_04	2.902	500	0.026
ESA15_04	2.860	500	0.033

v_o void ratio at the beginning of the wetting; ε_v^{coll} volumetric strain at the end of the wetting phase; *: step for the decrease of the suction.

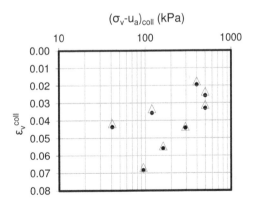

Figure 1. Experimental (dots) and numerical results (triangles) for the wetting-collapse tests of table 1.

Then, some selected experimental tests (Table 1) were reproduced referring to a generalized plasticity model formerly set up by Pastor et al. (1990) and extended to unsaturated conditions by Manzanal (2008) and Manzanal et al. (2011).

Focusing on the so-called "wetting-collapse-deformation", the numerical results match the experimental evidences reasonably well for all tests performed at different net vertical stress (Fig. 1).

The results achieved outlined a satisfactory capability to simulate the mechanical behaviour of the investigated pyroclastic unsaturated soil.

REFERENCES

Bilotta, E., Cascini, L., Foresta, V., and Sorbino, G. 2005. Geotechnical characterization of pyroclastic soils involved in huge flowslides. *Geotechnical and Geological Engineering*, 23: 365–402.

Bilotta, E., Foresta, V. and Migliaro, G. 2006. Suction controlled laboratory tests on undisturbed pyroclastic soil: stiffnesses and volumetric deformations. *Proc. International Conference on Unsaturated Soils*, 2–6 April, Carefree, Arizona USA, 1, 849–860.

Cascini, L., Cuomo, S., Pastor, M. and Sorbino, G. 2010. Modelling of rainfall-induced shallow landslides of the flow-type. *ASCE's Journal of Geotechnical and Geoenvironmental Engineering*, 1, 85–98.

Cascini, L., Cuomo, S., Pastor, M., Sacco, C., 2013. Modelling the post-failure stage of rainfall-induced landslides of the flow-type. *Canadian Geotechnical Journal*, Vol. 50, No. 9, pp. 924–934.

Manzanal, D. 2008. Modelo constitutivo basado en la teoría de la Plasticidad Generalizada con la incorporaciòn de paràmetros de estado para arenas saturadas y no saturadas, PhD thesis, Universidad Politecnica de Madrid (Spain).

Manzanal, D., Pastor, M., Merodo, J.A.F. 2011. Generalized plasticity state parameter-based model for saturated and unsaturated soils. Part II: Unsaturated soil modeling. *International Journal for Numerical and Analytical Methods in Geomechanics*, 35 (18), 1899–1917.

Pastor, M., Zienkiewicz, O.C. and Chan, A.H.C. 1990. Generalized plasticity and the modelling of soil behaviour. *International Journal for Numerical and Analytical Methods in Geomechanics*, 14 (1990), 151–190.

Sorbino, G. and Nicotera, M.V., 2013. Unsaturated soil mechanics in rainfall-induced flow landslides. *Engineering Geology*, Vol. 165, No. 24, pp. 105–132.

Volcanic Rocks and Soils – Rotonda et al. (eds)
© 2016 Taylor & Francis Group, London, ISBN 978-1-138-02886-9

Shear strength of a pyroclastic soil measured in different testing devices

Sabatino Cuomo, Vito Foresta & Mariagiovanna Moscariello
Laboratory of Geotechnics, Department of Civil Engineering, University of Salerno, Italy

EXTENDED ABSTRACT

The pyroclastic soils are produced by volcanic explosive eruptions, wind transportation and air-fall deposition (Cuomo, 2006) and they mantle steep slopes nearby the volcanoes. Thus, the pyroclastic soils may be involved in landslides of the flow-type, which frequently result in casualties and damage to property (Cascini et al., 2010).

This is the case of the pyroclastic soils which mantle steep carbonate reliefs in a large area ($3000\,\mathrm{km}^2$) surrounding the Vesuvius volcano (Southern Italy). These soils have a metastable structure (Bilotta et al., 2005; Lancellotta et al., 2012) and their strain behaviour is typical of transition soils composed of both silts and sands (Migliaro, 2008).

In the paper, the shear strength of a Vesuvian pyroclastic soil, sampled in the Sarno-Quindici area (Cascini et al., 2008), was investigated. Purposely, triaxial tests, direct shear tests and simple shear tests were carried out (Table 1).

The use of different testing devices was motivated by an extensive literature, which outlines some differences among the results achievable from simple

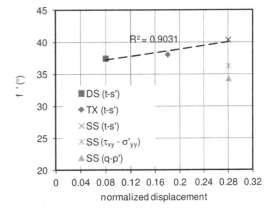

Figure 1. Values of the effective friction angle obtained from different devices (DS, TX, and SS) and through different interpretation of the results (the normalized displacement is assumed equal to δ_h/B at steady-state strength for DS tests, δ_a/H_c at critical state for TX tests, and δ_h/H at the final stage of SS tests).

shear, direct shear and triaxial tests, for instance due to different failure mechanisms (Randolph and Worth, 1981).

The interpretation of the experimental results was pursued using two different procedures: one method neglects the intermediate principal stress (Saada et al., 1983); another method takes into account the complete stress state to relate the results of simple shear and triaxial (axialsymmetric) tests (Wood et al., 1979; Budhu, 1984).

As main insight, some differences were pointed out for the effective friction angle (φ') (Fig. 1), as computed in the deviatoric ($q - p'$) plane for triaxial conditions.

Particularly, the results of the simple shear tests, whether interpreted in the ($t - s'$) plane, well agree the direct shear and triaxial tests, and the differences were related to the normalized displacement of the specimen at failure.

Table 1. Details of the tests carried out with different devices.

Test type	#	v_0	σ'_{yy} (kPa) consolidation	p'_{cons} (kPa)
DS	TAL01	2.564	20.0	–
DS	TAL02	2.611	49.0	–
DS	TAL03	2.586	79.0	–
DS	TAL04	2.545	34.8	–
DS	TAL05	2.552	65.4	–
DTX	BIS24_06	2.901	–	100
DTX	BIS26_06	2.934	–	50
DTX	BIS28_06	2.904	–	30
DTX	BIS29_06	2.916	–	30
UTX	BIS25_06	2.931	–	50
CLSS	SSP0115	3.018	–	100.0
CVSS	SSP0225	3.028	–	69.8*
CLSS	SSP0315	3.025	–	75.0

DS: Direct Shear test; DTX: drained triaxial test; UTX: undrained triaxial test; CLSS: simple shear test carried out at constant load; CVSS: simple shear test carried out at constant volume.

(*) p' varies during the test up to about $50\,\mathrm{kPa}$ at failure.

REFERENCES

Bilotta, E., Cascini, L., Foresta, V., and Sorbino, G. 2005. Geotechnical characterization of pyroclastic soils involved in huge flowslides. *Geotechnical and Geological*

Engineering, 23: 365–402. DOI:10.1007/S10706-004-1607-3.

Budhu M., 1984. On comparing simple shear and triaxial test results. J. Geotech. Engrg. 1984. 110: 1809–1814. DOI: 1984.110:1809-1814.

Cascini L., Cuomo S., Guida D., 2008. Typical source areas of May 1998 flow-like mass movements in the Campania region, Southern Italy. *Engineering Geology* 96, 107–125. DOI:10.1016/j.enggeo.2007.10.003

Cascini L., Cuomo S., Pastor M., Sorbino G. 2010. Modelling of rainfall-induced shallow landslides of the flow-type. *ASCE's Journal of Geotechnical and Geoenvironmental Engineering*, No. 1, pp. 85–98. DOI: 10.1061/ASCEGT. 1943-5606.0000182.

Cuomo S., 2006. Geomechanical modelling of triggering mechanisms for flow-like mass movements in pyroclastic soils. PhD Thesis,University of Salerno, pp. 274.

Lancellotta, R., Di Prisco, C., Costanzo, D., Foti, S., Sorbino, G., Buscarnera, G., Cosentini, R.M. and Foresta V., 2012. Caratterizzazione e modellazione geotecnica. In: Criteri di zonazione della suscettibilità e della pericolosità da frane innescate da eventi estremi (piogge e sisma)/ Leonardo Cascini. Composervice srl, Padova, pp. 266–319. ISBN 9788890687334 (in Italian).

Migliaro G., 2008. Il legame costitutivo dei terreni piroclastici per la modellazione di scavi in ambiente urbanizzato ed influenza della parziale saturazione. PhD thesis at University of Salerno (in Italian).

Randolph M. F., Wroth C. P., 1981. Application of the failure state in undrained simple shear to the shaft capacity of driven piles. Géothecnique Vol. 31, No. 1, pp. 143–157.

Saada A.S., Fries G., Ker C., 1983. An evaluation of Laboratory Testing techniques in Soil Mechanics. *Soils and Foundations, Japan Society of Soil Mechanics and Foundation Engineering*, Vol. 23, No. 2, 1983, pp. 381–395.

Wood D.M., Drescher A., Budhu M., 1979. On the Determination of Stress State in the Simple Shear Apparatus. *Geotechnical Testing Journal*, Vol. 2, No. 4, Dec. 1979 pp. 211–222.

Volcanic Rocks and Soils – Rotonda et al. (eds)
© *2016 Taylor & Francis Group, London, ISBN 978-1-138-02886-9*

Experimental study on the shear moduli of volcanic soil with various fines content on equivalent granular void ratio

T. Hyodo
Tokyo University of Science, Chiba, Japan

EXTENDED ABSTRACT

*a*In the south of Kyushu in Japan, widely distributed deposits of Shirasu resulting from pyroclastic flows are frequently used for soil structures and fills. It contains non plastic fines which was created by its fragment of coarse particles. Shirasu consists of rapidly cooled very rough glassy particles whose morphology is not clear.

In this study, a series of bender element tests on Shirasu was performed with varying its fines content. As a result, it was observed that the shear modulus of Shiarasu was decreased with increasing the fines content although void ratio decreased as fines content increased.

However, it was found that there is a unique relationship between shear moduli of Shirasu and the equivalent granular void ratio, in which the proper contribution factor b for fines was assumed. Based on the findings in the study, an empirical formula to express the shear moduls of Shirasu with various fines content was developed as the function of the fines content, the equivalent granular void ratio and the effective confining stress.

The maximum and minimum void ratios were determined using the JIS A 1224:2009 method without washing out the fines. Shirasu particles have a glassy pumice like structure with intra-particle voids which result in a low value of specific gravity and high maximum and minimum void ratios as compared to Toyoura sand.

The equivalent granular void ratio was introduced bt Thevenayagam (2000) to extend the concept of granular void ratio. In this case the solids include a proportion of the fnes. By using this concept it was possible to construct a unique e vs p' steady state line for various fines contents. If the specific gravity for the fine and coarse materials are assumed to be identical then the fines content by volume is defined as:

$$F_c = \frac{V_{sf}}{V_s}$$

and the effective granular void ratio is:

$$e_{ge} = \frac{e + (1-b)F_c}{1 - (1-b)F_c}$$

where: e is the void ratio and b is the fines contribution factor.

In the case of $b = 1$, $e_{ge} = e$ and in the case of $b = 0$, $e_{ge} = e_g$. b is assumed to vary between 0 and 1 for the fines.

The fines content was varied between 0% and 30% in order to evaluate an appropriate b value using the line for 0% fines as a benchmark.

Bender element tests on triaxial samples have been carried out to determine the influence of fines on the the shear modulus of Shirasu, a volcanic soil. It was concluded that:

1) The shear modulus of Shirasu decreases with increasing fines and decreasing void ratio.
2) The shear modulus was shown to be a function of effective stress raised to the power 0.5.
3) The contribution factor $b = 0.5$ was applicable to Shirasu with a fines content from 0% to 30%.
4) Using this factor a unique relation was found between shear modulus, void ratio and mean effective confining stress.

REFERENCE

Thevanayagam, S. (2000). Liquefaction potential and undrained fragility of silty soils, Paper No. 2383, Proceedings of the 12th World Conference on Earthquake Engineering, New Zealand, January.

Volcanic Rocks and Soils – Rotonda et al. (eds)
© *2016 Taylor & Francis Group, London, ISBN 978-1-138-02886-9*

A laboratory investigation on the cyclic liquefaction resistance of pyroclastic soils

V. Licata, A. d'Onofrio & F. Silvestri
DICEA, Università di Napoli Federico II, Napoli, Italy

L. Olivares
Seconda Università degli studi di Napoli, Napoli, Italy

V. Bandini
Università degli studi di Messina, Messina, Italy

ABSTRACT: The paper analyzes the influence of some peculiar factors on the cyclic liquefaction behavior of a pyroclastic silty sand. The results highlight that liquefaction resistance of the natural soil is significantly enhanced by fabric with respect to that of reconstituted samples tested at the same state conditions; on the other hand, the cyclic strength of the reconstituted soil appears poorly affected by the relative density and confining stress. Furthermore, the non-plastic fine ash was observed to increase the cyclic resistance of the pumice sand with respect to that of the same coarse matrix added with plastic clay. Finally, no appreciable evidences of grain crushing were detected on all tested soils containing pumice sand.

1 INTRODUCTION

The Campania region is widely covered by pyroclastic soils originated by the intense activity of Phlegraean fields and Somma-Vesuvius volcanic districts. Many historical and recent case studies show that such materials significantly affect the stability of the territory versus both seismic and hydrologic extreme events. In particular, loose saturated shallow covers of silty sands ('pozzolana') are potentially prone to liquefaction during seismic events.

Depending on the parent volcanic activity, pyroclastic silty sands are basically characterized by two peculiar lithological fractions: pumice sand with crushable grains and non-plastic fine ash.

The role of grain crushing on cyclic liquefaction has been investigated through field and laboratory tests in few cases, e.g. by Hyodo et al. (1998) and Orense & Pender (2013). The latter authors recently found that crushing of pumice particles results in an increase of non-plastic fine fraction that, in turn, leads to a reduction of the liquefaction potential.

Fine content and plasticity are traditionally recognized as constitutive factors increasing the liquefaction resistance (e.g. Ishihara, 1996). On the contrary, the progressive accumulation of laboratory experience on silty sands (e.g. Chang et al., 1982; Troncoso & Verdugo, 1985; Mominul et al., 2013; Noda & Hyodo, 2013) shows that the dependence of cyclic strength on plastic/non-plastic fine content can be ambiguous. Summarizing, it results from the complex interplay between the contrasting effects of permeability reduction and inter-particle forces, as well as on the granular void ratio of the sandy matrix (Silvestri, 2013).

2 MATERIALS AND EXPERIMENTAL PROCEDURE

In this paper the influence of fabric, grain crushability and non-plastic fine content on the behavior of a typical pyroclastic soil of the Neapolitan area was evaluated by a wide laboratory investigation.

Cyclic undrained triaxial tests were carried out in a hydraulic Bishop & Wesley cell on natural (PAN) and reconstituted loose (PARa) and dense (PARb) soil samples, in order to investigate the role of fabric, relative density and confining stress on the cyclic resistance. Furthermore, the likely effects of particle fragility were analysed by comparing the behaviour of the reconstituted pyroclastic soil to that of a material having the same grain size distribution, in which the pumiceous sand fraction was replaced with a hard-grained silica sand (SAR). Finally, the role of non-plastic fine was investigated comparing the cyclic resistance of the reconstituted pyroclastic material with that of samples prepared with the same grading and relative density, but replacing the fine ash content with a low plasticity clay (PCR).

Two preparation procedures were considered for the reconstituted soils: a moist-tamping (MT) method was adopted for the dense mixture (PARb), while water

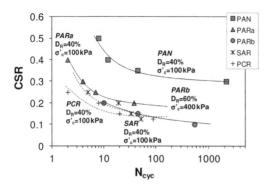

Figure 1. Comparison between the tested mixtures.

pluviation and freezing (WPF) was necessary for the loose PARa, SAR and PCR materials. The undistributed specimens were saturated by CO_2 flushing.

3 RESULTS

The results were analyzed in terms of stress-strain loops, effective stress-paths and evolution with the number of cycles of axial strain and pore pressure ratio (Licata, 2015). Cyclic mobility mechanisms were exhibited by the natural (PAN) and the dense reconstituted (PARb) soil, while flow liquefaction characterized by runaway failure was triggered in the reconstituted specimens of silica sand with ash (SAR) and pumice sand with plastic clay (PCR).

The experimental results are synthesized in Figure 1 in terms of cyclic resistance curves, showing the relationship between the number of cycles at liquefaction (Ncyc) and the cyclic stress ratio (CSR), pertaining to natural and reconstituted materials.

The results primarily showed that fabric has a key role on the cyclic resistance of pyroclastic soil, since, at the same number of cycles, the cyclic stress amplitude needed to trigger liquefaction on the natural soil is almost twice that relevant to the reconstituted material, tested at the same stress state and relative density.

Although relative density has some noticeable effect on the cyclic resistance of pumice+ash mixtures, it was not as significant when compared to that traditionally observed for hard-grained sands. As the confining pressure was increased, the liquefaction resistance curve of reconstituted pumice+ash specimens was shifted downward, consistently with the observations made on hard-grained sands.

The cyclic resistance was poorly influenced by the volcanic nature of the sandy particles. Negligible pumice grain crushing occurred during the tests, as highlighted by the comparison among the 'virgin' and 'after dead' grain size distributions, measured by a dynamic image analysis technique. This may be justified by the presence of a fine content as high as 30%, which might have increased the strength of greater particles against breakage.

The effect of non-plastic fine ash on liquefaction strength is still unclear. The experimental results showed a reduction of the cyclic resistance as the plasticity of the fine fraction increases. This behavior might be simply justified by the reduction of static frictional strength with plasticity.

REFERENCES

Chang N.Y., Yeh S.I., Kaufman L.P. (1982). Liquefaction potential of clean and silty sands, *Proc. Third International Earthquake Microzonation Conference*, Seattle, Washington, 2:1017–1032.

Hyodo M., Hyde A. F. L., Aramaki N. (1998). Liquefaction of crushable soils (1998). *Geotechnique*, 48(4):527–543.

Ishihara K. (1996). *Soil behavior in earthquake geotechnics*. Clarendon press, Oxford, 1–350.

Licata V. (2015). A laboratory and field study on cyclic liquefaction of a pyroclastic soil. *PhD Thesis in Geotechncial Engineering, University of Napoli Federico II*.

Mominul H.M., Alam M.J., Ansary M.A., Karim M.E. (2013). Dynamic properties and liquefaction potential of a sandy soil containing silt. *Proc. 18th International Conference on Soil Mechanics and Geotechnical Engineering*, Paris, 2:1539–1542.

Noda S., Hyodo M. (2013). Effects of fines content on cyclic shear characteristics of sand-clay mixtures. *Proc. 18th International Conference on Soil Mechanics and Geotechnical Engineering*, Paris, 2:1551–1554.

Orense R. P., Pender M. J. (2013). Liquefaction characteristics of crushable pumice sand. *Proc. 18th International Conference on Soil Mechanics and Geotechnical Engineering*, Paris, 2:1559–1562.

Silvestri F. (2013). Experimental characterization and analysis of soil behaviour under earthquake loads. 2nd General Report for TC203. *Proc. 18th International Conference on Soil Mechanics and Geotechnical Engineering, Paris*, 2:1399–1406.

Troncoso J.H., Verdugo R. (1985). Silt content and dynamic behavior of tailings sands, *Proc. 11th International Conference on Soil Mechanics and Foundation Engineering*, San Francisco, 3:131–134.

Volcanic Rocks and Soils – Rotonda et al. (eds)
© *2016 Taylor & Francis Group, London, ISBN 978-1-138-02886-9*

Experimental evaluation of liquefaction resistance for volcanic coarse-grained soil in cold region using temperature- and/or moisture-controlled triaxial apparatus

S. Matsumura
Foundations Group, Port and Airport Research Institute, Japan

S. Miura
Faculty of Engineering, Hokkaido University, Japan

ABSTRACT: The aim of this paper is to experimentally evaluate the effect of the freeze-thaw action and moisture condition on the cyclic mechanical property of a volcanic soil deposited in Hokkaido. To achieve this, using the cyclic triaxial test apparatus possible to control temperature and/or moisture conditions of specimens prior to cyclic shearing, the liquefaction resistance of the volcanic soil was verified under variable conditions of compaction, freeze-thaw and moisture content at shearing.

1 INTRODUCTION

The aim of this paper is to experimentally evaluate the effect of the freeze-thaw action and moisture condition on the cyclic mechanical property of a volcanic soil deposited in Hokkaido. To achieve this, using the cyclic triaxial test apparatus possible to control temperature and/or moisture conditions of specimens prior to cyclic shearing, the liquefaction resistance of the volcanic soil was verified under variable conditions of compaction, freeze-thaw and moisture content at shearing. The testing results show the freeze-thaw action has the great impact on the cyclic strength, but such tendency depends on the compaction degree at the freeze-thaw. In addition, it concludes that the cyclic strength can be more susceptible to the saturation degree than the freeze-thaw action.

2 SOIL MATERIAL, APPARATUS AND PROCEDURE

The soil material used was a coarse-grained volcanic soil, which is deposited in the Komaoka district of Sapporo City in Hokkaido (referred to as 'K soil' in this paper), and is classified into a pumice flow deposit provided by Shikotsu caldera. Table 1 shows the physical properties and compaction test result for the K soil.

The main features of the triaxial apparatus developed for this study are as follows (Matsumura et al., 2015): 1) to simulate a soil subjected to a freeze-thaw sequence by controlling the temperature, 2) to control the moisture content arbitrarily prior to loading

Table 1. Physical properties of K soil.

	Komaoka volcanic soil	
Density of soil particle, ρ_s (kg/m^3)		2512
Particle size distribution	D_{max} (mm)	9.5
	D_{50} (mm)	0.20
	U_c	48
	F_c (%)	31.0
Compaction	w_{opt} (%)	40.5
	ρ_{dmax} (kg/m^3)	1059

and 3) to conduct liquefaction tests by applying stress-controlled cyclic loads. Each process can be simulated in the apparatus, throughout.

Figure 1 indicates the flow for each process of the testing. The specimen conditions are divided into four categories depending on whether or not the freeze-thaw and moisture control were applied prior to the cyclic loading. After installing the specimens prepared under the desired compaction condition, whole process of the above, i.e. freeze-thaw, moisture-control and cyclic loading, was applied in the triaxial cell.

3 TEST RESULTS

Figure 2 describes the relations between the degree of compaction after consolidation D_{cc} and the cyclic stress ratio at $N_c = 20$, SR_{20}, to reach the double amplitude of axial strain, DA, of 1% and 5%. Some clear differences in the cyclic strength behavior caused by freeze-thawing are indicated: i.e. 1) freeze-thawing

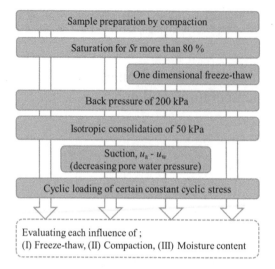

Figure 1. Test procedures of temperature- and/or moisture-controlled triaxial tests.

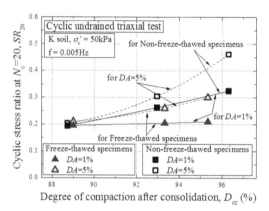

Figure 2. Relations between D_{cc} and SR_{20} at $DA = 1$, 5% for the freeze-thawed and the non-freeze-thawed specimens.

disturbs the increase in cyclic strength accompanied with the increase in degree of compaction, 2) the SR_{20} value for $DA = 1\%$ can hardly increase under the freeze-thawed condition, while compacted more densely, unlike the specimens not freeze-thawed. According to this study, it is emphasized that the cyclic strength characteristics of K soil can remarkably deteriorate when exposed to the freeze-thaw sequence. However, such effects of freeze-thawing ought to be incorporated into the relation with the initial degree of compaction prior to freezing.

Figure 3 shows the number of cyclic loads to cause $DA = 5\%$, shown as N_c, against saturation degree at cyclic shear, S_r (%). In this figure, no plot at the saturation degrees lower than 75% are shown, because the DA values for the specimens at the corresponding S_r conditions did not reached 5%. From the figure, the N_c value exponentially increases with the saturation

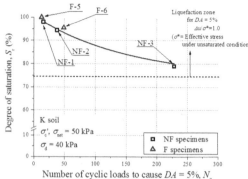

Figure 3. Relations of N_c to cause DA of 5% vs. saturation degree at cyclic shear with freeze-thawed specimens and non-freeze-thawed ones.

degree lower as indicated by the solid line, regardless of whether or not the specimens were freeze-thawed. Therefore, it is clear that the de-saturation leads to the increase of cyclic strength of K soil. Furthermore, in the figure, the dotted line shows the asymptotic line to the solid line. On the basis of the above, the saturation degree of the specimens leading to the liquefaction (that is, the axial strain of 5%) is higher than 75%. That implies the liquefiable potential can be found by the saturation degree boundary. On the other hand, comparing the freeze-thawed specimens with the non-freeze-thawed ones in the similar saturation degree, the difference of the cyclic strength seems to be not significant unlike the influence of saturation degree. That may indicate the cyclic strength can be more susceptible to the de-saturation than the freeze-thaw sequence, and the saturation degree at cyclic shear can be highly significant on evaluating the liquefaction resistance.

4 CONCLUSIONS

The paper mainly concludes as follows:

- Freeze-thaw action leads to significant deterioration of liquefaction resistance of the volcanic soil. However, the influence apparently seems to disappear in the case of loosely-compacted specimens.
- Desaturation of compacted volcanic soils prior to cyclic loading causes the liquefaction resistance to increase to a significant extent. The experimental results indicate a threshold of degree of saturation to be liquefiable in terms of axial strain.

REFERENCE

Matsumura, S., Miura, S., Yokohama, S. & Kawamura, S. 2015. Cyclic deformation-strength evaluation of compacted volcanic soil subjected to freeze-thaw sequence, *Soils and Foundations*, Vol. 55, No. 1: 86–98.

Volcanic Rocks and Soils – Rotonda et al. (eds)
© *2016 Taylor & Francis Group, London, ISBN 978-1-138-02886-9*

One-dimensional consolidation of unsaturated pyroclastic soils: Theoretical analysis and experimental results

F. Parisi
Civil Engineer, Italy

V. Foresta & S. Ferlisi
Department of Civil Engineering, University of Salerno, Italy

EXTENDED ABSTRACT

The prediction of settlement rates that simple geotechnical systems (e.g., shallow foundations) may undergo under live loads represents a fundamental step in design processes. To this aim, the theory of uncoupled consolidation earliest provided by Terzaghi (1923) and Rendulic (1936) for saturated soils with incompressible constituents still represents a valuable reference framework for civil engineers. However, natural soils are often in unsaturated conditions and solutions deriving from the above mentioned theory could not be adequately representative of the real trend of settlements with time and, in general, could lead to an excessive overestimation of maximum values of expected settlements.

In this regard, several contributions are provided by the scientific literature on the consolidation of unsaturated porous media under different geometrical and load conditions (e.g., Fredlund & Hasan 1979, Lloret & Alonso 1980, Alonso et al. 1988, Khalili & Khabbaz 1995, Conte 2004, Qin et al. 2008, Qin et al. 2010, Shan et al. 2012). The novelty of this paper is to highlight the role played by some relevant compressibility parameters (Biot & Willis 1957, Brown & Korringa 1975, Fredlund & Rahardjo 1993, Ferlisi et al. 1999) in the analysis of the one-dimensional consolidation specifically involving pyroclastic soils.

To this aim, the set of differential equations (including those describing the flow of the fluid phase, the flow of the air phase and the stress-strain relationship) which – in a coupled formulation – govern the phenomenon was first derived following a methodological approach provided by Khalili & Khabbaz (1995). In this regard, initial and boundary conditions were posed equal to those existing in the suction controlled oedometer (Aversa & Nicotera 2002) adopted to carrying out laboratory tests on a remoulded non-plastic ashy soil composed of 60% to 50% silt and 40% to 50% sand (Bilotta et al. 2008). In the undisturbed state, the soil is characterized by a high value of the initial void ratio (e_i) and a metastable behaviour. On the contrary, in the remoulded state, the e_i values are lower than in the undisturbed state; in such a case,

the remoulding technique adopted for the specimen's preparation was the one given by Bilotta et al. (2008).

The suction controlled oedometer tests consisted on applying a sequence of vertical net stress values (from 5 to 4200 kPa) to each specimen initially equalized to an imposed matric suction value ranging from 0 to 200 kPa. Test results were used for calibrating the compressibility parameters, defining the hydraulic conductivity function of tested soil specimens and validating the theoretical model.

As far as the volumetric compressibility is concerned, its values were estimated on the basis of the compressibility curves which link the applied vertical net stresses with the corresponding volumetric strains exhibited – for given values of the matric suction – by the ashy soil specimens tested in the suction controlled oedometer. On the other hand, data collected during the initial stages of the performed tests (dealing with the initial suction of soil specimens and the equalization to the target values of the matric suction) were used to derive both the variation of the saturation degree and the volumetric compressibility of solid skeleton under changes in matric suction. Furthermore, considering that the average void ratio of the tested soil is 1.975, data from saturation degree-matric suction relation were first rearranged in order to obtain the volumetric water content as a function of the matric suction. Then, assuming a mean value of the saturated hydraulic conductivity equal to $3.0 \cdot 10^{-9}$ m/s (this value was derived by analyzing the consolidation curves of the test carried out at an imposed matric suction equal to zero) and adopting the procedure proposed by Fredlund & Xing (1994), the unsaturated conductivity function of tested soil specimens was finally obtained.

Once the soil mechanical parameters were calibrated and the hydraulic conductivity function was defined, the set of differential equations which govern the one-dimensional consolidation of unsaturated soils was numerically solved via the implementation of the Finite Difference Method in the commercial software MatLab2013 (www.mathworks.com).

In order to check the capability of the adopted theoretical model to predict the behaviour of the tested ashy soil specimens during the entire stress-path, one

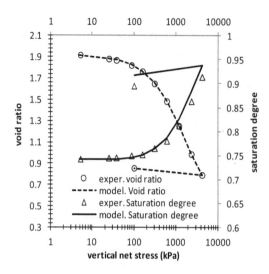

Figure 1. Experimental and modelled variation of void ratio and saturation degree with vertical net stresses applied to tested ashy soil specimens.

of the performed test was taken into account for the numerical simulation. The soil specimen was initially equalized to an imposed suction value of 200 kPa and later subjected to a sequence of load increments up to 4200 kPa of vertical net stress. As final stage of the test, an unloading was performed at the target vertical net stress value of 100 kPa. The obtained results show a good agreement between the evolution with time of the experimental volumetric strains and the modelled ones for both loading and unloading stages. Furthermore, changes in void ratio and saturation degree values as a consequence of applied vertical net stress were compared (Fig. 1). Outcomes highlights that *i*) the predicted compressibility curve perfectly matches the experimental results while *ii)* the predicted values of the saturation degree overestimate the experimental ones, mainly in correspondence of the highest values of applied vertical net stresses. The latter result can be explained considering that the predicted volume of water out-flowing is lower than the measured one and the computational time is not long enough to allow a complete dissipation of the overpressures of water phase. Of course, better results could be obtained by rearranging the values of the hydraulic conductivity function taking into account its dependence on vertical net stress values.

Bearing in mind that consolidation settlements of unsaturated soils are lower than those experienced by the same soils in saturated conditions (at a parity of all factors involved in the considered problem, including changes in boundary conditions), the proposed procedure – which strictly combines theoretical and experimental results – can turn out to be useful in the engineering design processes where the quantitative estimation of the expected consolidation settlements is a requirement of particular concern.

REFERENCES

Alonso, E.E., Battle, F., Gens, A. & Lloret, A. 1988. Consolidation analysis of partially saturated soils – Application to earth dam construction. In Swoboda G. (ed.) *Proc. of the 6th International Conference on Numerical Methods in Geomechanics*: 1303–1308. Rotterdam: Balkema.

Aversa, S., & Nicotera, M.V. 2002. A Triaxial and Oedometer Apparatus for Testing Unsaturated Soils. *Geotechnical Testing Journal* 25(1): 3–15.

Bilotta, E., Foresta, V. & Migliaro, G. 2008. The influence of suction on stiffness, viscosity and collapse of some volcanic ashy soils. *EUNSAT2008, Proc. of 1st Eur. Conf. on Unsaturated Soils, Durham, UK, 2–4 July 2008*, pp. 349–354.

Biot, M.A. & Willis, D.G. 1957. The elastic constants of the theory of consolidation. *ASME, Journal of Applied Mechanics* 24: 594–601.

Brown, R.J.S. & Korringa, J. 1975. On the dependence of the elastic properties of a porous rock on the compressibility of pore fluid. *Geophysics* 40: 608–616.

Conte, E. 2004. Consolidation analysis for unsaturated soils. *Canadian Geotechnical Journal* 41: 599–612.

Ferlisi, S., Federico, F. & Musso, A. 1999. *Pressioni efficaci e coefficienti di compressibilità di mezzi elastici a singola porosità, con costituenti comprimibili*. Research report, Department of Civil Engineering, University of Rome "Tor Vergata" (in Italian).

Fredlund, D.G., & Xing, A. 1994. Equations for the soil-water characteristic curve. *Canadian Geotechnical Journal* 31: 521–532.

Fredlund, D.G. & Hasan, J.U. 1979. One-dimensional consolidation theory: unsaturated soils. *Canadian Geotechnical Journal* 16(3): 521–531.

Fredlund, D.G. & Rahardjo, H. 1993. *Soil mechanics for unsaturated soils*. New York: John Wiley & Sons Inc.

Khalili, N. & Khabbaz, M.H. 1995. On the theory of three-dimensional consolidation in unsaturated soils. In E.E. Alonso & P. Delage (eds.). *Unsaturated soils, Proc. of the 1st Intern. Conf. on Unsaturated Soils, Paris, 6–8 September 1995*: Vol. 2, pp. 745–750. Rotterdam: Balkema.

Lloret, A. & Alonso, E.E. 1980. Consolidation of unsaturated soils including swelling and collapse behavior. *Géotechnique* 30(4): 449–477.

Qin, A.F., Chen, G.J., Tan, Y.W. & Sun, D.A. 2008. Analytical solution to one-dimensional consolidation in unsaturated soils. *Applied Mathematics and Mechanics* 29: 1329–1340.

Qin, A.F., Sun, D.A. & Tan, Y.W. 2010. Analytical solution to one-dimensional consolidation in unsaturated soils under loading varying exponentially with time. *Computers and Geotechnics* 37: 233–238.

Rendulic, L. 1936. Porenziffer und Porenwasserdruck in Tonen. *Der Bauingenieur*, 17, 51/52, 559–564.

Shan, Z., Ling, D. & Ding, H. 2012. Exact solutions for one-dimensional consolidation of single-layer unsaturated soil. *International Journal for Numerical and Analytical Methods in Geomechanics* 36: 708–722.

Terzaghi, K. 1923. Die Berechnung der Durchlassigkeitsziffer des Tones aus dem Verlauf der hydrodynamischen Spannungserscheinungen. *Akademie der Wissenschaften in Wien. Sitzungsberichte. Mathematisch-naturwissenschaftliche Klasse*, IIa, 132, 3/4, 125–138.

Volcanic Rocks and Soils – Rotonda et al. (eds)
© 2016 Taylor & Francis Group, London, ISBN 978-1-138-02886-9

A new rheometer for mud and debris flow

A.M. Pellegrino
Department of Engineering, University of Ferrara, Ferrara, Italy

A. Scotto di Santolo
Pegaso University, Napoli, Italy

A. Evangelista
University of Napoli "Federico II", Napoli, Italy

ABSTRACT: The paper presents an innovative laboratory technique aimed at determining the viscous proprieties of soil-water mixture involved in fast landslides. The knowledge of their rheological properties are necessary for assessing the spreading areas that may danger and damage persons and properties (hazard mitigation). The new equipment is a drag ball rheometer, called Sphere Drag Rheometer (SDR). Compared to standard rheometer this system uses much higher volume of mixtures of wider grain size distribution. The calibration procedure are illustrated with reference to Newtonian and Non-Newtonian standard fluids. The flow curve of some pyroclastic soil-water mixtures are presented. The experimental results on fine mixture are compared with those obtained by conventional rheometrical tools. The results show that the SDR rheometer is able to evaluate the rheological properties of soil-water mixture similar to those involved into pyroclastic debris flow.

1 THE SPHERE DRAG RHEOMETER (SDR) AND MAIN RESULTS

The SDR system, illustrated in Figures 1, is composed of a cylindrical container (i.e. having radius d_c equal to 130 mm, height h_c equal to 60 mm and sample volume equal to 0.5 l) and an eccentric sphere (i.e. with a variable eccentricity r, relative to the motor, varying from 20 to 46 mm) fixed to a thin vertical shaft. The experiment consists in measuring the drag force F_D at a specified rotational speed Ω, while the sphere makes one full rotation within the material sample. For each imposed velocity the value of the required drag force F_D (value at which the sphere starts moving through the material sample) is evaluated as the average between the measured points in a full rotation, of the sphere. The apparent flow curve was obtained by applying an increasing and decreasing rotational speed ramp.

The SDR experimental results reported are related to pyroclastic soils which are representative of real debris flows occurred in Southern Italy. The soil mixture analyzed were before tested with standard rheometer and with fall-cone penetrometer according to the experimental program illustrated in Table 1. The soil A and B derives from the most recent deposits produced by the volcanic activity of mount Somma/Vesuvius. Soil S derives from the volcanic deposits related to the Strombolian activity of the summit craters that covered the Sciara del Fuoco depression. All the experiments involved mixtures of dry soils with appropriate amount of distilled water

Figure 1. Picture of the SDR rheometer.

in order to obtained mixtures having different solid volumetric concentrations Φ. Fine-grained mixtures (i.e., mixtures composed by soil fraction with a particle diameter less than 0.5 mm) were tested in order to compare the data to those obtained using standard rheometer (Scotto di Santolo et al., 2012).

Figure 2 reports the experimental results of some sweep tests in standard rheometer for these materials varying the solid volumetric concentration Φ. The rheological behaviour of the water-soil mixtures tested are typical of a Non-Newtonian fluid for which no flow is possible under a critical value of the stress. At low value of shear rate $\dot{\gamma}$, the shear stress τ increases with the increasing of shear rate: this corresponds to the response of the material in the solid regime.

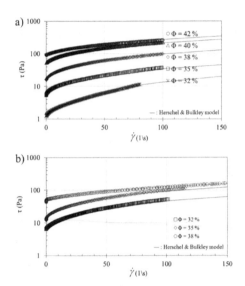

Figure 2. Results with standard rheometer sweep test at different solid volumetric concentration Φ: a) A mixtures; b) B mixtures.

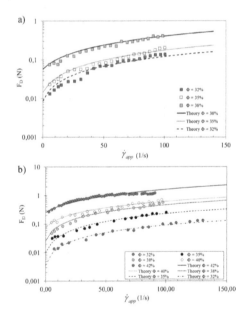

Figure 3. Comparison between SDR measured data (symbols) and theoretical curve (lines): a) A mixtures; b) B mixtures.

The transition to the liquid regime is associated with the rapid increase of the shear rate above some critical value of the stress. At larger stresses, the curve slope increases. The curve obtained for the stress increasing is associated with the value of stress at which the material ultimately flows in a liquid regime. It is especially interesting to look at the variations of the yield stresses with the solid fraction: the yield stress increases in a similar way with the solid fraction concentration, considering the fine-grained mixtures analyzed. Their

Table 1. Experimental program (i.e., SDR = Sphere Drag Rheometer; SR = Standard Rheometer; FCT = Fall-Cone Test).

Test #	Soil (–)	Φ (%)	SDR (–)	SR (–)	CFT (–)
1	A	32	x	x	–
2	A	35	x	x	–
3	A	38	x	x	–
4	A	40	x	x	x
5	A	42	x	x	–
6	B	32	x	x	x
7	B	35	x	x	–
8	B	38	x	x	–
9	S	55	x	–	x

(x = tested; – = not tested).

rheological behaviour were well described with the Herschel & Bulkley model reported in the same figure with continuous lines. Details was reported in Scotto di Santolo et al. (2012).

The A and B water-soil mixtures were then tested with SDR rheometer. The results are reported in Figures 3 at different solid volumetric concentrations Φ. in terms of the drag force F_D as a function of the apparent shear rate $\dot{\gamma}_{app}$. The water-soil mixtures behave like Non-Newtonian fluid and flow resistance increases with the increasing of shear rate regardless the value of solid concentration. The solid fraction clearly influence the rheological behaviour of the materials. The resistance force is strictly influenced by the solid volumetric concentration. The intrinsic strength of the mixtures increases as the particle fraction increases and there is a critical value of drag force below which the flow is possible.

Unlike traditional rheometers, the measurement system adopted for the SDR rheometer is not based on the classic shear flow between two parallel surfaces but on the flow regime around a sphere. For this reason it was very important to define of an appropriate theory of conversion that allow to relate the two schemes of flow. The analytical solution of Ansley & Smith (1967) was utilized.

There is a good match between the measured data and the theoretical ones and the obtained trend is in agreement with that obtained with standard rheometer tests on the same mixtures (Figure 2).

The obtained experimental results show that the SDR rheometer is able to evaluate the rheological properties of soil-water mixture similar to those involved into pyroclastic debris flow.

REFERENCES

Ansley, R.W. & Smith, T.N. 1967. Motion of spherical particles in a Bingham plastic. *AIChE Journal*, 13: 1193–1196.
Scotto di Santolo, A., Pellegrino, A.M., Evangelista, A., Coussot, P. 2012. Rheological behaviour of reconstituted pyroclastic debris flow. *Géotechnique Journal*, 62 (1): 19–27.

Volcanic Rocks and Soils – Rotonda et al. (eds)
© *2016 Taylor & Francis Group, London, ISBN 978-1-138-02886-9*

Hydraulic characterization of an unsaturated pyroclastic slope by in situ measurements

M. Pirone, R. Papa, M.V. Nicotera & G. Urciuoli
University of Naples 'Federico II', Naples, Italy

ABSTRACT: Full understanding of slope failure conditions in an unsaturated pyroclastic slope needs sound analysis of groundwater flow and hence proper hydraulic soil characterization. In order to explore the hydraulic hysteretic behaviour of pyroclastic soils and their ability to interpret site behaviour, in this paper pairs of water content and matric suction measurements recorded at the same depth in surficial layers of an instrumented unsaturated pyroclastic slope are compared to a number of retention curves obtained in the laboratory. A theoretical interpretation of the observed behaviour in situ and in the laboratory is proposed.

1 INTRODUCTION

Full understanding and analysis of triggering mechanisms of debris flows and mudflows in unsaturated slopes requires proper hydraulic soil characterization. Paths followed in situ, in terms of matrix suction and water content, show a hysteretic hydraulic soil behaviour which influence soil hydraulic conductivity and hence pore water pressures in the subsoil (Pirone et al., 2014).

Therefore, choosing to adopt the *main wetting curve* or a *scanning curve* to model the soil domain during rainfall infiltration is crucial to carry out a numerical analysis of groundwater flow in an unsaturated slope and to obtain reliable results. In this paper suction and soil water content measurements collected over three years at the pilot site of Monteforte Irpino (Southern Italy) are shown to derive hydraulic paths actually followed in situ throughout the pyroclastic cover. Moreover, the main drying path and wetting-drying cycles were derived in the laboratory on undisturbed samples taken from the pyroclastic cover of the pilot site. The results are compared with paths detected in situ and differences are discussed. In 2005 a pilot site was set up in the municipality of Monteforte Irpino in Southern Italy. The monitoring system installed on site provides suction and water content measurements in the pyroclastic cover and meteorological data (Pirone et al., 2015). Starting from the ground surface, eight layers are recognized, ashy soils (soil 1, 2, 4, 6, 8) spaced by pumice layers (soil 3, 5, 7). The main drying paths were already available from previous laboratory experiments using evaporation tests performed in a ku-pF apparatus on undisturbed samples collected at different depths (Nicotera et al., 2010). The main drying curves for soils 1–2 are plotted in Figure 1.

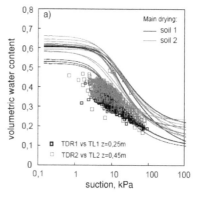

Figure 1. Soil retention curve in main drying, carried out in laboratory tests; suction-water content measured in situ in soils 1–2.

2 IN SITU MEASUREMENTS: SUCTION AND WATER CONTENT

On the basis of readings carried out regularly three times a month, suction measurements are available from 2006, water contents from 2008 (Pirone et al., 2015). In Figure 1 suction and water content collected at the same depth are overlapped to the main drying curves obtained in the laboratory for the same soils. The data collected in situ always lie below the main drying curves; the maximum volumetric water content is 0.46 for suction of 2 kPa in soils 1–2 (Fig. 1). The paired measurements of matric suction and water content collected along one vertical profile in soil 1 over the years from 2009 to 2011 are reported in Figure 2a (the numbered points shown represent a sampling date and indicate the path direction). In situ data identify a narrow and open hysteresis loop in all the soils.

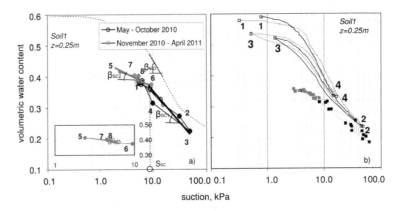

Figure 2. Suction and water content measured in situ over the years 2010–2011 at a selected vertical in soil 1 (a); suction and water content measured in situ and drying and wetting cycles carried out in the laboratory on undisturbed samples from soil 1 (b).

Regardless of the year, the path recorded in situ over drier periods, from May to October/November, detects a series of scanning curves without exhibiting a significant hysteresis. These lines seem to be almost straight in the semi-logarithmic water retention plane. All the measurements are interpolated by a logarithmic function with a coefficient of determination higher than 0.92. The mean slopes, β_{SC1}, of the logarithmic function interpolating the data over the drier periods are less than those calculated for the main dry retention curves, β_{MD}, in the range of suction from 5 to 80 kPa (Fig. 2a). During the wet season, from November to April, measurements in all the soils describe almost straight lines with no significant hysteresis; the suction value is always smaller than 5–6 kPa in soils 1–2, the straight lines interpolating data in the semi-logarithmic water retention plane has significantly smaller slope, β_{SC2}, than that detected in the dry period, β_{SC1}. In conclusion, it seems that the paths traced by the measurements available in the semi-log water retention plane can be simply described by two straight lines. The mean values of s_{sc} separating the ranges of validity of the two different lines are in all soil layers smaller than the air entry suction value.

3 COMPARISON BETWEEN LABORATORY AND FIELD WATER RETENTION CURVES

In Figure 2b the results of laboratory tests carried out on undisturbed samples from soil 1 are reported. Tests consisted of the following phases: (i) main drying, path 1–2, obtained by imposing forced evaporation on an initially saturated soil core in ku-pF apparatus up to a suction value lower than 70 kPa; (ii) a wetting-drying cycle, path 2–3–4 obtained by progressively wetting the same soil core and then drying it again. These results show that: (i) the hysteresis of the second cycle is lower than that of the first; (ii) the suction corresponding to the knee on the wetting path is lower than that on the main drying curve (entry air suction). The paths observed

in situ are in some sense qualitatively similar to the results of the laboratory tests but the range of water contents measured in situ (0.25–0.46 in soils 1–2) is rather small with respect to the range attained in the laboratory (0.20–0.60) in the same suction interval. The considerable difference between the water content measured close to saturation (i.e. 3 kPa) along the laboratory wetting branch (i.e. $\theta = 0.50$ see Fig. 2b) and on the path detected from field measurements (i.e. $\theta = 0.38$ see Fig. 2b) is due to a larger amount of entrapped air in the second case (Pirone et al., 2014). Hence the actual values of in situ water content and, in turn, of unsaturated permeability, cannot be merely estimated on the basis of laboratory-determined water retention curves alone.

4 CONCLUSION

Retention curves of pyroclastic soils are known to be hysteretic, with the main drying branch clearly distinct from the main wetting one. The pairs of matric suction and water content measured on site at the same point, depicting the drying/wetting cycle occurring in situ, display a unique, narrow and closed hysteresis loop. All the measurements are well interpolated by a logarithmic function with two different slopes over dry and wet periods. Differences between the paths from field measurements and the laboratory results were observed close to saturation: in the wetting phase, a higher fraction of air in situ has no possibility of escaping through the soil surface.

REFERENCES

Nicotera, M.V., Papa, R., Urciuoli, G. 2010. An experimental technique for determining the hydraulic properties of unsaturated pyroclastic soils. *Geotechn. Test. J.* 33(4): 263–285.

Pirone, M., Papa, R., Nicotera, M.V., Urciuoli, G. 2015. In situ monitoring of the groundwater field in an unsaturated pyroclastic slope for slope stability evaluation. *Landslides*, 12(2): 259–276.

Volcanic Rocks and Soils – Rotonda et al. (eds)
© *2016 Taylor & Francis Group, London, ISBN 978-1-138-02886-9*

The behaviour of Hong Kong volcanic saprolites in one-dimensional compression

I. Rocchi
DICAM Department, University of Bologna, Italy
City University of Hong Kong, Hong Kong SAR

I.A. Okewale & M.R. Coop
Department of Architecture and Civil Engineering, City University of Hong Kong, Hong Kong SAR

ABSTRACT: An extensive research programme focusing on the one-dimensional compression behaviour of reconstituted and intact volcanic saprolites was carried out. A number of block samples, including saprolites originating from different parent rocks or taken at different sites within a same formation, were tested. The intrinsic behavior of reconstituted samples indicated that some of the soils had very slow convergence to a unique instrinsic normal compression line (ICL) and/or rather shallow slopes of compression paths. The slope of the ICL (λ) changed with the sampling location, even if most samples belonged to the same weathering grade and were taken at shallow depths. The relationship between λ and the amount of fines was established and the effects of structure were studied comparing the compression tests of intact specimens to their respective ICL. The extent to which these tests crossed the ICL was variable and in some cases the curves simply converged to the ICL, showing no effect of structure.

1   MATERIALS AND METHODOLOGY

Many recently developing countries, for example in south east Asia or South America are in tropical or sub-tropical areas, where weathering is an important factor in the genesis of soils. The effects of weathering on the mechanical behaviour have been studied in sedimentary soils (e.g., Cafaro & Cotecchia, 2001). However, a systematic investigation of whether the critical state and sensitivity frameworks (Cotecchia & Chandler, 2000) can be applied to saprolitic and residual soils, originating solely from the chemical decomposition of a rock, is still missing.

Several samples covering saprolites originating from different parent rocks or taken at different sites within a same formation, were studied. The soils tested were taken from a number of block samples from Hong Kong Island. Despite being from several locations, they only covered two geological formations: the Ap Lei Chau Formation (ALCF) and the Mt. Davis Formation (MtDF), which are respectively a fine ash tuff having a vitric matrix and a layered structure and a coarse ash tuff. According to the GEO (1988) classification system, the soils tested were either Completely or Highly Decomposed Volcanic rocks, which correspond to grades IV and V on a scale where I represents a fresh rock and VI a residual soil.

After characterising the soils, the samples were tested in one-dimensional compression, using both reconstituted and intact samples to investigate the effects of structure. The particle size distributions were carried out by a combination of wet sieving and sedimentation, according to the BS1377 (1990), while the oedometer tests were carried out in conventional front loading frames. For the tests on reconstituted samples 50mm diameter fix rings were used, while most of the tests on intact specimens were carried out using 30mm diameter floating rings. It is likely that some of the scatter in the data is due to the heterogeneity that will arise from the presence of relic joints in rock mass and could be reduced by carrying out tests at a very much larger scale.

2   RESULTS

The results were analysed with respect to depth, but distinguishing between the two different weathering degrees. Figures 1a and g show the composition of the soils with depth for the CDV and HDV, respectively. For the CDV from the ALCF, where more depths were investigated, it is possible to see an increase in clay content at shallower depths. The sand content reduces steadily, while the silt content increases slightly with reducing depth. Based on the limited data, the parent rock seems to have considerable influence. However, the HDV is not always coarser than the CDV as would be expected, which cannot be easily explained.

The Atterberg limits are presented in Figures 1b and h together with the saturated water contents in situ.

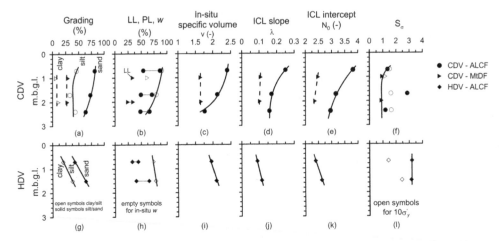

Figure 1. Profiles of the index properties and intrinsic and intact mechanical properties. Composition (a and g), plasticity (b and h), specific volume (c and i), ICL slope (d and j), ICL intercept (e and k), stress sensitivity (f and l).

There is an increase in the LL, PI and the in-situ water content with reducing depth. Again the values for the MtDF are significantly different to those of the ALCF. The HDV shows an increase in plasticity for increasing depth, which is an anomaly. The in-situ specific volume (Figs 1c and i) also reduces with depth in the CDV and it is lower for the MtDF.

For each location, a number of one-dimensional compression tests were carried out on the reconstituted soil, covering the widest possible range of initial specific volume. It was found that some soils had rather shallow slopes of their compression paths and a slow convergence to a unique intrinsic normal compression line (ICL). The slope and the intercept of the ICL changed with the sampling location, even if most samples belonged to the same weathering grade and all were taken at shallow depths. On Figure 1d, e, j and k the CDV of the ALCF show both the λ and N_0 increasing towards the surface where there is more weathering and a greater fines content, as expected. But for the CDV of the MtDF and the HDV the trends are less clear with too few data and too great a scatter, probably due to the heterogeneity of the weathering.

Intact specimens were tested to investigate any effect of structure. To provide a quantitative assessment, the stress sensitivity (S_σ) defined by Cotecchia & Chandler (2000) was used. This is defined as the ratio between the yield stress (σ'_y) and an equivalent pressure (σ'^*). The yield stress was identified using the Casagrande construction and the equivalent pressure was taken as that pressure on the ICL which has the same specific volume as that of the intact compression path at σ'_y. In addition, values of S_σ at a stress

10 times σ'_y was also calculated, as in some cases the soil yielded very gradually and subsequently diverged from the ICL. The S_σ values were then analysed with respect to depth (Figs 1f and l). It can be seen that for the HDV, $S\sigma$ is approximately constant with depth, for the small range investigated, but higher than the CDV, which is as expected. However, for the CDV there is no clear trend. Both positive and negative effects of structure can be observed, i.e. values larger and smaller than unity, respectively, which might possibly reflect the presence of relic fissures, which could be observed in the block samples, but might have been present in different amounts in the specimens tested. If the values at ten times the yield stress are considered, an approximately constant value slightly larger than unity is obtained, indicating that only one cycle of load on the log scale is sufficient to erase most of the effects of structure observed.

REFERENCES

BSI (1990) BS 1377:1990: Methods of test for soils for civil engineering purposes. British Standard Institution, London, UK.

Cafaro, F. & Cotecchia, F. 2001.Structure degradation and changes in the mechanical behaviour of a stiff clay due to weathering.*Géotechnique* 51(5): 441–453.

Cotecchia, F. & Chandler, R.J. 2000. A general framework for the mechanical behaviour of clays.*Géotechnique* 50(4): 431–447.

Geotechnical Engineering Office 1988. *Guide to rock and soil descriptions. Geoguide 3.* Geotechnical Engineering Office, Hong Kong.

Volcanic Rocks and Soils – Rotonda et al. (eds)
© 2016 Taylor & Francis Group, London, ISBN 978-1-138-02886-9

Microstructure insights in mechanical improvement of a lime-stabilised pyroclastic soil

G. Russo & E. Vitale
Department of Civil and Mechanical Engineering, University of Cassino and Southern Lazio, Italy

M. Cecconi & V. Pane
Department of Engineering, University of Perugia, Italy

D. Deneele
Institut des Matériaux Jean Rouxel (IMN), Université de Nantes, France

C. Cambi & G. Guidobaldi
Department of Physics and Geology, University of Perugia, Italy

ABSTRACT: A large experimental work has been developed on a pyroclastic soil from Southern Italy to show its suitability to lime stabilisation and the effectiveness of the treatment. The beneficial effects of the mechanical improvement induced after the addition of lime have been linked to the chemical–physical evolution of the system, with particular reference to the short term.

1 INTRODUCTION

Lime stabilization appears to be a very effective improvement technique for the reuse of soils that are not suitable for earthworks in their natural state. Pyroclastic soils are commonly not used for lime stabilization, even though it is well known that natural pozzolanas are very reactive with calcium oxide (CaO) provided by lime addition. This circumstance has motivated the present research on lime stabilisation of pyroclastic soils typical of Central and Southern Italy, considering also the lack of systematic studies on this topic in literature.

In this paper the mechanical improvement induced after lime addition has been detected on lime stabilised samples in terms of reduction of compressibility. The volumetric collapse behaviour of unsaturated not treated samples has been evidenced by performing wetting paths during one dimensional loading. Similar wetting paths have been performed also on lime stabilised sample, highlighting the beneficial effects of lime addition.

The improved behaviour has been interpreted in terms of microstructural evolution of the stabilised pyroclastic soil. At this end, some microstructure analyses as X Ray Diffraction (XRD), Differential Thermal Analysis (DTA) and Fourier Transform Infrared (FTIR) have been performed.

2 MATERIALS

The Monteforte (MF) pyroclastic soil is a weathered and humified ashy soil belonging to the stratigraphic succession of Somma-Vesuvius eruptions products.

The MF soil characterizes the upper layers of the succession in the test site of Monte Faggeto, about 40 km northwest of the volcano Somma-Vesuvius, where an ongoing experimental research project on mudflows in pyroclastic soils has been carried on (Evangelista et al., 2008). The relevant hydromechanical properties of MF soil are discussed in Papa (2007), Papa et al. (2008) and Nicotera et al. (2008).

The suitability of Monteforte soil for lime stabilisation has been assessed (Cecconi et al., 2010) by means of Initial Consumption of Lime measurements (ASTM D6276-99a). Lime treated reconstituted samples were prepared by thoroughly hand mixing the soil with 7% by soil dry weight of quicklime powder and distilled water, allowing the quicklime to hydrate for 24 h. The samples were sealed in plastic bags and cured for increasing curing times of 24h (referred to as 0 days), 7, 14, 28 days.

3 EXPERIMENTAL RESULTS

The lime addition induces a relevant modification of the observed behaviour during one dimensional

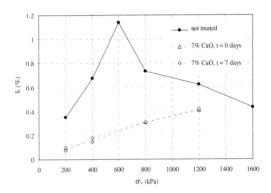

Figure 1. Comparison between collapse potentials for not treated and lime stabilised samples.

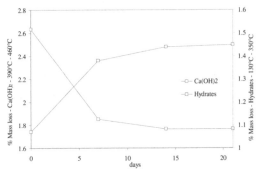

Figure 2. Lime stabilised samples: consumption of portlandite and development of hydrates with time.

loading of the sample. Under one dimensional loading the stabilised samples reduce the volume strains in the short term with relevant decrease within the first 7 days since the addition of lime.

The increase of yield stress is relevant, with an extension of the stress interval in which the stabilised soil exhibit a reversible behaviour in loading/ unloading paths. The wetting path was performed at 1200 kPa stress level, corresponding to the yield stress level of stabilised samples. The increase of the saturation degree does not trigger the collapse of the soil skeleton, differently from the observed behavior of natural ones.

The final observed settlement of samples after wetting paths was used to calculate the collapse potential I_c (I_c assumes a positive sign when the sample collapses). In Figure 1 the largest value of I_c corresponds to the yield stress of not treated soil, whereas a consistent reduction is calculated at pre-yield and post yield stress levels. The collapse potentials of lime stabilised samples at increasing curing time seem to be not dependent on curing time as is evidenced by the two curves relative to samples cured 0 days and 7 days. The reduction of collapse potential, considered as a relevant improvement of the mechanical behaviour of the stabilised pyroclastic soil, can then be assumed as a short term effect of lime addition.

The mechanical improvement induced by the addition of lime and detected at macroscopic level has been linked to the chemico–physical evolution of the system at microscopic scale. Since the pyroclastic soil is mainly constituted by amorphous phases, ionic exchange was not expected to take place in the system. The highly alkaline environment favored the dissolution of alumina and silica provided by the amorphous phases. The reaction of the available calcium ions with alumina and silica (pozzolanic reactions) induced the formation of secondary hydrates phases in the system. The evolution of the system is shown in Figure 2, where the results of several Differential Thermal Analyses (DTA) have been represented and interpreted. The consumption of portlandite (reduction of mass loss over the time) is very rapid reaching the minimum value in the short term as consequence of the

rapid development of pozzolanic reactions. The mineralogical nature of the pyroclastic soil, mainly formed by amorphous phases, makes the lime stabilization suitable also in absence of clay minerals. In fact, the results confirm that no ionic exchange is expected in the system after the addition of lime, as highlighted also at the macroscopic level by the reduced evolution of grain size distributions due to absent or not relevant flocculation of particles.

The mechanical behaviour of the stabilised soil is improved due to the formation of bonding compounds (hydrates) in the very short term, which is responsible of both the reduction of compressibility and the improved stability of the soil skeleton upon wetting.

REFERENCES

Cecconi M., Pane V., Marmottini F., Russo G., Croce P. dal Vecchio S. (2010). Lime stabilisa-tion of pyroclastic soils. Proc. Vth Int. Conf. on Unsaturated Soils, Barcelona, 6–8 september 2010, Balkema Editors, Vol. I, 537–541.

Croce P., Russo G. (2002). Reimpiego dei terreni di scavo mediante stabilizzazione a calce – Proc. XXI Geotechnical National Conference, L'Aquila, 2002, 387–394, Patron.

Delage, P. & Pellerin, F.M. (1984). Influence de la lyophilisation sur la structure d'une argile sensible du Québec. Clay Minerals, 19: 151–160.

Evangelista A., Nicotera M.V., Papa R., Urciuoli G., (2008). Field investigations on triggering mechanisms of fast landslides in unsaturated pyroclastic soils. Unsaturated soils: advances in geo-engineering. Toll et al. (eds.). Taylor & Francis Group, London, 909–915.

Nicotera M.V., Papa R., Urciuoli G., Russo G., (2008). Caratterizzazione in condizioni di parziale saturazione di una serie stratigrafica di terreni suscettibili di colata di fango. Incontro Annua-le dei Ricercatori di Geotecnica (IARG), Catania.

Papa R., (2007). Indagine sperimentale sulla coltre pироclastica di un versante della Campania. PhD Thesis, University of Napoli Federico II. Napoli, Italy.

Papa R., Evangelista A., Nicotera M.V., Urciuoli G., (2008). Mechanical properties of pyroclastic soils affected by landslide phenomena. Unsaturated soils: advances in geo-engineering. Toll et al. (eds.). Taylor & Francis Group, London, pp. 917–923.

Volcanic Rocks and Soils – Rotonda et al. (eds)
© 2016 Taylor & Francis Group, London, ISBN 978-1-138-02886-9

Dynamics of volcanic sand through resonant column and cyclic triaxial tests

A. Tsinaris, A. Anastasiadis & K. Pitilakis
Department of Civil Engineering, Aristotle University, Thessaloniki, Greece

K. Senetakis
School of Civil and Environmental Engineering, University of New South Wales, Sydney, Australia

EXTENDED ABSTRACT

The objective of this study was the determination of the dynamic properties of volcanic soil (pumice) through a series of tests carried out at the Laboratory of Soil Mechanics, Foundations and Geotechnical Earthquake Engineering of Aristotle University of Thessaloniki, Greece. A Long-Tor resonant column apparatus and a cyclic triaxial apparatus were used to cover a wide range of shear strain amplitudes, including both torsional excitation at resonance of the samples and frequency-controlled experiments at a desired number of loading cycles.

Three uniform granular pumice fractions of mean grain size, D_{50}, equal to 2.16 mm, 3.40 mm and 7.08 mm, respectively, were studied. Specific gravity of soil solids, G_s, was found equal to 1.97, 1.84 and 1.72, respectively, for the three pumice fractions. In total, six saturated specimens of pumice were tested in the present study. All specimens were prepared in dry conditions and compacted at the same number of layers of equal mass, as well as the same number of tips reaching high values of relative density, in the range of $D_r \approx 93\%-98\%$. The basic characteristics of the specimens are presented in Table 1.

In order to study the behavior of the volcanic sands from small to medium shear strain amplitudes ($10^{-4}\% < \gamma < 10^{-2}\%$), a resonant column (RC) device of free-fixed end (Drnevich, 1967) was used. Both, the dynamic small-strain shear modulus (G_0) and the damping ratio (DT_0), as well as the variation stiffness and material damping with shear strain were examined in the RC apparatus (denoted as

$G/G_0-\gamma$-DT curves). For greater levels of shear strain ($10^{-2}\% < \gamma < 1\%$) undrained cyclic triaxial (CTRX) tests were carried out. The results of the two set of tests were combined to form the complete dynamic behavior of the pumice sands by means of $G/G_0-\gamma$-DT curves in a wide range of strains. The amplitude of the applied radial stress, σ'_m, was 50, 100, 200 and 400 kPa, while the consolidation time applied between the four levels of mean effective stress was 60 minutes. The imposed loading frequency in the cyclic triaxial test was 0.5 Hz with an amplitude of the applied vertical deformation that ranged between 0.00001 and 2.0 mm which corresponded to axial strains in a range of $2 \cdot 10^{-6}$ to $2 \cdot 10^{-2}$ mm. The maximum number of cycles in the cyclic triaxial tests and for a given level of strain was equal to 10.

The experimental results of the study showed that pumice, like others volcanic materials, because of their high porosity, exhibit lower shear modulus and a higher torsional damping values compared to common sandy soils of similar particle size and density but stronger-massive grains (Figure 1 and 2).

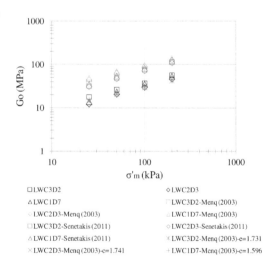

Figure 1. Small-strain shear modulus against confining pressure of saturated pumice specimens for different uniform fractions and corresponding values from empirical relationships.

Table 1. Specimens features study in Resonant Column and Cyclic Triaxial.

Specimen code	Type of Test	γ_d (kN/m³)	e_o	D_r (%)
LWC3D2	RC	7.29	1.704	97.6
LWC2D3	RC	6.71	1.741	92.8
LWC1D7	RC	6.62	1.596	93.1
LWC3D2	CTRX	7.22	1.727	93.4
LWC2D3	CTRX	6.74	1.731	95.2
LWC1D7	CTRX	6.68	1.576	97.7

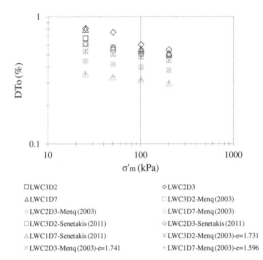

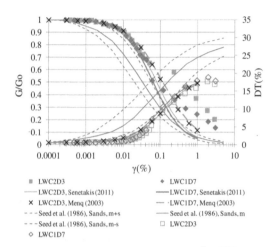

□ LWC3D2   ◇ LWC2D3
△ LWC1D7   □ LWC3D2-Menq (2003)
◇ LWC2D3-Menq (2003)   △ LWC1D7-Menq (2003)
□ LWC3D2-Senetakis (2011)   ◇ LWC2D3-Senetakis (2011)
△ LWC1D7-Senetakis (2011)   × LWC3D2-Menq (2003)-e=1.731
× LWC2D3-Menq (2003)-e=1.741   + LWC1D7-Menq (2003)-e=1.596

Figure 2. Small-strain damping ratio against confining pressure of saturated pumice specimens for different uniform fractions and corresponding values from empirical relationships.

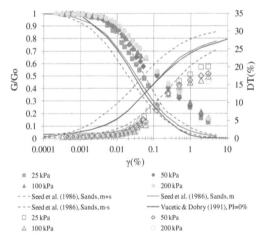

■ 25 kPa   ◆ 50 kPa
▲ 100 kPa   ● 200 kPa
– – – Seed et al. (1986), Sands, m+s ——— Seed et al. (1986), Sands, m
– – – Seed et al. (1986), Sands, m-s ——— Vucetic & Dobry (1991), PI=0%
□ 25 kPa   ◇ 50 kPa
△ 100 kPa   ○ 200 kPa

Figure 3. Effect of shearing strain amplitude and mean effective confining pressure on the non-linear G/G_o-log γ and DT-log γ curves of pumice (specimen LWC3D2).

In the range of small strains amplitudes the increase of the mean grain diameter affected the dynamic response of saturated samples as it led to higher values of small-strain shear modulus and lower values of torsional damping, with an increase of grain size (Figure 1 and 2). Particle-contact response may be one important factor for this observation.

■ LWC2D3   ◆ LWC1D7
——— LWC2D3, Senetakis (2011)   ——— LWC1D7, Senetakis (2011)
× LWC2D3, Menq (2003)   — ·LWC1D7, Menq (2003)
– – – Seed et al. (1986), Sands, m+s ——— Seed et al. (1986), Sands, m
– – – Seed et al. (1986), Sands, m-s □ LWC2D3
◇ LWC1D7

Figure 4. Effect of mean grain size on the non-linear G/G_o-log γ and DT-log γ curves of pumice at a pressure of $\sigma'_m = 100$ kPa.

In the range of medium to high strain levels the G/G_o–γ-DT(%) curves of pumice showed a more linear behavior than those of typical soils with the increase of the implemented shear strain (Figure 3). This trend was more pronounced for specimens with smaller mean grain size since they presented higher γ_{ref} values for the same σ'_m magnitude (Figure 4).

The increase of the number of loading cycles led to a reduction of the shear strain and the damping ratio for all the tested specimens. Once again the diameter of the grains along with the relative density had important effect on the rate of reduction or increase of shear modulus and damping ratio.

Session 4:   Geotechnical aspects of natural hazards

Volcanic Rocks and Soils – Rotonda et al. (eds)

Geological evolution of the Ischia volcanic complex (Naples Bay, Tyrrhenian sea) based on submarine seismic reflection profiles

G. Aiello & E. Marsella
CNR IAMC Sede di Napoli, Naples, Italy

EXTENDED ABSTRACT

The geological evolution of the Ischia active volcanic complex (Naples Bay, Southern Tyrrhenian sea) has been reconstructed based on recently acquired submarine seismic reflection data. Implications on submarine slope stability both in the northern and western submerged flanks of the island, characterized by thick submarine slide deposits and southern flank of the island, characterized by active erosion of the coastal systems due to submarine canyons, will be discussed.

A densely-spaced grid of single-channel seismic profiles has been recently acquired and interpreted in the frame of research programs on marine cartography (CARG Project) financed by the Campania Region (Sector of Soil Defence, Geothermics and Geotechnics) during the mapping of the marine areas of the geological sheet n. 464 "Isola d'Ischia" at the scale 1:25.000 (Aiello et al., 2009; 2012; Aiello and Marsella, 2014). New geological data on the Ischia volcanic complex are here presented based on integrated geologic interpretation of Multibeam bathymetric and single channel seismic data to constrain the Ischia geologic evolution during Quaternary times.

Marine geophysics around the Ischia island has shown a great improvement during the last ten years, due to the bathymetric surveys acquired by the GNV and CARG Projects. These bathymetric surveys resulted in a Multibeam coverage all around the Ischia island (Aiello et al., 2010). Swath bathymetry was acquired in a wide depth range, with various echosounders, each characterized by its own frequency (from 60 to 300 kHz) and spatial resolution (from 0.3 to 10 m the footprint). The merging of these surveys with the island subaerial topography has allowed the construction of a DTM in which the elementary cell size has been averaged to 20 m obtaining a good compromise among coverage depth and resolution in shallow waters. The sea floor structure has been determined through the analysis of the DTM. The study grid is composed of dip lines perpendicular to the shoreline and tie lines parallel to the shoreline, interpreted accordingly to criteria of seismic stratigraphy and volcanic geomorphology. The classification of volcanic landforms has been recently improved taking into account the complexity in the generation

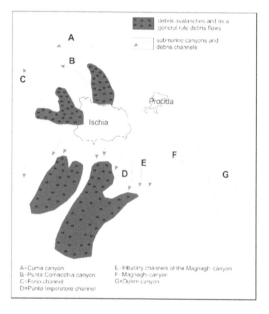

Figure 1. Volcanic and tectonic lineaments at Ischia.

of volcanic structures and their control factors, such as the magmatic systems, the styles of eruption and the erupted materials. The DEM datasets are often available, making possible the morphometric characterization of large composite volcanoes at a global scale.

Main volcanic and tectonic lineaments on the Ischia shelf and slope have been represented (Fig. 1).

New insights into seismo-stratigraphy and Ischia geologic evolution are here discussed. The Ischia offshore is characterized by alkali-potassic volcanic rocks (trachytes, latites, alkali-basalts) and pertains to a volcanic complex emplaced during the last 55 ky. Four main phases have been distinguished in the eruptive activity of the Ischia volcanic complex from 150 ky B.P. to 1302 A.D. The geological interpretation of marine DEM and Sparker data has allowed the identification of important submarine instability processes, both catastrophic (debris avalanches) and continuous (creep and accelerated erosion along canyons). Debris avalanches are mainly controlled by the volcano-tectonic uplift of the Epomeo block, related to a caldera resurgence during the last 30 ky. The most important one is the IDA corresponding to a large scar of southern Ischia; on the contrary, the large slides occurring off Casamicciola and Forio do not appear related to evident slide scars onshore. The Casamicciola slide is characterized by two distinct episodes of slide emplacement suggested by seismic interpretation. The Southern Ischia canyon system engraves a narrow continental shelf from Punta Imperatore to Punta San Pancrazio and is limited to the SW by the relict Ischia bank volcano. It consists of 22 drainage axes, whose planimetric trending has been reconstructed through morpho-bathymetric analysis. The eastern margin of the canyon system is tectonically-controlled, being limited by a NE-SW (counter-Apenninic) normal fault. Its western margin is controlled by volcanism, due to the growth of the Ischia volcanic bank. Important implications exist regarding the coastal monitoring and beach nourishment of southern Ischia, involved by strong erosion and shoreline retreatment, mainly in correspondence to Maronti and Barano coastal systems.

REFERENCES

Aiello G., Marsella E. 2014. The Southern Ischia canyon system: examples of deep sea depositional systems on the continental slope off Campania (Italy). *Rendiconti Online della Società Geologica Italiana* 32:28–37.

Aiello G., Marsella E., Passaro S. 2009. Submarine instability processes on the continental scope off the Campania region (southern Tyrrhenian sea, Italy): the case history of Ischia Island (Naples Bay). *Bollettino di Geofisica Teorica Applicata* 50 (2):193–207.

Aiello G., Budillon F., Conforti A., D'Argenio B., Putignano M.L., Toccaceli R.M. 2010. *Note illustrative alla cartografia geologica marina. Foglio geologico n. 464 Isola d'Ischia*. Regione Campania, Settore Difesa Suolo, III SAL, Geologia Marina, Preprints.

Aiello G., Marsella E.., Passaro S. 2012. Stratigraphic and structural setting of the Ischia volcanic complex (Naples Bay, southern Italy) revealed by submarine seismic reflection data. *Rendiconti Lincei* 23 (4):387–408.

Biagio G. 2007. Studio vulcanologico delle porzioni sommerse del Distretto Vulcanico Flegreo attraverso indagini di sismica marina ad alta risoluzione e correlazione delle facies alle interfacce terra-mare. *PhD Thesis Earth Sciences*, University of Pisa, 126 pp.

Volcanic Rocks and Soils – Rotonda et al. (eds)
© *2016 Taylor & Francis Group, London, ISBN 978-1-138-02886-9*

Investigation on the hydraulic hysteresis of a pyroclastic deposit

L. Comegna, E. Damiano, R. Greco, A. Guida, L. Olivares & L. Picarelli
Dipartimento di Ingegneria Civile, Design, Edilizia e Ambiente,
Seconda Università degli Studi di Napoli, Aversa, Italy

ABSTRACT: The paper describes the results coming from the automatic monitoring of the annual hydrological response of a deposit in loose pyroclastic soils located in a mountainous area of Cervinara, Campania Region. The collected field data, consisting in rainfall, soil moisture content and suction measurements, allow to estimate the soil water retention features, putting into evidence some hysteretic nature of the wetting/drying processes determined by the weather events.

1 MONITORING SITE

In order to collect useful information about the hydrological processes related to meteorological forcing, an automatic monitoring station has been installed in the area of Cervinara, Campania Region, where a catastrophic flowslide was triggered by a rainstorm totaling 320 mm in 50 hours (Olivares & Picarelli 2003; Damiano et al. 2012). According to geological surveys and geotechnical investigations, the pyroclastic cover along the slope reaches a highest thickness of about 2.5 m and consists of alternating layers of volcanic ashes and pumices laying upon a fractured calcareous bedrock. Automatic monitoring started on 2009 (Comegna et al. 2011; Guida et al. 2012). Hourly precipitations are recorded by a rain gauge having a sensitivity of 0.2 mm. Soil suction is measured by "*Jet-fill*" tensiometers, equipped with tension transducers. Soil volumetric water contents are measured by probes for Time Domain Reflectometry (TDR). Suction and moisture sensors have been vertically installed within volcanic ashes at different depths from 0.60 m to 1.70 m and are connected to a Data Logger, that allows the automatic acquisition and storage of data with a time resolution of two hours.

Next section reports data obtained from January, 2011, to January, 2012, at the depths $z = 0.60$ m and $z = 1.00$ m, where the tensiometer tips are located closely to TDR probes, allowing to couple recorded volumetric water content and suction values.

2 FIELD DATA

The data obtained by coupling matric suction and moisture content values measured at the depths $z = 0.60$ m and $z = 1.00$ m are plotted in Figures 1 and 2. The observed scattering is probably due to some error in the measured volumetric water content of

± 0.02 m^3/m^3, mainly caused by the imperfect matching of suction and water measurements. The data are interpolated by curves whose slope is strongly governed by the actual weather conditions: the most gentle observed paths (curves *AB*, *EF*, *HI*) are the result of alternating wetting/drying stages, while the steepest drying paths (curve *CD*) correspond to prolonged stages occurred during the dry season (that started on June and continued until the end of October). In particular, curves *EF* and *HI* (Figs. 1b and 2b) show that different values of the volumetric water content for

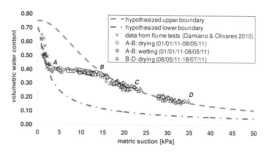

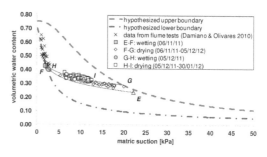

Figure 1. Volumetric water content and matric suction measured at depth $z = 0.60$ m from January to July, 2011 (a) and from November, 2011, to January, 2012 (b).

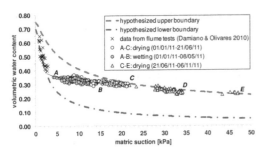

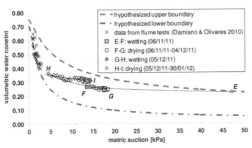

Figure 2. Volumetric water content and matric suction measured at depth $z = 1.00\,m$ from January to November, 2011 (a) and from November, 2011, to January, 2012 (b).

the same matric suction have been monitored at both instrumented depths.

Such a hydrological response seems to be an indication of some hysteretic behaviour of the deposit, that could depend on energy dissipative effects occurring during infiltration and/or evaporation processes, as revealed by empirical evidences and theoretical considerations reported in the literature (Li 2005, Tarantino 2009, Yang et al. 2012, Pirone et al. 2014). In particular, different hysteretic loops should be found in the water retention plan. Such paths, named *scanning curves*, are usually located between two main curves, known as the *main drying curve* and the *main wetting curve*, but may also partly or fully coincide with one of them. Furthermore, the scanning curves are less inclined than the two primary curves. In our case, the less inclined envelopes (*AB*, *EF*, *HI*) are located below the most inclined curve (*CD*), that might represent the uppermost boundary (possibly coinciding with the main drying curve). Figures 1 and 2 show also the results of some laboratory infiltration tests performed on small-scale slopes reconstituted with the same Cervinara ashes in an instrumented flume (Damiano & Olivares 2010): such data are well interpolated by a curve that seems to represent a reliable lowermost boundary (or main wetting curve) for the field data.

3 CONCLUSIONS

The setting up of an automatic field station with high temporal data resolution is allowing to investigate the annual cyclic hydrological response of a sloping deposit of loose pyroclastic soils. Obtained data highlight that the effects of weather on soil moisture changes are strongly influenced by the initial conditions. In particular, different values of the volumetric water content for the same matric suction can be observed, depending on preceding wetting or drying history. Such evidences raise some questions on the reliability of the models commonly used to simulate transient infiltration and evaporative processes, that usually bear on a unique water retention curve simulating the hydraulic soil behaviour. Further field data are being collected to check the reliability and repetitively of the observed relations. Moreover, a laboratory testing program is starting in order to verify through a different approach the obtained results.

REFERENCES

Comegna, L., Guida, A., Damiano, E., Olivares, L., Greco, R., & Picarelli L. 2011. Monitoraggio di un pendio naturale in depositi piroclastici sciolti. *Proc. XXIV Convegno Nazionale di Geotecnica, Napoli, 22–24 June 2011*, Vol. 2: 681–686. Edizioni Associazione Geotecnica Italiana, Roma.

Damiano, E. & Olivares, L. 2010. The role of infiltration processes in steep slope stability of pyroclastic granular soils: laboratory and numerical investigation. *Natural Hazards* 52: 329–350.

Damiano, E., Olivares, L. & Picarelli L. 2012. Steep-slope monitoring in unsaturated pyroclastic soils. *Engineering Geology* 137–138: 1–12

Guida, A., Comegna, L., Damiano, E., Greco, R., Olivares, L. & Picarelli L. 2012. Soil characterization from monitoring over steep slopes in layered granular volcanic deposits. In Luciano Picarelli, Roberto Greco and Gianfranco Urciuoli (eds.), *Proc. The second Italian Workshop on Landslides, Napoli, 28–30 September 2011*, 147–153. Cooperativa Universitaria Editrice Studi, Fisciano.

Li, X.S. 2005. Modelling of hysteresis response for arbitrary wetting/drying paths. *Computers and Geotechnics* 32: 133–137.

Olivares, L. & Picarelli, L. 2003. Shallow flowslides triggered by intense rainfalls on natural slopes covered by loose unsaturated pyroclastic soils. *Geotechnique* 53(2): 283–288.

Pirone, M., Papa, R., Nicotera, M.V. & Urciuoli, G. 2014. Evaluation of the hydraulic hysteresis of unsaturated pyroclastic soils by in situ measurements. *Procedia Earth and Planetary Sciences* 9: 163–170.

Tarantino, A. 2009. A water retention model for deformable soils. *Géotechnique* 59(9): 751–762.

Yang, C., Sheng, D. & Carter, J.P. 2012. Effect of hydraulic hysteresis on seepage analysis for unsaturated soils. *Computers and Geotechnics* 41: 36–56.

Volcanic Rocks and Soils – Rotonda et al. (eds)
© 2016 Taylor & Francis Group, London, ISBN 978-1-138-02886-9

An investigation of infiltration and deformation processes in layered small-scale slopes in pyroclastic soils

E. Damiano, R. Greco, A. Guida, L. Olivares & L. Picarelli
D.I.C.D.E.A., Second University of Naples, Aversa, Italy

ABSTRACT: In recent years, a number of flowslides and debris flows triggered by rainfall affected a wide mountainous area surrounding the "Piana Campana" (southern Italy). The involved slopes are constituted by shallow unsaturated air-fall layered deposits of pyroclastic nature, which stability is guaranteed by the contribution of suction to shear strength. To understand the infiltration process and the soil suction distribution in such layered deposits, strongly affecting slope stability, infiltration tests in small-scale layered slopes reconstituted in a well-instrumented flume have been carried out. The results highlight that the presence of a coarse-textured pumiceous layer interbedded between two finer ashy layers, in unsaturated conditions, delays the wetting front advancement, thus initially confining the infiltration process within the uppermost finer layer. However, when high hydraulic gradients establish across the pumices, water infiltration into the deepest layer starts. In sloping deposits, under high applied rainfall intensity, a seepage parallel to the slope seems to occur through the uppermost finer layer when it approaches saturation. At the same time, water infiltrates through pumices. Thus, in presence of long lasting intense rainfall, the presence of a coarse-gained layer does not necessarily impede the progressive saturation of the lowest soil layer.

1 INTRODUCTION

In Campania, a wide mountainous area, mantled by pyroclastic soils in unsaturated conditions, experienced rainfall-induced landslides during the last decades. Along the Apennines, the pyroclastic mantles are usually constituted by alternating ashy and pumiceous layers few meters thick, which lay upon a fractured limestone bedrock. It is widely reported in literature that a layered soil profile may deeply affect rainfall infiltration and water content distribution, in some cases conditioning slope failure (Yang et al. 2006; Mancarella et al. 2012). However this is still an open topic since contrasting results are reported in the literature. To this aim, infiltration tests in small-scale layered slopes reconstituted in a well-instrumented flume (Olivares et al. 2009) have been carried out, which allowed to evaluate the effect of slope inclination on the infiltration and deformation processes during the pre-failure stage, investigating the role of an interbedded pumiceous coarser layer.

2 EXPERIMENTAL RESULTS

The soil deposit, reconstituted in the flume at porosities as high as the in situ ones, was constituted of three soil layers: from top to bottom, an ashy layer, a pumiceous layer and again an ashy layer, with a total thickness of 0.2 m.

The investigated soils were sampled at the Cervinara slope (Damiano et al. 2012; Greco et al. 2013).

The volcanic ash is a silty sand, the pumiceous soil is a gravel with sand. Both soils are characterized by a very low dry unit weight (13–14 kN/m^3) as a consequence of their nature and of their high porosities. A synthesis of the mechanical and hydraulic properties of the investigated soils is reported in Olivares & Picarelli 2003.

The deposit was monitored by means of minitensiometers, placed at three depths within the two ashy layers, laser transducers to measure the settlements at the ground surface, and a vertical TDR probe 19 cm long, buried into the soil to retrieve profiles of volumetric water content (Greco 2006).

The layered deposit was subjected to two stages. During the first one (stage I), the deposit was horizontal and subjected to a rainfall intensity of 45 mm/h for 60 minutes. During the stage II, the slope was tilted to an angle of 40° and a rainfall intensity of 83 mm/h was applied for about 100 min.

In both cases a wetting front develops from the ground surface towards the base. However, the presence of the intermediate pumice layer causes a marked delay in the infiltration process.

The water content profiles retrieved during stage I (Fig. 1a) reveal that, for more than 30 minutes the infiltration process remains confined into the upper ashy layer which progressively increases its water content. On the contrary, the measured values of θ_w in the pumices do not change significantly, indicating that the hydraulic conductivity of the pumices is so low that nearly no flow crosses the interface under the established hydraulic gradient. The wetting front enters the

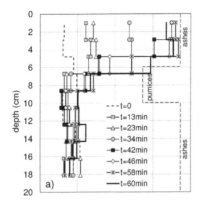

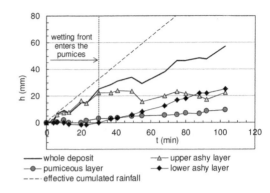

Figure 2. Cumulated stored water heights during stage II.

Figure 1. Volumetric water content profiles during: a) stage I; b) stage II.

middle layer about 42 minutes after the start of rain. At this stage, a water potential difference of about 45 kPa establishes across the pumices, allowing a significant water flow despite their still low conductivity.

The soil deposit undergoes a settlement in the order of 0.4 cm as measured by laser sensor transducers, corresponding to a volumetric deformation of 2%. Settlements continue even after the stop of rain, as an effect of progressive saturation of the lower soil layer. The volumetric deformation is also accompanied by formation of cracks.

During stage II, an almost complete saturation of the whole deposit is reached. The water content profiles (Fig. 1b) reveal that the complete saturation of the upper ashy layer occurs ($t = 30$ min) and that this layer saturates from the top whereas the lower one saturates from the bottom. The cumulated stored water heights illustrated in Figure 2 add some more information about the process. During a first time, as long as the upper ashy layer does not reach saturation, the rate of water height accumulation into the upper ashy layer equals the effective rainfall inten-sity applied over the inclined surface. Afterwards, the rate of the water height accumulating within the whole deposit (bold line in Fig. 2) decreases to a value of about 27 mm/h, becoming lower than the applied rainfall intensity. As there is no evidence of surface water run-off nor of leakage at the flume bottom, the drainage of the deposit can occur only by means of a subsurface downslope seepage through the saturated upper ashy layer. This subsurface drainage continues until the end of the test. The wetting front enters the pumices after nearly 30 min. Infiltration penetrates into the lower ashy layer few minutes later. During this last stage, the rate of water accumulating within the lower ashy layer becomes greater than that of pumices.

3 CONCLUSIONS

The experiences performed highlight that the pumiceous layer delays the entering of the wetting front, until an high hydraulic gradient establishes across it. This result shows that the presence of a coarser layer below a finer one does not necessarily imply the formation of a capillary barrier, since infiltration is governed not only by the hydraulic properties of the soils, but also by the hydraulic gradients which establish during the process, in turn related to the rate and duration of infiltration. The test on the small-scale slope inclined at 40° reveals that, under the applied rainfall intensity, a seepage parallel to the slope inclination occurs within the upper ashy layer. At the same time water infiltrates into the lowermost ashy layer: pumices, although in unsaturated conditions, let infiltrate an amount of water greater than that accumulating within them.

REFERENCES

Damiano E., Olivares L., Picarelli L. 2012. Steep-slope monitoring in unsaturated pyroclastic soils. *Eng Geol* 137–138: 1–12.

Greco, R. 2006. Soil water content inverse profiling from single TDR waveforms. *Journal of Hydrology* 317: 325–339.

Greco R., Comegna L., Damiano E., Guida A., Olivares L., Picarelli L. 2013. Hydrological modelling of a slope covered with shallow pyroclastic deposits from field monitoring data. *Hydrology and Earth System Sciences* 17: 4001–4013.

Mancarella D., Doglioni A., Simeone V. 2012. On capillary barrier effects and debris slide triggering in unsaturated soil covers. *EngGeol* 147–148: 14–27.

Olivares, L. & Picarelli, L. 2003. Shallow flowslides triggered by intense rainfalls onnatural slopes covered by loose unsaturated pyroclastic soils. *Geotechnique* 53(2), 283–288.

Olivares L., Damiano E., Greco R., Zeni L., Picarelli L., Minardo A., Guida A., Bernini R. 2009. An instrumented flume for investigation of the mechanics of rainfall-induced landslides in unsaturated granular soils. *ASTM Geotechnical Testing Journal* 32 (2): 108–118.

Yang H., Rahardjo H., Leong E-C., 2006. Behavior of unsaturated layered soil columns during infiltration. *J HydrolEng* 11(4): 329–337

Volcanic Rocks and Soils – Rotonda et al. (eds)
© 2016 Taylor & Francis Group, London, ISBN 978-1-138-02886-9

Rainfall-induced slope instabilities in pyroclastic soils: The case study of Mount Albino (Campania region, southern Italy)

G. De Chiara, S. Ferlisi & L. Cascini
Department of Civil Engineering, University of Salerno, Italy

F. Matano
Istituto per l'Ambiente Marino Costiero, Consiglio Nazionale delle Ricerche, Naples, Italy

EXTENDED ABSTRACT

Rainfall-induced slope instabilities later propagating as flow-like phenomena are widespread all over the World and, generally, cause catastrophic consequences in urbanised areas due to their imperceptible premonitory signals, long run-out distances, high velocities and huge mobilised volumes (Sorbino et al. 2010). Features of slope instabilities/flow-like phenomena depend on predisposing factors as well as on initial and boundary conditions. Therefore, carrying out susceptibility and hazard analyses of slope instabilities/flow-like phenomena that might occur in a defined geological context first requires the definition of a deepen cognitive framework embracing the physical processes leading to either triggering or propagation stage (Corominas et al. 2014).

In this paper susceptibility and hazard analyses at the source areas of slope instabilities in shallow deposits of pyroclastic soils were carried out with reference to the case study of Mount Albino, a carbonatic relief – located in the municipality of Nocera Inferiore (Campania region, southern Italy) – covered by pyroclastic soils deriving from the explosive activity of Somma-Vesuvius volcanic complex. The adopted procedure included two level of analysis (Cotecchia et al. 2014): the first one focused on an in-depth phenomenological interpretation of the mechanisms accompanying the prevailing slope instabilities later propagating as flow-like phenomena; whereas the second level was aimed at objectivising the previously identified mechanisms on the basis of deterministic models.

In order to detect the mechanisms characterising the slope instabilities which could be triggered over the Mount Albino slopes and to collect the data to be used within procedures aimed at pursuing susceptibility and hazard purposes (first level of analysis) a number of activities (historical and geomorphological analyses, field surveys and in-situ tests) were carried out following a multidisciplinary approach.

Results of above activities, revealed that Mount Albino hillslopes are prone to different kind of rainfall-induced slope instabilities namely *i*)

hyperconcentrated flows, *ii*) debris flows and *iii*) debris avalanches, which substantially differ in terms of triggering mechanisms, mobility (in terms of run-out distance) and intensity (Hungr et al. 2001).

Features of both boundary and initial conditions leading to above mentioned slope instabilities (which can be profitably used as starting point to perform more advanced susceptibility and hazard analysis) were derived from a historical analysis, widely discussed in De Chiara et al. (2015), dealing with 121 municipalities – including Nocera Inferiore – of the Campania region belonging to the a geological context where pyroclastic soils cover a carbonate bedrock. In particular, the authors observed that the occurrence of hyperconcentrated flows – which mainly concentrates in a period of the year between September and November months – relates to meteorological processes associated to rainfall of high intensity and short duration. Since in such a period of the year the shallow deposits of pyroclastic soils are characterised by high values of matric suction – and, therefore, low hydraulic conductivities – the infiltration of rain water is inhibited and the runoff prevails; this, in turn, leads to the occurrence of erosion phenomena and/or small-size slope instabilities which mainly concentrate along the channel's sides of gullies. On the other hand, debris flows or debris avalanches mainly occur from December to June when the continuous sequence of rainy days leads to a progressive decrease of soil suction values and to the consequent occurrence of first-time shallow slides at source areas later propagating as debris flows or debris avalanches (Cascini et al. 2014).

Among the investigated rainfall-induced slope instabilities, this paper focused on susceptibility and hazard analyses at source areas of hyperconcentrated flows.

The susceptibility analysis was carried out at large scale (1:5,000) by adopting a heuristic procedure. The latter allowed a classification of the gullies on Mount Albino hillslopes taking into account the combined role played by two predisposing factors, namely the maximum slope angle (in degrees) of the gullies' sides and the thickness (in meters) of coarser ashy soil covers. The combination of these parameters allowed the estimation of the susceptibility to erosion phenomena and to local instability of the gullies' sides on the basis of four levels (VL = very low, L = low, M = medium, H = high). The results of the analysis were summarised in a susceptibility zoning map showing that slope portions with the highest susceptibility are located in the western side of Mount Albino which is less evolved (in morphological terms) and characterized by stream channels with cross section having a typical "U-like" shape.

In order to carry-out a hazard analysis, the frequency (F) of hyperconcentrated flows was estimated via a hydrological analysis which resulted in three different return period (T = 50, 100 and 200 years) of the triggering rainfall while the magnitude (M) of potentially mobilised soil volumes was assessed by separately considering both the surface erosion processes and localised slope instabilities.

The analysis of surface erosion processes was carried out through the use of the physically-based openLISEM model (Baartman et al. 2012) while small-localised slope instabilities, resulting by erosion processes at the toe of the gullies' sides, were computed through slope stability analyses performed via the Morgenstern and Price's limit equilibrium method implemented in the SLOPE/W code (GeoSlope 2005) and integrated with the results of the TRIGRS-unsaturated (Savage et al. 2004) physically based model.

The performed analyses allowed the generation of F-M curves for each mountain catchment of Monte Albino hillslope. In this regard, the frequency was normalised with respect to the areal extent of each mountain catchment while the relative magnitude was posed equal to the total mobilised volume of soil at source areas of hyperconcentrated flows. These obtained results reveal that, for each mountain catchment, as the frequency of occurrence decreases, the mobilised volume increases. Moreover, for an equal value of frequency per unit area, the catchments B1, B3, B6 are those potentially involving the highest soil volumes.

Finally, in order to zones the hazard of hyperconcentrated flows at their triggering stage the potentially mobilised soil volumes were normalised with reference to the length of the gullies' stretches for which a given susceptibility level was recognised. The obtained magnitude values were grouped into four classes and combined with the return period of the triggering rainfall so that the hazard can be distinguished into four levels (VL = very low, L = low, M = moderate, H = high).

The results obtained for each considered return period were finally represented in three distinct hazard maps. These maps show that the maximum value of hazard level lies in different gullies' stretches, with reference to triggering rainfall with a return period T equal to 50 years.

It is worth to observe that the portions of Mount Albino hillslopes where the highest hazard level was recognized are those which prioritarily may require structural mitigation measures (of active type). Their design, in turn, implies a suitable prediction of the potentially mobilized soil volume which, as highlighted in the case study at hand, could be largely overestimated if results of susceptibility (instead of hazard) analyses are taken into account.

In this regard, further efforts are needed to properly carry out susceptibility and hazard analyses dealing with further kind of slope instabilities potentially affecting Mount Albino hillslopes and which may propagate as debris flows/avalanches, usually causing huge consequences in terms of loss of life and properties (Cascini et al. 2014).

REFERENCES

Baartman, J.E.M., Jetten, V.G., Ritsema, C.J. & de Vente, J. 2012. Exploring effects of rainfall intensity and duration on soil erosion at the catchment scale using openLISEM: Prado catchment, SE Spain. *Hydrological Processes* 26:1034–1049.

Cascini, L., Sorbino, G., Cuomo, S. & Ferlisi, S. 2014. Seasonal effects of rainfall on the shallow pyroclastic deposits of the Campania region (southern Italy). *Landslides* 11(5):779–792.

Corominas, J., van Westen, C., Frattini, P., Cascini, L., Malet, J.P., Fotopoulou, S., Catani, F., Van Den Eeckhaut, M., Mavrouli, O., Agliardi, F., Pitilakis, K., Winter, M.G., Pastor, M., Ferlisi, S., Tofani, V., Hervàs, J. & Smith, J.T. 2014. Recommendations for the quantitative analysis of landslide risk. *Bulletin of Engineering Geology and the Environment* 73:209–263.

Cotecchia, F., Ferlisi, F., Santaloia, F., Vitone, C., Lollino, P., Pedone, G. & Bottiglieri, O. 2014. La diagnosi del meccanismo di frana nell'analisi del rischio. In: *La geotecnica nella difesa del territorio e dalle infrastrutture dalle calamità naturali*, Proc. of the XXV Italian Congress of Geotechnics, Vol. 1:167–186. Edizioni AGI, Roma.

De Chiara, G., Ferlisi, S. & Sacco, C. 2015. Analysis at medium scale of incident data dealing with flow-like phenomena in the Campania region (southern Italy). In *16th ECSMGE: European Conference on Soil Mechanics and Foundation. Edinburgh, 13–17 September 2015* (accepted for publication).

Ferlisi, S., De Chiara, G. & Cascini, L. 2015. Quantitative Risk Analysis for hyperconcentrated flows in Nocera Inferiore (southern Italy). *Natural Hazards*, DOI: 10.1007/s11069-015-1784-9.

GeoSlope 2005. *User's guide*. GeoStudio 2004, version 6.13. Geo-Slope Int. Ltd, Calgary, Canada.

Hungr, O., Evans, S.G., Bovis, M.J. & Hutchinson, J.N. 2001. A review of the classification of landslides of the flow type. *Environmental & Engineering Geoscience*, 7(3):221–238.

Savage, W.Z., Godt, J.W. & Baum, R.L. 2004. Modeling time-dependent areal slope stability. In W.A. Lacerda, M. Erlich, S.A.B. Fontoura & A.S.F. Sayao (eds.) *Landslides – Evaluation and Stabilization, Proc. of the 9th International Symposium on Landslides*, Vol. 1:23–36. Rotterdam: Balkema.

Volcanic Rocks and Soils – Rotonda et al. (eds)
© *2016 Taylor & Francis Group, London, ISBN 978-1-138-02886-9*

Geotechnical characterization and seismic slope stability of rock slopes in the Port Hills during the New Zealand 2011 Canterbury Earthquakes

F.N. Della Pasqua, C.I. Massey & M.J. McSaveney
GNS Science, Lower Hutt, New Zealand

ABSTRACT: Field information from four rock-slope sites was used to study the seismic response and amplification effects induced in the Port Hills of Christchurch, New Zealand, by the 2011 Canterbury Earthquakes. The dynamic response of each slope was characterised in terms of input free-field seismic acceleration A_{FF}, slope yield acceleration K_Y, maximum crest acceleration A_{MAX} and the maximum average acceleration K_{MAX}. The measured permanent slope displacements at each site were used in back analysing numerical models of the stability of each slope. The amounts of permanent co-seismic slope displacement at the cliff sites were found to result from a combination of lithology, amplification, degradation and preconditioning prior to each event, and were unique for each site.

1 INTRODUCTION

During the 2010–2011 Canterbury earthquake sequence, extensive ground cracking and failure occurred onmany cliffs ofthe Port Hills (Figure 1), (Massey et al. 2014a, b, c and d). In this study we show how the permanent co-seismic slope displacements can be accounted for by a combination of the local site characteristics such as lithology and slope shape, and their effects on amplification.

The extent of permanent slope displacement was quantified using a combination of cadastral and crack-aperture surveys. Slope stability was assessed with reference to procedures outlined in Eurocode 8 and permanent displacements simulated using the Quake/W software program (Slope/W, 2012).

70m

Figure 1. Aerial view of a damaged Port Hills rock slope at Redcliffs.

2 RESULTS

The maximum crest amplification and average maximum amplification were measured in terms of A_{MAX}/A_{FF} and K_{MAX}/A_{FF}, respectively. K_{MAX} was used to represent the average acceleration response of the slope.

The crest amplification at low $A_{FF} < 0.3$ g was variable and ranged up to >4 times. At higher $A_{FF} > 0.5$, the degree of crest amplification are less and tended to converge, ranging from 1.5 to 3, depending on the slope (Figure 2).

The slope amplification in terms of K_{MAX}/A_{FF} varied from 1 to 2, (Figure 2). Thus the average maximum acceleration experienced by the slope was about half that at the slope crest. The higher slopes also appeared to show a higher degree of amplification than the lower slopes.

The back analysis of measured displacements showed that the higher slopes developed permanent slope displacements (>0.1 m) at A_{FF} values above about 0.2 g, whereas the lower slopes developed permanent slope displacements at A_{FF} values of about >0.4 g.

3 CONCLUSIONS

For a given earthquake, the extent of slope deformation islargely site-specific and dependent on geometry and the rock-mass strength.

In this study, the simulated amplifications for slope-crest response with respect to the input free-field acceleration (A_{MAX}/A_{FF}) were found to vary up to 7 times, at horizontal accelerations of less than 0.2 g.

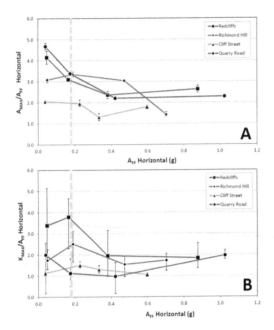

Figure 2. Variation in amplification ratios for A) peak crest acceleration A_{MAX} and B) average maximum acceleration K_{MAX}.

However, this amplification value decreases rapidly with increasing A_{FF}, above a threshold of 0.2 g. This threshold representst he minimum yield accceleration of the assessed slopes and correspondingto the minimum horizontal acceleration above which permanent rockslope deformation was observed in the field.

K_{MAX} was found to provide a better representation of the acceleration response of the slope, compared to a single point measurement at the cliff crest (A_{MAX}).

At A_{FF} values greater than about 0.2 g, the amplification ratios showed a tendency to merge, yielding relatively constant horizontal amplification ratios (K_{MAX}/A_{FF}) for all four slopes of between 1 and 2. This suggestedthat there might bea threshold acceleration above which the amplification becomes linear. In all cases, this linearity appears to develop above the acceleration value where cracks develop, i.e. the yield acceleration.

The choice of amplification factor used for assessing the dynamic response of slopes should be addressed on a site-specific basis as it can vary mainly as a result of site geology site geometry and the nature of the earthquake.

REFERENCES

Eurocode 8.EN1998-5. 2004. Design of structures for earthquake resistance Part 5: *Foundations, retaining structures and geotechnical aspects.*

Massey, C. I., Della Pasqua, F., Lukovic, B., Ries W. & Heron, D. 2014a. *Canterbury Earthquakes 2010/11 Port Hills Slope Stability: Risk assessment for Cliff Street.* GNS Science Consultancy Report 2014/73.

Massey, C. I., Della Pasqua, F., Taig, T., Lukovic, B., Ries, W. & Heron, D. 2014b. *Canterbury Earthquakes 2010/11 Port Hills Slope Stability: Risk assessment for Quarry Road.* GNS Science Consultancy Report 2014/75.

Massey, C. I., Della Pasqua, F., Taig, T., Lukovic, B., Ries, W.; Heron, D. & Archibald, G. 2014c. *Canterbury Earthquakes 2010/11 Port Hills Slope Stability: Risk assessment for Redcliffs.* GNS Science Consultancy Report 2014/78.

Massey, C. I., Taig, T., Della Pasqua, F., Lukovic, B., Ries, W. & Archibald, G. 2014d. *Canterbury Earthquakes 2010/11 Port Hills Slope Stability: Debris avalanche risk assessment for Richmond Hill,* GNS Science Consultancy Report 2014/34.

Slope/W, 2012. Stability modelling with Slope/W. An engineering methodology. November 2012 Edition. *GEO-SLOPE International Ltd.*

Volcanic Rocks and Soils – Rotonda et al. (eds)
© *2016 Taylor & Francis Group, London, ISBN 978-1-138-02886-9*

High-resolution geological model of the gravitational deformation affecting the western slope of Mt. Epomeo (Ischia)

M. Della Seta, C. Esposito, G.M. Marmoni, S. Martino & C. Perinelli
Dipartimento di Scienze della Terra e Centro di Ricerca CERI, Sapienza Universitá di Roma, Roma, Italia

A. Paciello
Agenzia Nazionale per le Nuove Tecnologie, l'Energia e lo Sviluppo Economico Sostenibile (ENEA-Casaccia), Roma, Italia

G. Sottili
CNR, Istituto di Geologia Ambientale e Geoingegneria (IGAG), Roma, Italia

EXTENDED ABSTRACT

The recent geological history of the Ischia Island, emerged portion of the Phlegraean Volcanic District, is closely related to the volcano-tectonic dynamic of the Mt. Epomeo resurgent caldera. The presence of a relatively shallow magmatic body strongly conditioned the geological evolution of the island that is responsible for volcanic activity, crustal deformations and strong seismicity. This magmatic body, located at about 2000 m depth, produces the development of a stable hydrothermal system characterised by high heat flow (200–400 mW/m^2) (Cataldi et al. 1991) and several thermal springs and gas vents. In addition, diffuse gravitational processes involved the edge of the resurgent block. The present study focuses on the gravitational deformation of Mt. Nuovo, located in the western portion of Mt. Epomeo, which involves alkali-trachytic pyroclastic flow deposit (Mt. Epomeo Green Tuff (MEGT); Brown et al., 2008) and trachytic and phonolitic lavas (133 ka; Vezzoli, 1988). Historical and archaeological documents led some authors (Del Prete & Mele, 2006) to consider that the deep gravitational phenomenon connected with the formation of Mt. Nuovo is related to a catastrophic volcano-tectonic event which took place around 460–470 BC during the last stages of Mt. Epomeo uplift. Based on a multidisciplinary approach, this study proposes a new high-resolution engineering-geological model of the Mt. Nuovo slope, that provides constraints to the ongoing gravitational slope deformation. By means of geomorphological surveys and terrain analyses through GIS, fault scarps, terraces, saddles, trenches and morphological counter-slopes were also identified, thus providing more details and constraints to the geometry of the slope deformation. A geomechanical survey was also carried out to derive the geomechanical properties of the outcropping jointed rock mass and to evaluate their influence on the landslide process.

In order to define the dynamic properties of the rock masses involved in the slope deformations and to derive further constraints to the engineering-geological model of the slope, a seismic geophysical campaign was performed in the Mt. Nuovo by 17 measurements of ambient noise. The here reported results were processed in agreement with the horizontal to vertical spectral ratios (HVSR) (Nakamura, 1989). This technique was used to focused on possible resonance frequencies and indirectly assess the thickness of the softer layer constituted by the highly jointed MEGT involved in the landslide respect to the MEGT and to the trachitic lavas that compose the landslide substratum.

The surveyed data revealed the complexity of the ongoing slope deformation and led to the identification of a multiple compound mechanism of the landslide process. This model allow us to constrain the shear zones that presently drive the gravitational slope deformations and to infer the mechanisms and the volumes associated to this gravity-induced process. New field observations, combined with literature geological data, allowed to identify the geological framework in which the Mt. Nuovo slope deformation is taking place. Based on the collected geological data it was possible to obtain a cross-section of the Mt. Nuovo from the top of Mt. Epomeo to the Falanga plain. This geological section intercepts three distinct gravitational surfaces: the main one is at about 200–250 m b.g.l., while the two minor ones are located in correspondence of Mt. Nuovo and downslope, respectively. Based on this geometry, the main gravitational surface originates in correspondence with the main fault that displaces Mt. Nuovo from the top of the Falanga with a main bi-planar structural setting according to a "type D" structurally-defined compound slide, (Hungr and Evans, 2004). The geometry of the secondary failure surfaces, the deformed mass structure, as well as the counter-slope terraces indicate a marked roto-translational mechanism. The so reconstructed model of the Mt. Nuovo landslide mass allowed to identify a rock mass volume involved in the slope deformation of about 0.36 km^3, i.e. lower than the one of the greatest debris avalanche surveyed in the island (1.5 km^3). The seismic noise measurements provided an useful contribution to the reconstruction of such a geological setting as they pointed out a resonance peak of 0.8 Hz and revealed that the azimuthal distribution of these peaks has not directivity. Taking into account the geological setting resulting in the cross-section reconstructed along Mt. Nuovo, the observed resonance could be related to approximately 200 m of both MEGT and landslide-involved (i.e. intensely jointed) trachytic lavas that overlay stiffer lavas constituting the landslide substratum. The absence of directivity validates the hypothesis that the resonance peak is related to an impedance contrast resulting from an almost 1D condition. At all the other sites the noise measurements do not show significant HVSR peaks: the absence of resonance that can be ascribed to the presence of major faults or shear zones and to the consequent increase of the rock mass jointing conditions and reduces the impedance contrast between the two media avoiding resonance effects to be measured.

The obtained geological model shows a significant control of the pre-existing tectonic pattern on slope deformation mechanisms, since the main morphological evidence of sliding surfaces partly reflect the tectonic pattern.

Furthermore, the here proposed geological model highlights the correspondence between the gas vent location and the basal breccias outcropping at the base of the two MEGT flow units. Such a correspondence is evident in both the Donna Rachele (Rione Bocca)

fumarole field and in the Mt. Nuovo area and could be related to the permeability contrast between the MEGT breccia levels and the massive lavas, which represent the local hydrothermal reservoir. Based on the here reported geological section the fluid emergences are justified as they rise from a complex net of faults and fractures, mostly related to the structural setting of the slope and the ongoing gravitational deformation. The presence of extinct fumaroles upslope the main surface seems to confirm that both lithological and structural elements control the emergence of ascent fluids from the deep reservoir. In this framework the tectonic discontinuities represent, because of the induced fracturing, preferential escape routes for the ascent of fluids, while the rupture surfaces may have interrupted the lateral continuity of the reservoir, preventing the ascent of fluids in the higher areas and leading to the extinction of gas vents.

The here proposed geological model will be enriched by further mechanical features (i.e. integrating new geomechanical surveys and providing specific laboratory tests on samples) for future stress-strain numerical simulations aiming at reconstruct the evolution of the gravitational slope instability and evaluate the role of the rock mass rheology, thermomechanical features and destabilizing actions possibly related to the hydrothermal system. In this regard, the results reported in this paper constitute a start point for future studies that will be focused on the dynamics and the mechanisms of the gravity-induced slope evolution affecting the SW sector of Mt. Epomeo in the Mt. Nuovo area, so driving to landslide hazard assessment through the identification and the analysis of scenarios for a possible generalized collapse of the slope.

Volcanic Rocks and Soils – Rotonda et al. (eds)
© 2016 Taylor & Francis Group, London, ISBN 978-1-138-02886-9

Geo-engineering contributions to improve volcanic rock and soil slopes stabilization

C. Dinis da Gama
Geomechanics Laboratory, IST, Lisbon University, Portugal

EXTENDED ABSTRACT

Long-term needs of mankind must involve ambitious research methods, as well as innovative practical applications, which may provide a comprehensive investigation framework for managing Earth resources and mitigate natural hazards, as well as global changes.

Within the scope of these aims, Geo-engineering may help to implement and identify new large scale methods and processes that are required, as the means of applying scientific principles to attain practical objectives such as the design, construction, and operation of efficient and economical structures, equipment and systems, both natural and man-made.

Landscape stabilization and control are certainly one of those mankind values that must be kept and improved, thus justifying a joint effort in which stabilization of rock and soil slopes is essential, particularly in volcanic environments such as islands. A short contribution is attempted with this presentation.

Nowadays, the broad concept of Geo-engineering involves the deliberate modification of Earth's features to suit consensual human needs. It started with the formulation of proposals to deliberately manipulate the Earth's climate in order to counteract the effects of global warming from greenhouse gas emissions, but has been expanded to many other topics of interest (The Royal Society, 2009).

Early Geo-engineering techniques were based on carbon sequestration seeking to reduce the liberation of greenhouse gases in the atmosphere. These include direct methods (such as carbon dioxide air capture) and indirect methods (like ocean iron fertilization).

These processes were regarded as mitigations forms of global warming. Alternatively, solar radiation management techniques do not reduce greenhouse gas concentrations, and can only address the warming effects of carbon dioxide and other gases; they cannot address problems such as ocean acidification, which are expected as a result of rising carbon dioxide levels.

Examples of proposed solar radiation management techniques include the production of stratospheric sulfur aerosols, space mirrors, and cloud reflectivity enhancement. To date, no large-scale Geo-engineering projects have been undertaken. Some limited tree planting and cool roof projects are already underway, and ocean iron fertilization is at an advanced stage of research, with small-scale research trials and global modeling having been completed.

Field research into sulfur aerosols has also started. Some commentators have suggested that consideration of Geo-engineering presents a moral hazard because it threatens to reduce the political and popular pressure for emissions reduction.

Typically, scientists and engineers proposing those strategies do not suggest that they are an alternative to earth problems, but rather an accompanying strategy.

In this presentation, a review of the methods commonly utilized to stabilize slopes is conducted so that long-term strategies are proposed with the intention of identifying potentially stronger and weaker schemes which may be valuable to future generations, particularly for those living in volcanic islands.

This analysis is presented in two parts: the large scale events (under Geo-engineering criteria) and the detailed stability conditions (using geomechanical analysis methods).

Current world problems of deforestation, degradation and collapse of productive environments are commonly reported to humans. They are considered in the context of natural processes of formation and modification of the environment's aspects of climate, ecology and geomorphology. These aspects have profound influence on resource availability and utilization, thus shaping the culture and social organization of millions of people, mostly those living in mountainous regions.

Besides the normal and cyclic changes of temperature between day and night, as along the year, there are periods of drought and floods which may cover millions of years, even including the formation of mountain chains and valleys.

In addition to this constant evolution, many low-frequency events can also happen, such as earthquakes, tectonic uplifts, major floods and glacial phenomena that may shape ground and slopes.

This complex dynamics leads to the formation of diverse environments where short differences in altitude, slope angle and lithology can have important effects on vegetation, soils, erosion (especially caused by snow melt and heavy rain fall) and land use potential. If in addition there are frequent earthquakes and volcanic activity, then the region is characterized by great levels of instability and consequently more difficulties for human occupation.

This is why classification of a mountain region in terms of its stability is a complex subject which requires a reliable characterization of its changing mechanics, either in the form of cyclic, or elastic, or even constant performances.

Therefore, it is essential to define that the scale of observation in time or space affects our perception of stability. In particular, the factor of altitude provides to the slopes an image of enhanced activity and hazard which is relevant in most cases.

The well-known minimum potential energy principle states that natural systems, in order to reach a situation of equilibrium, continuously change through a process that is characterized by a sequence of events always decreasing its potential energy. This is the main cause for earth surface unrest, under the influence of distinct natural phenomena, including the topic of slope stability. Therefore, it is quite understandable that mankind tries to incessantly resist those effects, with all possible available tools and utilizing innovative technologies and equipment whenever possible. Geo-engineering is one of these tools, leading to a permanent search for the best integrated solutions.

It is thus justifiable that we may change the classic statement of Francis Bacon issued in 1620, with an additional idea: Nature to be commanded must be obeyed, *except when humans, fauna and flora are in danger.*

REFERENCES

Coates, D.F. 1981. *Rock Mechanics Principles.* Mines Branch Monograph 874, Ottawa.
Doebrich, J.F. & Theodore, T.G. 1996. *Geologic history of the Battle Mountain mining district, Nevada.* U.S. Geological Survey publication, Washington.
Kanji, M.A., Gramani, M.F., Massad, F., Cruz, P.T. & Araujo, H.A. 2000. *Main factors intervening in the risk assessment of debris flows.* International Workshop on the Debris Flow Disaster of Dec 1999, Caracas.
NaturalHazards.org website. 2005.
The Royal Society, 2009. *Geo-engineering the climate: science, governance and uncertainty.* London.
Willie, D.C. & Mah, C.W. 2004. *Rock slope engineering – civil and mining* (4th edition). New York. Spon Press.

Volcanic Rocks and Soils – Rotonda et al. (eds)
© 2016 Taylor & Francis Group, London, ISBN 978-1-138-02886-9

Wave erosion mechanism of volcanic embankment subjected to cyclic loadings

S. Kawamura
Muroran Institute of Technology, Muroran, Japan

S. Miura
Hokkaido University, Sapporo, Japan

EXTENDED ABSTRACT

In the 2011 off the Pacific coast of Tohoku Earthquake, serious damage due to great seismic loadings and subsequent tsunami occurred widely in Japan. It has been also reported that road embankments in Tohoku expressway play an important role to evacuate from tsunami (see Photo 1). After that, many evacuation places have been constructed in other expressways in Japan. This paper aims at clarifying wave erosion mechanism and failure of volcanic embankments subjected to cyclic loadings such as seismic loadings. A series of model tests was performed on model volcanic embankments by using a cyclic loading and a wave paddle systems; especially the effects of initial water content and compaction conditions on wave erosion mechanisms were presented herein.

Volcanic soil which was sampled from the Shikotsu caldera in Hokkaido was used in this study. It is estimated that the eruption age for Komaoka volcanic soil belonging to Shikotsu primary tephra was 31,000–34,000 years and was pyroclastic flow deposits (the notation is *Spfl*; Shikotsu pumice flow deposits).

Photo 2 shows typical deformation behavior of a model embankment during wave loadings after cyclic loadings. The embankment was eroded due to wave loadings. Thereafter, wave-induced failure was caused by the development of notch. Figure 1 summarizes the relationship between elapsed time at failure T_f normalized by that of wave-induced failure of embankment without cyclic loadings ($T_{ffor\ case\ without\ cyclic\ loadings}$) and shear strain γ due to cyclic loadings for all tests. It is apparent that the elapsed time increases until $\gamma = 1\%$ and decreases for more than $\gamma = 3\%$ in this study. Specifically, slope failure seems to differ depending strongly on the stress-strain history due to cyclic loadings such as seismic loadings.

From the model tests presented, failure mechanisms of volcanic embankments subjected to pre-deformation due to cyclic loadings such as seismic loadings were clarified. It has been indicated that estimation of shear strain after cyclic loadings is important for stability of volcanic embankments.

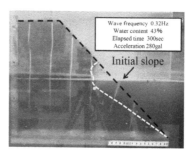

Photo 2. Typical deformation behavior of a model embankment of Dc = 85%, $w_0 = 43\%$, 280 gal.

Photo 1. Serious damage around Tohoku Expressway due to tsunami in the 2011 off the Pacific coast of Tohoku Earthquake (provided by Nippon Expressway Research Institute Co., Ltd.).

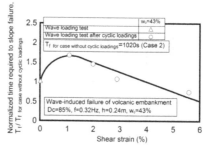

Figure 1. Changes in elapsed time at failure due to difference in shear strain.

Volcanic Rocks and Soils – Rotonda et al. (eds)
© *2016 Taylor & Francis Group, London, ISBN 978-1-138-02886-9*

Earthquake-induced flow-type slope failures in volcanic sandy soils and tentative evaluation of the fluidization properties of soils

M. Kazama, T. Kawai, J. Kim & M. Takagi
Tohoku University, Sendai, Japan

T. Morita
Gunma Institute of Technology, Gunnma, Japan

T. Unno
Utsunomiya University, Utsunomiya, Japan

ABSTRACT: Japan has many experiences of mudflow-type failures of volcanic sandy soil slopes during earthquakes. Firstly in this paper, several major cases of the flow-type failures are outlined. In general, undisturbed pyroclastic deposits have high strength against to fluidization because of its cementation effect. However, once the deposits are disturbed to use as an earth fill materials, the shear strength of disturbed soils cannot be expected. Rather, the disturbed pyroclastic soils are weaker than ordinary sandy soils. Therefore, it is necessary to evaluate the fluidization property of disturbed pyroclastic sandy soils for mitigating mudflow disaster. In this study, we introduce a drum type testing machine to evaluate the fluidization property. Two kinds of volcanic sandy soils are used as a specimen sampled from volcanic area in Japan. In addition to this, the wireless multi sensor which can be installed in the movable soils is introduced.

1 INTRODUCTION

When volcanic sandy soils liquefy during an earthquake, mudflow-type failure can occur. It is important to evaluate the fluidization property of volcanic sandy soils under both disturbed and undisturbed conditions because mud flow-type failures occur in both natural volcanic sandy soil deposits and ground artificially filled with disturbed volcanic sandy soil material.

2 MAJOR CASE HISTORIES OF THE MUD FLOW TYPE FAILURE IN JAPAN

Liquefaction occurs with unsaturated soil, and particularly volcanic ash soils characterized by porous soil particles. These soils are comparable to soils with high water content in their peacetime condition even in the absence of rainfall. When subjected to an earthquake of large magnitude with a strong motion of

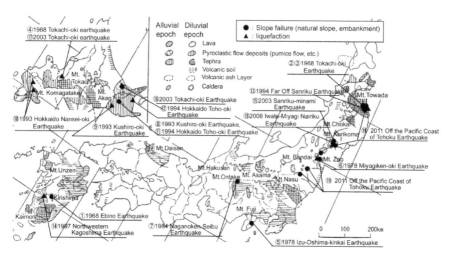

Figure 1. Liquefaction damage and mudflow-type slope failure composed of volcanic sandy soils in Japan.

Photo 1. Drum-type flow testing apparatus developed for evaluating the fluidization property of soils.

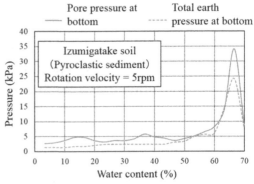

Figure 2. Variation of earth pressure and pore water pressure with water content of Izumigatake soil.

long duration, these volcanic sandy soils are prone to liquefaction.

The liquefaction damage and mudflow-type slope failure associated with volcanic sandy soils throughout Japan are shown in Figure 1. Besides being a volcano country, Japan is located in one of the world's highest seismicity zones. There are volcanos in Hokkaido, eastern Japan and Kyushu, and 40% of the country is covered with volcanic ash soil deposits, as shown in Figure 1. There are two distinct types of mudflow failures of volcanic soil deposits: the liquefaction of soils used as artificial fill material and natural slope failure. While both type of mudflow occur with almost the same frequency, the magnitude of damage is much larger in the case of the latter type.

3 EVALUATION OF FLUIDIZATION PROPERTY BY DRUM TYPE EQUIPMENT

3.1 Testing apparatus

An attempt has been made to evaluate the fluidization property of disturbed pyroclastic soils by using a drum type testing apparatus. The testing equipment used was a drum-type machine as shown in Photo 1 with an inner diameter of 1 m and depth of 50 cm. An earth pressure gauge and a pore pressure meter are embedded in the cylindrical surface. Data is acquired by the wireless data acquisition system.

3.2 Testing material and testing procedure

Two kinds of pyroclastic volcanic sandy soils were used as specimens. Testing procedure is as follows: First, 30-40 kg of the dried sample is placed into the drum shown in Photo 1 through the opening. The rotation velocity of the drum is gradually increased from the static position to the prescribed value 5 rpm/ 10 rpm. Once the velocity becomes constant, the behavior of soils is observed and both the earth pressure and pore water pressure are measured. When rotation is stopped, 1 kg of water is added to the soil mass, and the operation is repeated until complete fluidization occurs. When the surface angle of the soils mass is horizontal, we understand that complete fluidization taken place.

3.3 Test results

Figures 2 shows the variation of pore water pressure and total earth pressure with water content. The pore water pressure and total earth pressure are indicated by the solid line and broken line respectively. Since the maximum soil thickness about 20 cm, the pressure under the static condition must be in the range of several kilo Pascals. The sudden rise in the pore water pressure and total earth pressure indicated in this figure suggests that the soil particle cannot maintain water beyond a certain water content value. As a result, amount of free water increases, and the dynamic pressure of the falling soil mass is increased. After that, pressure falls rapidly and converges to a small value. This is because the total earth pressure and pore water pressure becomes the same as the hydrostatic pressure in saturated fluidized soil.

4 CONCLUSIONS AND REMARKS

In this paper, we first outline the major cases of the flow-type failures experienced in Japan. We can classify mudflow-type slope failures during earthquake composed of volcanic sandy soils into two major categories. One is the artificial fill flow failure type, which is caused by liquefaction due to long period seismic motion from relatively large magnitude subduction zone earthquake. The other is the natural slope failure type, which is caused by inland earthquakes and involves soils composed of volcanic ash in highly mountainous areas.

An attempt was made to evaluate the fluidization property of soils. Using a drum type testing apparatus, we evaluated the fluidization property of two volcanic pyroclastic sandy soils. It was found that the dynamic impact water and earth pressure increase rapidly at a certain water content. This may depend on the water retention property of particular type of soil. In this paper, we also introduce a wireless multi sensor, which can be installed in movable soil mass. Further study is necessary to confirm the effectiveness of the apparatus developed here.

Volcanic Rocks and Soils – Rotonda et al. (eds)
© 2016 Taylor & Francis Group, London, ISBN 978-1-138-02886-9

Rock fall instabilities and safety of visitors in the historic rock cut monastery of Vardzia (Georgia)

C. Margottini & D. Spizzichino
ISPRA, Italian Institute for Environmental Protection and Research, Rome, Italy

G.B. Crosta & P. Frattini
Department of Earth and Environmental Sciences, University of Milano-Bicocca, Milano, Italy

P. Mazzanti, G. Scarascia Mugnozza & L. Beninati
Nhazca S.r.l. Roma, Rome, Italy

ABSTRACT: This paper reports the main results of a project developed in cooperation with National Agency for Cultural Heritage Preservation of Georgia, and aimed at envisaging the stability conditions of the Vardzia Monastery slope (rupestrian rock cut city cave in the south-western Georgia). The site has always been affected by instability processes along the entire slope, including small block falls from the upper breccia layer and large collapses from the middle layer. The study involves: rock mechanics characterization, geo-engineering survey, geo-structural and kinematic analysis, rockfalls modelling, geomatic acquisitions and elaboration. The geomechanical characteristics of volcaniclastic and pyroclastic rocks were determined by means of geomechanical field surveys, rock mass classification through scan lines techniques, and laboratory tests on rock blocks and cores. In order to carry out a semi-automatic detection of discontinuities and to implement mitigation activities, a 3D Terrestrial Laser Scanning survey has been carried out. Potential rockfalls have been simulated by the 3D modelling code HY-STONE. The model, used both with a downward and backward approach, allowed the recognition of most critical sectors belonging to the upper part of the cliff (volcanic breccia) and to provide a support for designing both short and long-term mitigation measures. A general master plan for landslide risk reduction and mitigation measures is actually under development for the entire rock cut city of Vardzia.

1 INTRODUCTION

The rock-cut city of Vardzia is a cave monastery site in south western Georgia, excavated from the slopes of the Erusheti mountain on the left bank of the Mtkhvari river. The main period of construction was the second half of the twelfth century.

The caves stretch along the cliff for some 800 m and up to 50 m within the rocky wall (Figure 1).

The monastery consists of more than 600 hidden rooms spread over 13 floors. The main site was carved from the cliff layer of volcanic and pyroclastic rocks, Gillespie & Styles (1999) at an elevation of 1300 m above sea level. The cave city included a church, a royal hall, and a complex irrigation system. The earthquake that struck Samstkhe in 1283 AD destroyed about two thirds of the city cave, exposing the majority of the rooms to view outside. The site was largely abandoned after the Ottoman takeover in the sixteenth century. Now part of a state heritage reserve, the site has been submitted for future inscription on the UNESCO World Heritage List. The site is by the time affected by frequent slope instability processes along the entire volcanic cliff (Margottini et al, 2015). Similarly, the rapid onset falls and instabilities can become

Figure 1. The rock-cut city of Vardzia (Georgia).

extremely dangerous also for the many visitors that are interested to discover the site. Due to these phenomena, the National Agency for Cultural Heritage Preservation of Georgia (NACHPG) has promoted, with the support of ISPRA, a landslide hazard assessment for the entire area through rock mechanics characterization, geotechnical engineering survey, geo-structural and kinematic analysis, rockfalls 3D model, terrestrial

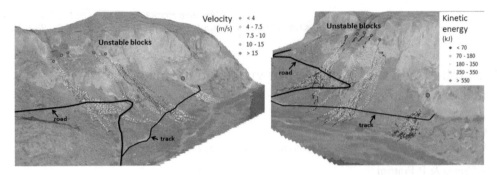

Figure 2. Results of potential rockfalls from unstable blocks recognized on field (s2). a) velocity; b) kinetic energy. Each point corresponds to the position of the block sampled along each trajectory every 4 meters.

laser scanner acquisitions for the identification of the most hazardous areas.

2 THE SITE

The main geomorphological, geo-structural and geomechanical evidences obtained after field missions suggest the followings:

- the potential instability processes and mechanisms observed for the entire rock cliff can be referred to different failure modes (or their combination): rock fall; planar rock slide; roto-translational rock slide; wedge and toppling failure;
- actual and/or potential instability processes at the Vardzia monastery are the result of a combination of different predisposing factors such as: lithology, presence, frequency and orientation of discontinuities vs. slope orientation, physical and mechanical characteristics of materials, morphological and hydrological boundary conditions as well as human activities;
- relevant factor in Vardzia slope stability is the reduction of UCS and tensile strength, from laboratory tests, when saturated; such drop can reach up to 70% of original values, then suggesting an important role in rainy period;

3 3D ROCKFALL SIMULATION

During field activities, it has been recognized that the rocky cliff is frequently affected by small-volume fragmental rockfalls, especially from the upper breccia layer (Figure 2).

A set of rockfall simulations was performed by means of the 3D numerical model HY-STONE (Crosta et al., 2004, Frattini et al., 2012).

4 CONCLUSIONS

The Vardzia Monastery is one of the most important Georgian Cultural Heritage.

- the coupling of different survey techniques (e.g. 3D laser scanner, engineering geological and geomechanical field surveys) is the best strategy to be adopted in the interdisciplinary field of Cultural Heritage protection and conservation policies;
- the area is seriously affected by rockfall risk that could affect the safety of both the monks and the visitors.
- The future development for the Vardzia monastery provide mitigation measures both structural and non-structural for the next two years in order to reduce risk and increase safety of the site and visitors.

REFERENCES

Crosta, G.B., Agliardi, F., Frattini, P. & Imposimato, S. 2004. A three-dimensional hybrid numerical model for rockfall simulation. Geophysical Research Abstracts, 6, 04502.

Frattini, P., Crosta, G.B., & Agliardi, F. 2012. Rockfall characterization and modeling. In Clague J.J., Stead. D. (eds) Landslides Types, mechanisms and modeling, 267–281, Cambridge University Press.

Gillespie, M., & Styles, M. 1999. BGS rock classification scheme, Volume 1. Classification of igneous rocks. Keyworth, UK: British Geological Survey.

Margottini, C., Antidze, N., Corominas, J., Crosta, G.B., Frattini, P., Gigli, G., Giordan, D., Iwasaky, I., Lollino, G., Manconi, A., Marinos, P., Scavia, C., Sonnessa, A., Spizzichino, D., & Vacheishvili, N. (2015). Landslide hazard, monitoring and conservation strategy for the safeguard of Vardzia Byzantine monastery complex, Georgia. Landslides, 12: 193–204.

Volcanic Rocks and Soils – Rotonda et al. (eds)
© *2016 Taylor & Francis Group, London, ISBN 978-1-138-02886-9*

Integration of geotechnical modeling and remote sensing data to analyze the evolution of an active volcanic area: The case of the New South East Crater (Mount Etna)

M. Martino, S. Scifoni, Q. Napoleoni, P.J.V. D'Aranno & M. Marsella
Dipartimento di Ingegneria Civile, Edile e Ambientale, Sapienza University, Roma, Italy

M. Coltelli
Etna Observatory National Institute of Geophysics and Volcanology, Catania, Italy

ABSTRACT: A combined approach based on remote sensing techniques and geotechnical modeling has been adopted to assess the hazard linked the instability a cinder cone recently formed on the summit area of the Mount Etna as a consequence of a very active volcanic phase started in January 2011 from the New South East Crater (NSEC). A multisensor and multitemporal approach was adopted to update the topographical surface at the top of the volcano: Digital Elevation Models (DEM) and maps from interferometric SAR data and high-resolution photogrammetric images processing were integrated. A probabilistic analysis has been conducted to determine the stability of the NSEC, quantify the possible unstable volumes and define the covered area from the feasible collapse. The work is aimed at assessing a consistent procedure that, starting from the capability of rapidly mapping the new volcanic structure and from a preliminary knowledge of the geotechnical characteristics of the material forming its slopes, may be adopted for a first evaluation of the occurrence and the effects of potential instabilities.

1 INTRODUCTION

Mount Etna is the largest volcano in Europe and one of the most active in the world, as testified by the frequent morphological changes of its summit area (at about 3300 m a.s.l.) and along its flanks. Between January 2011 and mid-2015 a sequence of about 50 paroxysms, begun within a pit crater placed in the proximity of the old Southeast Crater, has formed a cinder cone, named the New South East Crater (NSEC), that grew quickly by the deposition of pyroclastic materials (Benckhe et al. 2013). In this paper, after having remodeled the morphology of the summit area of the volcano by using remote sensing techniques, a two step geotechnical analysis has been conducted. First, by adopting literature data on the geotechnical characteristics (Rotonda et al. 2010), a probabilistic analysis was implemented to determine the stability of the NSEC and to quantify the potential unstable volume; secondarily a preliminary dynamic modeling was used to evaluate the areas potentially interested by the propagation of the collapsed masses. This approach may represent a first contribution towards the definition of a more rigorous procedure for slope stability analysis that would have required the collection of in-situ samples for estimating the geotechnical characteristics of the materials forming the cone and it may present also a non conventional study about the effects of a landslide to develop a risk analysis.

2 THE RECENT ACTIVITY AT MOUNT ETNA AND THE REMOTE SENSING DATA

Eruptive events that led to the birth of the NSEC are classified as "paroxysms" (i.e. "set of explosive phenomena that constitute the most violent and dangerous phase of an eruption"). These events are producing, obviously, deeply changes on the morphology of the Mount Etna volcano. The first paroxysm was recorded on the evening of 12th January 2011, following a vigorous Strombolian activity from the pit crater started 10 days before. This event marked the resumption of Etna eruptive activity after the last lava flow eruption lasted from 13th May 2008 to 6th July 2009. It was followed by 47 paroxysmal episodes occurred until to 31th December 2014 that resulted in a constant growth of the new crater up to an elevation of about 3230 m a.s.l. on the north side of the crater rim. By processing a multi-sensor and multi-temporal dataset formed by COSMO-SkyMed SAR data, WorldView-2 and Pleiades stereo-pairs and helicpter digital images, 3D and 2D mapping products were extracted and validated by using aerial photogrammetric maps at sub-meter level accuracy. The analysis provided data that permitted to define the geometry of the NSEC and its evolution between 2011 and 2014 (Martino et al., in press). The DEM extracted at the different data, permitted to follow the growth of volumes and the maximum height (Table 1) of the cinder cone.

Table 1. Summary of the data extracted from the DEMs to evaluate the growth of the NSEC come.

NSEC cumulative volume [m^3*10^6]					
	06.06. 11	25.07.12	07.08.12	12.05.13	27.10.13
SAR	5.7	20.3	–	–	33.9
VIS	–	–	16.8	28.2	–
NSEC cumulative height [m]					
	06.06.11	25.07.12	07.08.12	12.05.13	27.10.13
SAR	87	168	–	–	313
VIS	–	–	159	181	–

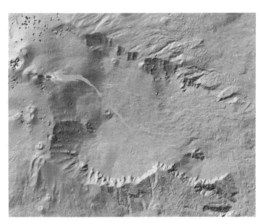

Figure 2. Results of the dynamical model showing the area invaded by the simulated flow.

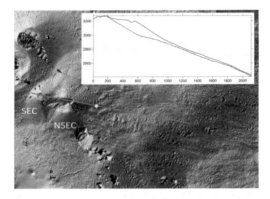

Figure 1. The direction of the A-A' section; the yellow point indicates the point where the February 2014 dyke emerged derived from the thermal images acquired before the avalanche; the inset shows the comparison of the topographic 2007 and 2013 profiles along the section A-A'.

3 GEOTECHNICAL MODELING

To carry out the slope stability analysis from a probabilistic point of view, the modeling has been developed associating to the involved parameters a normal distribution. Given the limited information available on the geotechnical characterization and assuming a constant unit volume equal to $16.00\,kN/m^3$, it was decided to analyze the results of direct shear tests in order to define a range of variability within which the peak values of the angle φ oscillate and determine its probability distribution.

The analysis is conducted on the eastern side of the NSEC that is considered more disposed to dyke formation and consequent instability that, in case of new magma intrusions. Therefore, the model adopted to evaluate the stability conditions was applied to the west-east longitudinal section AA' cut along the eastern side of the New South East Crater as shown in Figure 1.

The model was initially applied without considering the intrusion of magma as a triggering factor, providing a null probability of collapse. These results confirm that the distribution of pyroclastic material along the flanks of a typical cinder cone, like the NSEC, can hardly determine instabilities and the occurrence external triggering factors should be considered. In order to evaluate the distance covered by the potential sliding material and the relative area of invasion, the methodologies implemented for dynamic analysis in the DAN 3D tools was adopted (Marsella et al. 2015). The model is capable of providing a three-dimensional simulation of the evolutionary phase of debris flows, avalanches and landslides. The predicted parameters include the maximum distance reached by the flow, the speed, thicknesses and distributions of deposits and the behavior during the run. This approach provided an envelope of the covered areas from the possible landslides that may reach a maximum distance of about 4.0 km from the originating point and a thickness of the order of the meter within the Valle del Bove (Figure 2).

4 CONCLUSIONS

In relation to hazard assessment issues, the methodological approach implemented in this work can contribute to the understanding of the causes of collapses that may affect the investigated area.

REFERENCES

Behncke B., S. Branca, R. A. Corsaro, E. De Beni, L. Miraglia, and C. Proietti, 2013 The 2011–2012 summit activity of Mount Etna: Birth, growth and products of the new SE crater, Journal of Volcanology and Geothermal Research, vol. 270, pp. 10–21.

Marsella, M., P.J.V. D'Aranno, S. Scifoni, A. Sonnessa, and M.Corsetti 2015 Terrestrial laser scanning survey in support of unstable slopes analysis: the case of Vulcano Island (Italy) doi.org/10.1007/s11069-015-1729-3 Journal of Natural Hazards.

Rotonda T., Tommasi P., Boldini D., 2010 Geomechanical characterisation of the volcaniclastic material involved in the 2002 landslide at Stromboli. Journal of Geotechnical and Geoenvironmental Engineering (ASCE), vol. 136(2): 389–401.

Volcanic Rocks and Soils – Rotonda et al. (eds)
© 2016 Taylor & Francis Group, London, ISBN 978-1-138-02886-9

New mapping techniques on coastal volcanic rock platforms using UAV LiDAR surveys in Pico Island, Azores (Portugal)

A. Pires & H.I. Chaminé
Laboratory of Cartography and Applied Geology, DEG, School of Engineering (ISEP), Polytechnic of Porto, Porto, Portugal
Centre GeoBioTec (Georesources, Geotechnics, Geomaterials Research Group), University of Aveiro, Aveiro, Portugal

J.C. Nunes & P.A. Borges
Department of Geosciences, University of Azores, Ponta Delgada, Azores, Portugal
Centre GeoBioTec, University of Aveiro, Aveiro, Portugal

A. Garcia & E. Sarmento
Pico Island Delegation of the Regional Secretary for Tourism and Transport (Azores Regional Government), Pico Island, Azores, Portugal

M. Antunes
Cartography and Geographical Information System Services, Regional Secretary for Tourism and Transport, Pico Island, Azores, Portugal

F. Salvado
CartoGalicia: Geomatic Services and Unmanned Services Systems, Technical Department, A Coruña, Spain

F. Rocha
Centre GeoBioTec (Georesources, Geotechnics, Geomaterials Research Group), Department of Geosciences, University of Aveiro, Aveiro, Portugal

EXTENDED ABSTRACT

This work describes a preliminary methodological framework for the assessment of coastal volcanic rocky platforms in Pico Island (Azores). This study also deals with the importance of GIS-based mapping and unmanned aerial vehicles (UAVs) which were tested in two studied areas (Lajes and Madalena sites). This research gives an overview about the UAVs work flow and its application for photogrammetric assessment of coastal volcanic rocky platforms. The main purpose of this study is to explore the influence of basaltic lava coastal platforms and associated boulder strewn, which are well differentiated in Lajes and Madalena sites.

In general, this research describes the image acquisition process and the UAV LiDAR technology and the aerial surveys to acquire the geodatabase which will enable the design of geoscience and geotechnical maps on coastal volcanic rock platforms. The images analysis will allow: (i) boulder geotechnical description and evaluation; (ii) shore and coastal platform assessment; (iii) propose monitoring coastal plans (short to long-term); (iv) rock boulders (basaltic blocks) mobility, movement trend, imbrication, indicating flow and source direction; (v) obtain DTM and contour lines to generate 3D models of the coastal area. All these data is crucial to understand the coastal dynamics of the sites and develop an applied mapping which couples GIS technologies and UAV based spatial platform. It was also developed a GNSS (Global Navigation Satellite System) datasheet which incorporates the entire database of ground control points, using high resolution GPS system for the georeferencing and differential correction process.

This research presents the preliminary results of the coastal geotechnics mapping for volcanic environments. Furthermore, high resolution image acquisition, georeferencing and differential correction of the ground control points using high accuracy GPS, were also thorough analysed to improve the general methodology presented herein. Finally, it was proposed a preliminary integrated coastal engineering study for the rocky platform zoning and short to long-term monitoring in selected sites on Pico Island.

The study was partially financed by FEDER-EU COMPETE Funds and FCT (GeoBioTec\UA: UID/GEO/04035/2013) and LABCARGA\ISEP re-equipment program: IPP-ISEP\PAD'2007/08.

Keywords: UAV; LiDAR surveys; coastal geotechnics; coastal volcanic rock platforms.

Volcanic Rocks and Soils – Rotonda et al. (eds)
© 2016 Taylor & Francis Group, London, ISBN 978-1-138-02886-9

Cyclical suction characteristics in unsaturated slopes

M. Pirone & G. Urciuoli
University of Naples 'Federico II', Naples, Italy

ABSTRACT: The paper deals with the hydro-mechanical behaviour observed in an unsaturated pyroclastic slope instrumented at Monteforte Irpino, in southern Italy. Suction measured at different depths for four years is presented and interpreted. According to the pattern of data, three different periods are detected: (i) a *wet period* from January to April; (ii) a *dry period* from May to September; (iii) an *intermediate period* from September to December. For each period the mean profile of data is identified and limit equilibrium analysis is performed by modelling the test site as an infinite slope.

1 INTRODUCTION

Flowslides in partially saturated pyroclastic slopes are induced by rainwater infiltration, which is the main factor regulating pore pressure in shallow soil layers. Therefore in situ monitoring of rainfall, suction and water content in the subsoil may be used to investigate seasonal effects in the groundwater regime. In this regard, a test site was built at Monteforte Irpino (southern Italy) to monitor meteorological conditions, matric suction and water content in the subsoil. In the following meteorological data and measurements of volumetric water content and suction collected at field are shown. Using this database, three different periods are detected with different characteristics in terms of suction: (i) a *wet period* from January to April; (ii) a *dry period* from May to September; (iii) an *intermediate period* from September to December. For each period the profile of measurements is shown and the hydraulic response in the subsoil is analysed. Finally, the results of a limit equilibrium analysis are reported by modelling the test site as an infinite slope and using Bishop's effective stress approach to define the stress variables in unsaturated soils.

The pilot site of Monteforte Irpino (AV) was set up in 2005. The partially saturated pyroclastic cover resting on fractured limestone is a few metres thick (3–5.5 m); the slope is quite regular with an average slope angle of 25°–30°. The stratigraphic profile consists of a series of layers parallel to the ground surface. Starting from the ground level, it includes: the topsoil (soils 1–2), a pumice layer (soil 3), a palaeosoil consisting of weathered volcanic ash (soil 4), some pumiceous layers (soil 5), an older palaeosoil of pyroclastic nature (soil 6), a volcanic sand (soil 7), and highly weathered fine-grained brownish ash (soil 8). The Monteforte Irpino pilot site was monitored using the following equipment (Pirone et al., 2015; Papa et al., 2013): (i) 94 traditional vacuum tensiometers,

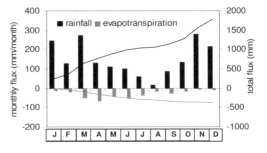

Figure 1. Mean values of monthly rainfall and monthly crop evapotranspiration averaged over four years.

(ii) 40 TDR probes; (iii) six Casagrande piezometers, (iv) a weather station.

2 FIELD DATA

In Figure 1 monthly mean values of rainfall and crop evapotranspiration calculated from meteorological data collected in four years are reported. Although most of the total rainfall falls from October to March, a monthly value of about 90–100 mm occurs from May to July. The mean monthly evapotranspiration is high from April to July (90–60 mm/month) while it is very low in the rest of the year (15–25 mm/month). According to the pattern of suction data, three different periods may be detected: (i) a *wet period* from January to April; (ii) a *dry period* from May to September; (iii) an *intermediate period* from September to December. Figures 2a–c show vertical profiles of matric suction measured along one of the vertical sections instrumented at the site. Suction measurements assume more or less constant values at all the depths over the four months from January to April: those always range between 2 and

183

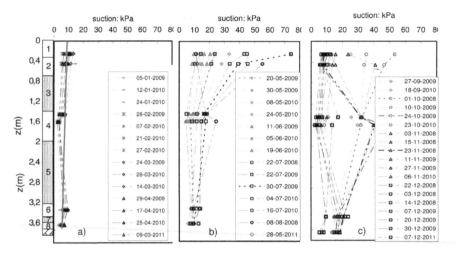

Figure 2. Profiles of suction data grouped by period: wet (a), dry (b), intermediate (c).

8 kPa at all depths. Therefore, steady hydraulic conditions for all the soils examined are established (Figure 2a). In the dry season the matric suction increases to maximum values due to evapotranspiration phenomena. The matric suction profile remains practically linear up to soil 4 but progressively increases toward the ground surface. Because of major rainfall events, from September to November matric suction was quite low in shallow and intermediate layers with values typical of the wet season.

3 LIMIT EQUILIBRIUM

Adopting the Mohr-Coulomb failure criterion, and applying the limit equilibrium method and the Bishop effective stress approach for $\chi = S_r$, the safety factor, FoS, of an infinite slope is given as:

$$FS = \frac{\tau_f}{\tau} = \frac{(S_r s + \sigma)\tan\varphi'}{\tau} = \frac{(S_r s + \gamma z \cos^2\alpha)\tan\varphi'}{\gamma z \sin\alpha\cos\alpha} \quad (1)$$

where the degree of saturation, S_r, is determined from water content measurements and mean soil porosity, n and ϕ' is the friction angle from tests on saturated samples. Therefore, FoS was calculated from the measurements of matric suction and water content available at different depths, i.e. 0.25 m (soil 1), 0.45 m (soil 2), 1.7 m (soil 4), 1.8 m (soil 4), 3.3 (soil 6), 3.5 (soil 8). In Figure 3 the profiles of FoS at fixed days in the wet period, are reported. The value of FoS corresponding to the condition of a nil matric suction is presented in the same figure and it takes on the minimum value of 1.2 in soil 4. The figure confirms that the minimum FoS occurs at soil 4 and even if the suction vanishes, the slope continues to be safe. Thus, the slope angle of the tested area being lower (30°) than that of the unstable area in western Campania (>35°), failure at the tested site could occur when positive pore pressures are established in soil 4.

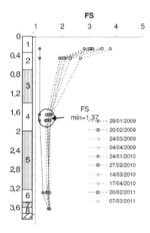

Figure 3. Profile of FoS over wet periods.

4 CONCLUSIONS

Using the monitoring data of a test site, three different periods are detected: (i) a *wet period* from January to April; (ii) a *dry period* from May to September; (iii) an *intermediate period* from September to December. The results of a limit equilibrium show that, for the experimental site (30° slope), the sliding surface should be placed in soil 4 and the FoS reaches unity when the pore water pressure is positive in soil 4.

REFERENCES

Papa, R., Pirone, M., Nicotera, M.V., Urciuoli, G. 2013. Seasonal groundwater regime in an unsaturated pyroclastic slope. *Geotechnique* 63(5):420–426

Pirone, M., Papa, R., Nicotera, M.V., Urciuoli, G. 2015a. In situ monitoring of the groundwater field in an unsaturated pyroclastic slope for slope stability evaluation. *Landslides*, 12(2): 259–276.

Session 5:   Geotechnical problems of engineering structures

Volcanic Rocks and Soils – Rotonda et al. (eds)
© *2016 Taylor & Francis Group, London, ISBN 978-1-138-02886-9*

The application of grouting technique to volcanic rocks and soils, to solve two difficult tunnelling problems

V. Manassero
Underground Consulting s.a.s., Pavia, Italy

G. Di Salvo
Icop s.p.a., Basiliano (UD), Italy

ABSTRACT: Permeation and fissure grouting techniques were applied to volcanic rocks and soils within Rome and Naples metro projects to solve two difficult tunnelling problems. Two case studies are illustrated, both relevant to the widening of station platform tunnels starting from twin TBM tunnels, in the presence of improved soil by grouting. The first is relevant to *Giglioli* and *Torre Spaccata* stations at the new Metro Line C in Rome, where four platform tunnels were excavated through an alternation of pyroclastite and tuff, with a hydrostatic head of 5 to 8 m. The second case study is relevant to *Duomo* Station at the extension of Metro Line 1 in Naples, where four platform tunnels were excavated within Neapolitan Yellow Tuff, under a hydrostatic head of up to 30 m; at the same station four further tunnels for pedestrian access were excavated under the same conditions.

1   INTRODUCTION

Permeation and fissure grouting are techniques commonly used to improve soil and rock characteristics, in terms of both reduced permeability and increased strength and modulus. The soil improvement is achieved by filling the natural voids and cracks with suitable grouts, without any essential change to the original soil/rock volume and structure. The technique was successfully used to solve two difficult tunnelling problems within volcanic soils at Rome and Naples metro stations.

In both case studies, grouting activity was monitored by a computerized system, which automatically measured and recorded a set of parameters, allowing reliable analysis of the results. Furthermore, displacements and groundwater level were constantly monitored, to control possible accidental induced effects on adjacent structures, buildings and utilities by grouting activities and excavation works.

2   GIGLIOLI AND TORRE SPACCATA STATIONS AT ROME METRO LINE C

Line C is the third metro line of Rome. It crosses the entire city from south-east to north-west.

The twin tunnels, 5.8 m I.D., have been bored by TBM. The underground stations have been built partly by cut and cover and partly by tunnelling.

Giglioli and *Torre Spaccata* stations are both composed of a "T" shaped shaft, and two platform tunnels, each about 110 m long. The shafts were excavated by an outer hydromill diaphragm wall and a jet grouting bottom blanket. The 4 platform tunnels, totalling 440 m, were excavated by widening the twin TBM tunnels, previously installed.

Grouting was applied as temporary soil improvement to form a U-shaped thick shell around the tunnel section to be excavated (Figure 1).

The geology of the site at *Giglioli* and *Torre Spaccata* stations is characterized by volcanic soils: an alternation of pyroclastite and volcanic tuff.

Pyroclastite is a volcanic sand with gravel and a silty and clayey fraction ranging between 5 and 25%. The average values of hydraulic conductivity are 5×10^{-5} m/s for pyroclastite and 1×10^{-6} to 5×10^{-5} m/s for tuff, as a function of the type of tuff.

The water-table is located 17 and 20 m below g. l., respectively at *Giglioli* and *Torre Spaccata* stations; this means that the water heads above the invert of tunnels to be excavated were respectively 8 and 5 m.

The job was preceded by an extensive trial field during the design stage at *Torre Spaccata* site, to demonstrate the suitability of permeation grouting to improve in situ soil and to select the most appropriate grouts to be applied. Further to cement-bentonite and silicate-based grouts, an innovative chemical mix was used, the colloidal nano-silica grout, allowing a further reduction of permeability. The standard *tube à manchettes* method was adopted and grouting was carried out by a repeated and selective procedure.

The results in terms of permeability reduction were gathered from comparative permeability tests carried out first on the natural soil and then on the improved soil: in situ Lefranc tests and pumping tests were performed. The original average permeability coefficient was reduced from $k = 1.6 \times 10^{-5}$ m/s to $k = 8.7 \times 10^{-8}$ m/s (without nano-silica grout) and to $k = 3.3 \times 10^{-8}$ m/s (with nano-silica grout), i.e. more than 2 and 2.5 orders of magnitude respectively.

To widen the four platform tunnels from 5.8 m to the final dimension of 10.2 m a fan-shaped grouting treatment was carried out, with 38 to 48 grouting pipes per fan (Figure 1) and a fan every 0.94 m.

The soil porosity was permeated and sealed by first injecting high-penetrability cement-bentonite grout and then Silacsol grout, through conventional TAMs, using a repeated and selective grouting procedure. Each grouting pipe was used first for injection of cement grout and then for chemical grout.

The results in terms of permeability reduction were gathered from comparative permeability tests carried out first on the natural soil and then on the improved soil: free discharge tests through holes drilled from the TBM tunnels were performed. The free flow from each borehole previously drilled through the natural soil ranged from 85 to 130 l/min.

After grouting treatment, the free flow through boreholes drilled from the TBM tunnels decreased to between 0 and 1.5 l/min, thus leading to a reduction of at least two orders of magnitude (well in accordance with the results achieved in the preliminary trial field) and allowing an almost dry tunnelling.

3 DUOMO STATION AT NAPLES METRO LINE 1

The 5 new stations of the Line 1 extension in Naples are all located downtown, in a deeply urbanized area, and are composed of one access shaft, 4 platform tunnels 50 m long and 4 pedestrian access tunnels 30 m long.

To allow for safe tunnelling, at *Duomo* station permeation grouting was chosen for soil improvement. This was performed from the twin TBM tunnels already in place. The 8 tunnels, totalling 320 m, were excavated entirely within the improved soil.

The geological history of the Naples soil formation was strongly influenced by two close volcanic systems: *Vesuvio* and *Campi Flegrei*. The geology of the site is characterized by a pozzolana layer lying on top of the Neapolitan Yellow Tuff (TGN); the underground works were carried out mainly within this deeper formation, under a hydrostatic head of up to 30 m (water-table is few meters below ground level). TGN, is a soft rock with a widespread presence of cracks, both vertical (*scarpine*) and horizontal (*suoli*). These cracks, are the reason of the significant secondary permeability of the tuff, which may release a huge amount of water during excavations.

A thick shell all around the tunnels to be excavated (Figure 2) was achieved by temporary soil improvement with fissure grouting. The platform tunnels were

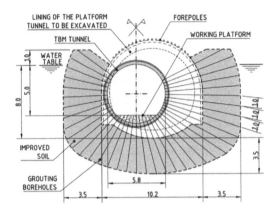

Figure 1. *Giglioli* station: typical cross section.

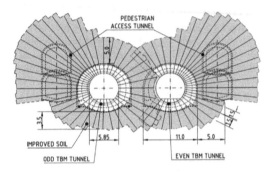

Figure 2. *Duomo* station: typical cross section.

widened from the initial TBM diameter of 5.85 m (ID) to the final dimension of 11.0 m, while the access tunnels, 6 m wide, were fully excavated through the improved soil.

The treatment was fan-shaped with 31 to 38 grouting pipes per fan (Figure 2) and fan spacing 1 m.

Due to the type of soil to be improved, the MPSP (Multi Packer Sleeved Pipe) method was selected. Grouting was carried out by a selective procedure. Every grouting pipe injected only one grout mix: through the primary-fan pipes cement-based mixes were injected (high-penetrability grout made from either standard fine cement or microfine cement), whilst the silicate-based mix Silacsol was injected through the secondary fans.

The combination of the different grout mixes solved the function which they were selected for, confirming the effectiveness of the soil improving method chosen. During tunnels excavation a systematic actual filling of the widespread crack system was observed, sometimes by the cement mix, where crack width permitted, sometimes by the chemical mix, the latter able to permeate a large number of thinner cracks. The residual permeability of the improved TGN, reduced by two orders of magnitude, allowed a safe, fast and almost dry tunnelling.

Volcanic Rocks and Soils – Rotonda et al. (eds)
© *2016 Taylor & Francis Group, London, ISBN 978-1-138-02886-9*

Study on volcanic sediment embankment collapse in the 2011 Earthquake off the Pacific Coast of Tohoku

S. Ohtsuka
Nagaoka University of Technology, Nagaoka, Japan

K. Isobe
Hokkaido University, Sapporo, Japan

Y. Koishi
Central Japan Railway Company, Nagoya, Japan

S. Endou
Shinwasekkei Co. Ltd, Yamagata, Japan

ABSTRACT: A large-scale slope failure of embankment, which consists of volcanic sediment, in Fukushima Prefecture of Japan, occurred due to the 2011 off the Pacific Coast of Tohoku Earthquake. Some signatures of liquefaction have been observed in the site. When investigating the embankment slope failure mechanism, it is necessary to study the effects of particle breakage. In this study, some monotonic and cyclic triaxial tests were conducted for the sampled soil to investigate the change of the particle size distribution caused by applying various stress and the influence on the shear strength characteristics due to the particle breakage. From the test results, it was revealed that the sampled soil, subjected to the effective isotropic consolidation stress of 100 kPa or less, increases the particle breakage amount rapidly and the shear resistance angle is significantly reduced with the increase of the particle breakage amount.

1 INTRODUCTION

Around Ikki junior high school of Aizu-Wakamatsu City in the Fukushima Prefecture, a large embankment collapse occurred. The seismic intensity observed in this region was 5 upper. According to the terrain analysis diagram, the collapsed embankment is classified as an embankmentwhich is filling valley. Figure 1 shows the plan view relating to the disaster situation. The liquefaction-induced slope failure, starting from the central part of the ground in Ikki junior high school located in the old valley terrain, has reached the sidewalk under the ground.

Photo 1 shows the boring survey results which were carried out in the main collapse point. The cores sampled at a range of depths 3–8 m, indicated by the red arrows, is very loose sandy soil with gravel and much water. The N-value is 1–2. Liquefaction-induced soil flow is also confirmed.

Figure 2 shows the geological sectional view assumed based on the boring survey results. The ground consists of three layers. The basement is pumice tuff which ispyroclastic flow sediment. The middle layer is clay with humic soil which is the former topsoil. The current topsoil is sandy soil with gravel which is pyroclastic flow sediment. N-value for the surfaceground around the collapse area is 10–15. N-value of the ground from the depth of 2 m to the basement is very small and 1–5.

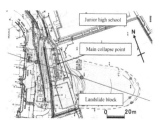

Figure 1. Disaster situation plan view.

Photo 1. Boring survey results of the affected areas.

The sampled soilmainly consists of volcanic ash soil, the particle size varieswidely andit contains much fine-grained fraction as shown in Table 1.

In this study, some isotropic consolidation tests, monotonic and cyclic shear tests using a triaxial compression testing apparatus were conducted for the sampled soil to investigate changes in the particle size distribution and shear strength due to particle

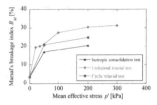

Figure 2. Sectional view of the site.

Table 1. Soil physical property of the sampled sand.

Name	Fukushima	Toyoura
ρ_s [g/cm^3]	2.542	2.640
ρ_{max} [g/cm^3]	1.659	1.645
ρ_{min} [g/cm^3]	0.831	1.335
D_{50} [mm]	0.33	0.18
U_c	6.57	1.60
F_c [%]	10.9	0.0
Remarks	Size control	For comparison

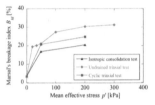

Figure 3. Impact on particle breakage B_M due to difference in test methods.

breakage. The characteristics of particle breakage for the fill materialare clarified. Subsequently, the results of seismic deformation analysis (Newmark, 1965) of the target embankment are reported based on the testing results.

2 PARTICLE BREAKAGE CHARACTERISTICS AND SHEAR STRENGTH PROPERTIES

The particle breakage rate due to differences in test methods is compared, focusing the results for D_r 60%. Figure 3 shows the impact on particle breakage rate of Marsal B_M due to the difference in test methods. B_M rapidly increases in the range of 100 kPa or less, and moderately increases in the range of 100 kPa or more. The particle breakage rate B_M generated by monotonic undrained shear tests is the greatest among three tests. This is because a large deviator stress is applied at the time of shearing. Comparing with the isotropic consolidation test results and the cyclic triaxial test results, the particle breakage rate B_M due to cyclic triaxial tests is 2–3% higher than that due to the isotropic consolidation tests. Hence, it is also revealed that particle breakage generates due to cyclic loading.

3 SEISMIC DEFORMATION ANALYSIS

In this chapter, seismic deformation analysis (Newmark's method) of the affected embankment is

Table 2. Analysis parameters.

Peak shear resistance angle ϕ' [deg]	45.3
Peak cohesive c' [kPa]	0.0
Residual shear resistance angle ϕ' [deg]	18.0
Residualcohesive c' [kPa]	0.0
Wet unit weight γ_{sub} [kN/m^3]	15.8
Saturation unit weight γ_{sat} [kN/m^3]	17.9

Table 3. Reduction in peak shear resistance angle due to particle breakage.

Effective stress of monotonic undrained shear test p' [kPa]	Peak shear resistance angle ϕ' [deg]
20	50.0
35	47.5
50	46.3
100	45.3
200	42.7
300	41.6

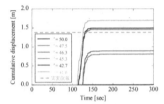

Figure 4. Comparison between cumulative displacement and the actual amount of displacement.

performed by the commercial software (Power SSA), based on the physical properties obtained by the above-mentioned tests.In this paper, the residual strength is assumed to be liquefactionstrength and calculated based on the cyclic triaxial test results.

The ground motion observed by K-net (National Research Institute for Earth Science and Disaster Prevention, 2011) in the Aizu-Wakamatsu city on March 11, 2011 is used as the input ground motion. Tables 2 and 3 show the analysis parameters and a reduction in peak shear resistance angle due to the influence of the particle breakage.

Figure 4 shows the analysis results. The total displacement amount during the earthquake increases with decrease of the shear resistance angle. The more shear resistance angle decreases, the faster the displacement occurs. From the field survey, the actual amount of displacement that occurred during the earthquake was 1.38 m. Comparing with the cumulative displacement and the actual amount of displacement, the cumulative displacement obtained using the shear resistance angle for $p' = 100$ kPa is 1.45 m and very similar to the actual amount of displacement. The overburden pressure in the deepest part of the circular slip line is 87 kPa and similar to $p' = 100$ kPa. Therefore, the reasonable analysis results are obtained by considering the reduction in the shear resistance angle due to the particle breakage.

Volcanic Rocks and Soils – Rotonda et al. (eds)
© 2016 Taylor & Francis Group, London, ISBN 978-1-138-02886-9

Numerical analysis of effects of water leakage with loss of fines on concrete tunnel lining

G. Ren, C.Q. Li & Y.Q. Tan
School of Civil, Environmental and Chemical Engineering, RMIT University, Melbourne, Australia

ABSTRACT: One of the undesirable effects of water leaking into a tunnel is the loss of fines from the ground behind the tunnel lining. The loss of fines creates erosion voids around the tunnel which alter the stress distribution on the lining. Excessive loss of fines due to water leakage would lead to serious stability issues about the tunnel structural stability. This study looks into a case where multiple tunnels constructed in weathered basaltic rocks and evaluates the effects of fine loss on circumferential stresses and bending moments in the tunnel lining using finite element modeling techniques. Analytical results from the finite element modeling show that the extents and the void locations as well as the combination of multiple voids will have significant impacts on the tunnel lining, and hence the engineering performance of the tunnel.

1 INTRODUCTION

Long term performance of tunnel depends on a number of factors, among which the stability of tunnel lining is the most important consideration (Wu et al. 2011). The leakage of groundwater through the cracks and joints of the lining commonly carries fines in the form of silt and clay from the surrounding ground. The muddy water leaking into the tunnel is normally an indication of loss of fines from the geological materials around the tunnel (Figure 1).

Leaking water into the tunnel can have multiple effects on the tunnel, including damages to service facilities, power supply lines, and corrosion to the rails, local flooding and causing instability to the tunnel structure. All of which will have impact on the service life span of the tunnel. This paper reports the findings of numerical analysis for a railway tunnel project. The subject tunnel networks can be summarized as four individual railway tunnels running parallel. The section of tunnels that was selected for modelling contains the most water leaking cracks. The leaking water was diverted onto the plinths and walkways and subsequently into the drainage channels beneath the railway tracks The geology in the project area can be characterized by 5 Zones of Silurian silt/mud stones of different weathering with a mixed sequence of sedimentary and volcanic materials (Basalt). Zone 1 is generally complete weathered material and is usually yellow-brown, sometimes mottled with grey silty or sandy clay. Zone 2 contains soft rock, which can be shattered easily with light hammer blow, and is somewhat friable. The colour is usually yellow-brown, though it may be pale grey or leached almost colourless. Within this zone, clay seams usually consist of individual decomposed silt stone beds or joint deposits. Zone 3 is of moderately hard, and requires a moderately strong siltstones of typically pale grey-brown, mottled pale grey colour. Zone 4 material is generally strong jointed siltstone. Zone 5 is relatively fresh rock with laminated bands of weak materials.

Tunneling in such weathered rocks will lead to the opening of discontinuities in the fractured rock mass, which will increase the hydraulic conductivity of the rock surrounding the tunnel. The highly fragmented rock debris and fines in the rock fractures is mobile in groundwater movement, especially when groundwater leaking into the tunnel.

2 FINITE ELEMENT MODELLING

A finite element model is set up to investigate the effects of fine loss in the vicinity of tunnel perimeter on the concrete lining. The basic finite element model comprises the four tunnels running parallel under the influence of gravity. A two-dimensional finite element model is established using a commercially available FE codes PLAXIS.

In the finite element analysis, the modeling sequence adopted is as follows:

1) Initial stage: generate water pressure and gravity loading within the model;
2) Install lining;
3) Excavate tunnel;
4) Excavate void 1 to simulate ground fine loss;
5) Excavate void 2 and 3 to simulate subsequent fine losses.

The effect of fine loss on the tunnel lining can be conveniently modeled by "switch off" the elements within the voids. The locations of the voids are taken as revealed in the observation (see Figure 1). At the chosen section, two leaking locations in Tunnel 1 and Tunnel 3 as illustrated in Figure 4 are modeled.

3 RESULTS AND DISCUSSION

The analysis revealed that the voids create stress redistribution in the vicinity of tunnels as illustrated in Figure 6.

The following observations are made based on results presented in the Tables:

1) Fine loss (void size) has significant impact on the concrete lining of tunnels. With the increase of void size in the perimeter of tunnel, there is an obvious increase in the maximum bending moment in the tunnel lining (see Figure 8a). The shear force increase is more significant (Figure 8b), however the axial force is almost unchanged.
2) The effects were found to be more concentrated near the location of void.
3) As in the case of this study, the local void appears to only affect the immediate tunnel. The adjacent tunnels are not substantially affected. For intance, Tunnel 2 and Tunnel 4 remain almost unchanged in both maximum bending moment and shear force.

It should be noted that the structural behaviour of tunnel lining depends not only on the tunnel structure integrity and strength but also the interaction with the surround soil and rock. The above analysis is based on a simplified two dimensional finite element modeling; in reality however, the effect of void is essentially a three-dimensional problem. The locations of voids can also be an influential factor in relation to stress distribution on the tunnel linings. For instance, if the leaking point is located between tunnels, the results in terms of bending moments, shear force distributions could be rather different from the above presented. Nevertheless, this paper presents a methodology to demonstrate that the effects of erosion of the geotechnical material behind tunnels can be numerically modeled, with quantitative results showing the changes in bending moment, shear and axial forces in the tunnel linings. These results can be used as input for further structural analysis to determine the safety margin of the tunnel lining.

Furthermore, this numerical analysis can be incorporated into a Quantitative Risk Assessment (S. Rosin 2005) to provide guidance on planning of tunnel inspection and maintenance. Further work will expand to link the effects on fine loss to the service life expectancy of tunnels (Li 2004), as it clear that the fine loss will contribute to the deterioration of the serviceability of the tunnels.

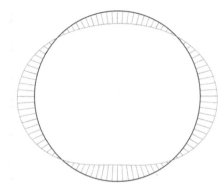

Tunnel 1 Max bending moment 41 KNm/m no void

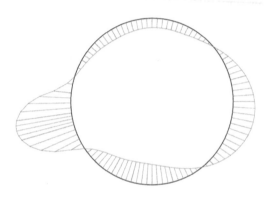

Tunnel 1 Max bending moment 71 KNm/m at 2%

REFERENCES

Wu, H.N., Xu, Y.S., Shen, S.L. and Chai, J.C., Long-term settlement behaviour of ground around shield tunnel due to leakage of water in soft deposit of Shaghai. Front. Archit. Civ. Eng. China 2011, 5(2): 194–198.

Peng, S.S., Surface Subsidence Engineering, Society for Mining, Metallurgy and Exploration inc., Littleton, Colorado 1992.

Meguid, M.A. and Dang, H.K., The effect of erosion voids on existing tunnel linings. Tunneling and Underground Space Technology 24 (2009), 278–286.

Rosin, S., Geotechnical risk assessment for maintenance of water conveyance tunnels in south eastern Australia, AGS AUCTA Mini-Symposium: Geotechnical Aspects of Tunneling for Infrastructure Projects, Oct. 2005.

Li, C.Q., Reliability Based Service Life Prediction of Corrosion Affected Concrete Structures. Journal of Structural Engineering © ASCE/Oct. 2004/1573.

Volcanic Rocks and Soils – Rotonda et al. (eds)
© *2016 Taylor & Francis Group, London, ISBN 978-1-138-02886-9*

Excavations in the Neapolitan Subsoil: The experience of the Toledo Station service tunnel

G. Russo & S. Autuori
University of Naples Federico II, Naples, Italy

F. Cavuoto & A. Corbo
Studio Cavuoto, Naples, Italy

V. Manassero
Underground Consulting S.a.s., Pavia, Italy

ABSTRACT: A Metro Line 1 extension is being executed in Naples from the terminal of Dante Station. The new underground section is composed of five new stations and two twin rail tunnels with a length of 5 km. Toledo Station has the shaft located laterally and a large size service tunnel (13 m wide, 17 m high and 40 m long) connects it to the rail and pedestrian access tunnels, starting from the access shaft. The tunnel is located in the historical centre, under a deeply urbanized area in volcanic soil (loose pozzolana silty sand and tuff) with a hydrostatic head of 27 m. For stabilizing the tunnel crown in loose volcanic soil, it was decided to adopt the Artificial Ground Freezing method among the various options of ground improvement techniques. In this paper, the observed behaviour is presented and discussed, allowing to throw a light on the effects of the rather complex execution steps of the underground excavations.

1 INTRODUCTION

This paper is concerned with the analysis of a 13 m wide, 17 m high and 40 m long service tunnel execution at Toledo Station within the Metro Line 1 extension project in Naples (Italy). From the terminal of Dante Station, the new stretch of the underground is composed of five new stations and two twin rail tunnels with a length of 5 km. Four stations out of five have the access shaft centered with the twin rail tunnels. On the contrary, Toledo Station (Fig. 1), situated in the deeply urbanized historical centre of the city, has the shaft located laterally and a large size service tunnel connects it to the rail and pedestrian access tunnels, starting from the access shaft.

The subsoil concerned can be divided schematically into two main layers: a top loose silty sand overlying the soft and sometimes fractured Neapolitan Yellow Tuff, with a hydrostatic head of 27 m above the tunnel invert. The first step of the service tunnel construction was the excavation of the small drift above it. Its transverse section is reported in Figure 2. It was excavated in loose to medium dense sandy soil immediately above the groundwater table at +4.75 m a.s.l. and only 8 m below the foundations of an existing building.

The excavation was carried out via NATM under the protection of injected forepoling at the crown and along the vertical side, using fibreglasses at the front face and advancing with steps of 1 m followed by the

installation of temporary steel ribs and shotcrete. Once finished the drift, from there both cement and chemical grouting were carried out around the underneath tunnel shape, through sub-vertical grouting pipes. In particular, the grouting method adopted was the Multi Packer Sleeved Pipe (MPSP). A low permeability jet-grouting end plug completed the first part of the consolidation works. Later on, directional drilling from the main shaft of the station and sub-vertical drilling from the drift were carried out to install freeze

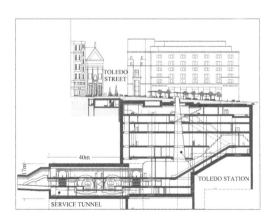

Figure 1. A longitudinal section of the Toledo Station.

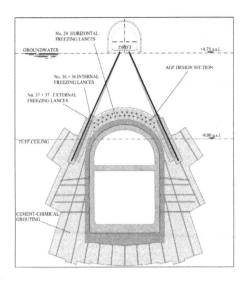

Figure 2. Schematic transverse section of the large service tunnel and soil treatments adopted.

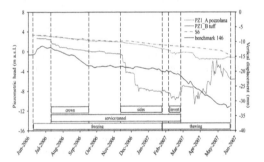

Figure 3. Settlements of benchmark 146 compared to the piezometric head variation during the construction progress.

lances. The Artificial Ground Freezing (AGF) technique was thus adopted to create a frozen arch in the ceiling of the large service tunnel to be excavated in the sandy soil. The frozen arch in the sandy soil and the Yellow Tuff injected with both micro-cements and silicates, together with the end plug in jet-grouting, formed a strong and waterproof contour around the large service tunnel to be excavated.

All the rather complex construction process was essentially monitored by survey on benchmarks diffused on the overlying buildings and by piezometers installed at various depths around the main shaft of the station and the service tunnel. In Figure 3 piezometric levels monitored by some of the installed piezometers, compared to the settlements of one of the most significant benchmarks, are reported versus time starting from June 2006 until June 2007. In the same plot, on the time axis, the main construction steps of the service tunnel execution are recalled.

In Figure 4 the contour lines of the surface subsidence related to the final stage (after the thawing) of the construction of the service tunnel are plotted.

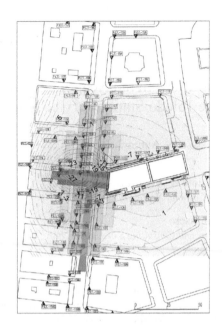

Figure 4. Ground surface subsidence due to the Toledo Station construction.

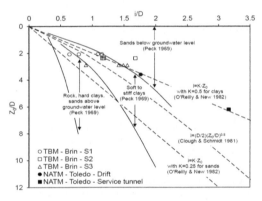

Figure 5. Empirical relations between settlement trough width and tunnel depth – Metro Line 1, Naples.

In the plan view also the existing buildings and the track of the underground works are superimposed. As we can see, the ground movements related to the tunnel execution are spread over a small distance from the tunnel centre line, probably due to the particular tunnel geometry and its depth from the ground surface.

In particular, the settlement trough width for both the tunnels is in substantial agreement with previous field data recorded during the excavation of the initial stretch of Metro Line 1 in Naples carried out by a TBM (Fig. 5). In fact, according to the original suggestions by Peck (1969) the range of the measurements falls in the area which is intermediate between *sands above the groundwater table* and *sands below the groundwater table*. This result is consistent with the intermediate nature of the tunnel cover.

194

Volcanic Rocks and Soils – Rotonda et al. (eds)
© 2016 Taylor & Francis Group, London, ISBN 978-1-138-02886-9

Partial reactivation of a DGSD of ignimbrite and tuff in an alpine glacial valley in Northern Italy

L. Simeoni
University of L'Aquila, L'Aquila, Italy

F. Ronchetti & A. Corsini
University of Modena and Reggio Emilia, Modena, Italy

L. Mongiovì
University of Trento, Trento, Italy

ABSTRACT: This paper describes the surveys, site investigations, field measurements and laboratory tests carried out to analyze the stability of a slope in the Isarco alpine glacial valley, in Northern Italy, where rocks and debris of ignimbrite and tuff outcrop. This valley is crossed by the major transport infrastructures connecting Italy to the central Europe: SS12 State Road, A22 (E45) Motorway, Verona-Brennero railway, high-speed railway network TNT-T5. Recently, large deformations of the pads of a viaduct were identified and were supposed to be caused by the downslope movement of the piers. In effect, subsequent investigations and measurements revealed the existence of sliding surfaces with movements smaller than 1 cm per year. Surveys identified that movements occur at the toe of a DGSD and back-analyses revealed that the residual shear strength estimated with shearbox tests is mobilized on the sliding surfaces.

1 GEOLOGICAL MODEL

The DGSD, affecting a slope characterized by Permian ignimbrites and tuffs, is a large and deep mass movement, that involves a slope for a length of 800 m, between the quotes 750 and 336 m a.s.l., and a width of 600 m (Figure 1). The deep movement has divided the slope in deep distinct morphological elements, which are composed by the original ignimbrite formation, and that are separated each other by rotational detachment zones. According field surveys and the deep monitoring systems (inclinometers),

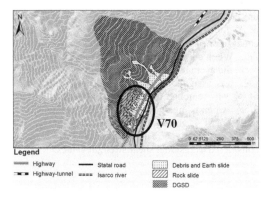

Figure 1. Landslide units.

Legend

Highway	Statal road	Debris and Earth slide
Highway-tunnel	Isarco river	Rock slide
		DGSD

this phenomenon can be considered non-active. The morphology of the deep mass movement has been re-shaped by other landslides, such as: rock slides, earth and debris slides, rock falls.

Rock slides are frequent in the south and lower part of the DGSD (V70 in Figure 1), and most of them are characterized by retrogressive movements, that are responsible of the development of trenches in the slope. According to field surveys and monitoring systems, most of these movements are active.

Inside the rock slides, from the vertical scarps, rock falls are frequent and they are responsible of the development of cones. The drilling of boreholes in these deposits have shown a stratigraphy characterized by the alternation between slope debris and alluvial gravels (deposits of the River Isarco).

2 SITE INVESTIGATIONS AND FIELD MEASUREMENTS

Since 1993 a number of open-hole or core-drilling borings have been drilled at the toe of the slope, close to the viaduct piers or the caissons built around them (Figure 2). Two boreholes (T8P and T9P) equipped with four piezometers and 10 inclinometers (I2, I3, I6, T1I, T2I, T3I, T4I, T5I, T6I, T10I) are currently used to collect the field measurements of pore water pressure and horizontal displacements.

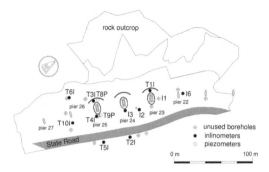

Figure 2. Location of boreholes and instrumentation.

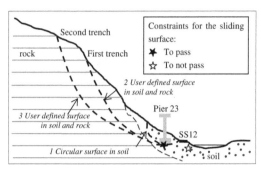

Figure 3. Example of the three different hypothesis of sliding surface location investigated at the pier 23.

The bedrock is of ignimbrite and was found at a depth greater than 18 m (333 m asl). Is is overlaid by debris, consisting of weathered rock fragments and blocks of tuff and ignimbrite accumulated at the foot of the cliffs (talus and rockfall deposit) and mixed with glacial sandy to clayey deposits and alluvial lens at the base of the slope.

Piezometers T8P and T9P revealed that the total heads always remain lower than the normal river elevation (336 m), and that the seepage is generally downslope.

Total Station have been used to measure the movement of the piers since 2004, and it was found that that all piers move downslope at constant rates ranging from a minimum of 1.3 mm/year at the pier 21 to a maximum of 10.6 mm/year at the pier 24.

Inclinometers identified two sliding surfaces in the Northern part of the landslide at the inclinometers T1I, I2, I3 and I6: one surface, with the major displacements, passes at the base of the pier foundations, the other, with the minor displacements, is about 4 m deeper. The former was called "foundation" sliding surface, the latter "deep" sliding surface. At the inclinometers T3I, T4I and T6I (Southern part of the landslide) only the "deep" sliding surface was recognized, while at the inclinometers T2I and T5I, close to the river Isarco and whose tops are located about 4 m upper than the pier foundations, none of the sliding surfaces was identified.

The redundancy between Total Station and inclinometer measurements, in terms of displacement rates, was very satisfactory, with differences generally less than 20%.

3 LABORATORY TESTING

The debris resulted well-graded soils, mainly gravelly at the top and sandy below, the rocks fragments were highly weathered. The alluvial soil buried at the toe of the slope contained levels of uniform sand.

Shearbox tests gave effective angles of shearing resistance at constant volume varying between 26° and 34°. The angles of residual strength varied between 20° and 34°.

4 BACK ANALYSES

2-D back analyses with the Limit Equilibrium Method by using Slope/W of Geostudio (2007) were carried out assuming a null cohesion and seeking the effective angle of shearing resistance (ϕ') corresponding to a factor of safety equal to one. Nine cross sections were investigated: 5 passing through piers 22, 23, 24, 25 and 26 and 4 intermediate to the previous. At each section three different sliding surfaces were assumed since it was not known at what altitude the sliding surface outcrop. It was then supposed that it develops 1) circular and entirely in the soil deposit, 2) partly in the rock up to the first trench, or 3) partly in the rock up to the second trench (Figure 3).

Each sliding surface was constrained to pass through the points recognized in the inclinometers close to the piers and to not involve the SS12 State Road. The "deep" sliding surfaces were investigated from pier 24 to pier 26 and the "foundation" sliding surfaces from pier 22 to pier 24. The pore water pressures were assumed to be null.

When adopting the sliding surface 1 (circular and entirely in the soil) only one material, and then one failure criterion, was assumed and the greater value ϕ'_{S1} of ϕ'_S for the soil corresponding to a factor of safety equal to one was searched by varying the sliding surface. When adopting the sliding surfaces partly in the rock, the average value of ϕ'_{S1} from the previous analyses was assigned to the soil and the greater value of ϕ'_R was searched (ϕ'_{R2} when using sliding surface 2, ϕ'_{R3} when using sliding surface 3). Average values of 21.8°, 31.4° and 29.9° were estimated for ϕ'_{S1}, ϕ'_{R2} and ϕ'_{R3}, respectively. These values can be reasonably considered as residual shear strength values.

Volcanic Rocks and Soils – Rotonda et al. (eds)
© 2016 Taylor & Francis Group, London, ISBN 978-1-138-02886-9

Vertical bearing mechanism of pile foundation in volcanic ash soil

K. Tomisawa, T. Yamanishi & S. Nishimoto
Civil Engineering Research Institute for Cold Region, PWRI, Sapporo, Japan

S. Miura
Hokkaido University, Sapporo, Japan

EXTENDED ABSTRACT

Pile foundations for bridge piers and abutments constructed in volcanic ash ground are mainly designed based on specifications for sandy soil ground due to the similar density and internal fiction angle of volcanic ash as those of normal sandy soil. However, mechanical characteristics of volcanic ash, such as friability against confining pressure, differ strongly from those of normal sandy soil. To confirm the vertical bearing capacity and deformation characteristics of piles in volcanic ash ground, vertical loading tests using actual piles for bridge foundations and static cone penetration tests were conducted at 14 construction sites.

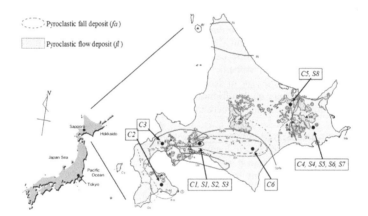

Distribution of volcanic ash soil in Hokkaido and the locations of field test sites

Push-in test for CCP

Impact test for SSP

As a result of testing and verification, the following information was obtained as basic data on the vertical bearing mechanism of piles in volcanic ash soil:

1) While skin friction of piles in pyroclastic fall deposit (fa) developed in a manner that was equivalent to that of sandy soil, it tended to decrease in the cases of cast-in-place piles and driven steel-pipe piles with an N value of less than 30 in pyroclastic flow deposit (fl), which was assumed to be in a welded state, and uniform relations shown in the table below were obtained.

	Cast-in-place pile	Driven steel-pipe pile
Pyroclastic fall deposit (fa)	$f = 5N(\leqq 200)$ (equivalent to sandy soil)	$f = 2N(\leqq 100)$ (equivalent to sandy soil)
Pyroclastic flow deposit (fl)	$f = 3.8N$ (decreased)	$f = 1.4N$ ($N < 30$) (decreased when N value was less than 30)

2) The decrease in skin friction in the welded pyroclastic flow deposit (fl) was thought to be due to the decrease in lateral pressure and grain size of soil around the piles as pile construction caused disturbances in the welded state of volcanic ash soil and grain fragmentation. In the case of driven steel-pipe piles, however, recovery of skin friction up to 20 kN/m² was achieved by curing for a certain period of time.

3) A certain correlation was observed between skin friction f of piles measured in the vertical loading test of piles and the mean end resistance q_t found by the cone penetration test, indicating the potential application of the cone penetration tests in the future.

	Cast-in-place pile	Driven steel-pipe pile
Pyroclastic fall deposit (fa)	$f = 0.0136q_t$	$f = 0.0050q_t$
Pyroclastic flow deposit (fl)	$f = 0.0024q_t$	$f = 0.0015q_t$

The authors intend to establish pile design and construction methods for volcanic ash soil through the further accumulation and evaluation of field data.

REFERENCES

Akai, K. et al. 1984. Bearing mechanism of steel-pipe piles in Shikotsu volcanic ash layer, Tsuchi-to-Kiso, JGS(1442): 41–46.
Hasegawa, K. et al. 1995. Bearing capacity characteristics of driven steel-pipe piles in volcanic ash ground, Proceedings of 21st Japan Road Conference, Bridge Section: 852–853.
Hokkaido Branch of Japanese Geotechnical Society 2004. Volcanic ash soil for engineers: 1–14.
Japan Road Association 2002. Specifications for highway bridges and instruction manual IV, sub-structures: 348–432.
Japanese Geotechnical Society 2000. Soil test methods and interpretations – first revised edition: 563–600.
Japanese Geotechnical Society 1974. Special Soils in Japan: 203–261.
Miura, S. & Yagi, K. 1997. Particle breakage of granular volcanic ash caused by compression and shearing and its evaluation, Journal of the Japan Society for Civil Engineers III-36(561): 257–269.
Machida, H. & Arai, F. 1992. Volcanic ash atlas, University of Tokyo Press: 6–162.
Meyerhof, G.G. 1976. Bearing capacity and settlement of pile foundation, Proc. of ASCE 102(GT3):197–228
Railway Technical Research Institute 2000. Design criteria for railway structures and instruction manual, foundation structures/retaining structures: 201–264.
Takada, M. et al. 1997. Evaluation of dynamic properties of secondary Shirasu ground, Journal of the Japan Society for Civil Engineers III-38(561): 237–244.
Tomisawa, K. & Miura, S. 2005. A study of the vertical bearing mechanism of bridge foundation piles in volcanic ash soil, Proceedings of the 50th Geotechnical Symposium: 303–310.
Uto, K. et al. 1982. Method for summarizing the results of a vertical loading test of piles, Foundation Work 10(9): 21–30.
Vesiæ, A.S. 1963. Bearing capacity of deep foundation in sand, Highway research record 39: 112–153.
Wakamatsu, M. & Kondo, T. 1989. Types of soil in Hokkaido – Volcanic ash soil in Hokkaido, Soils and Foundations 37(9): 24–29.
Yagi, K. & Miura, S. 2004. Evaluation of dynamic properties of crushable volcanic ash ground, Journal of the Japan Society for Civil Engineers III-66(757): 221–234.

Volcanic Rocks and Soils – Rotonda et al. (eds)
© 2016 Taylor & Francis Group, London, ISBN 978-1-138-02886-9

Author index